中国特色楼盘（I）

广州佳图文化传播有限公司　主编

CHINESE
CHARACTERISTICS ESTATE（I）

深圳出版发行集团
海天出版社

图书再版编目（CIP）数据

中国特色楼盘（I）/ 广州佳图文化传播有限公司主编.
深圳：海天出版社，2009.2
　（佳图建筑系列）
　ISBN 978-7-80747-474-6

　I. 中…　II. 广…　III. 住宅-建筑设计-中国　IV. TU241

　中国版本图书馆 CIP 数据核字（2008）第 198696 号

中国特色楼盘（I）
ZHONGGUO TESE　LOUPAN　（I）

出 品 人：陈锦涛
出版策划：毛世屏
责任编辑：王　颖（0755-83460593　E-mail:6021@sina.com）
责任校对：周　强
责任技编：钟愉琼
策　　划：佳图文化
装帧设计：杨先周

───────────────────────────────

出版发行　海天出版社
地　　址　深圳市彩田南路海天大厦（518033）
网　　址　www.htph.com.cn
电　　话　0755-83460137（批发）　83460397（邮购）
印　　刷　深圳市彩美印刷有限公司
版　　次　2009 年 2 月第 1 版
印　　次　2009 年 2 月第 1 次印刷
开　　本　889mm×1194mm　1/12
印　　张　22
总 定 价　560.00 元（I、II 两册）

───────────────────────────────

前言

　　中国房地产开发规模至今为止仍是空前巨大的，其规划、建筑、景观设计市场之大，实属全球罕见。据统计，中国人口众多，内需庞大，楼盘的开发数量在一定时期内仍将持续领先于全世界。

　　本书精选了全国各大地区68个独具特色的最新楼盘，包括贝尔高林、美国易道设计公司、SWA GROUP、澳洲U&A设计国际集团、优山美地·布里斯亚太联合规划设计公司、美国凯斯设计有限公司、美国RTKL、美国JWDA、美国道林建筑与城市规划设计集团、美国EEK建筑事务所、新加坡SCDA建筑师事务所、加拿大UDS国际建筑事务所、澳大利亚五合国际、德国（WSP）建筑设计、BDCL国际建筑设计有限公司、加拿大AEL建筑景观设计有限公司、香港泛亚国际、梁黄顾建筑师（香港）事务所、香港王董建筑师事务有限公司、许李严建筑师有限公司、罗麦庄马香港有限公司、深圳市城脉建筑设计有限公司、翰时国际、华森建筑与工程设计顾问有限公司、ZPLUS 普瑞思建筑规划设计咨询有限公司、北京都林国际工程设计咨询有限公司、新西林园林景观有限公司等国际国内一流设计公司的近期力作，其设计引领潮流，精美绝伦。

　　全书资料详尽，配合耗时费力拍摄的丰富的实景图片，实为楼盘设计不可多得的案头参考工具书。

目录
CONTENTS

别墅 ①

低密度住宅 ⑨⑤

小高层住宅 ⑱①

别墅

Villa

別墅

1

总居住户数：160
总居住人口：560
户均人数：3.5
建筑密度：20.2%
容积率：0.39
绿化率：50.9%
机动车位：298

1. 门楼实景图
2. 天井实景图
3. 聚落结构

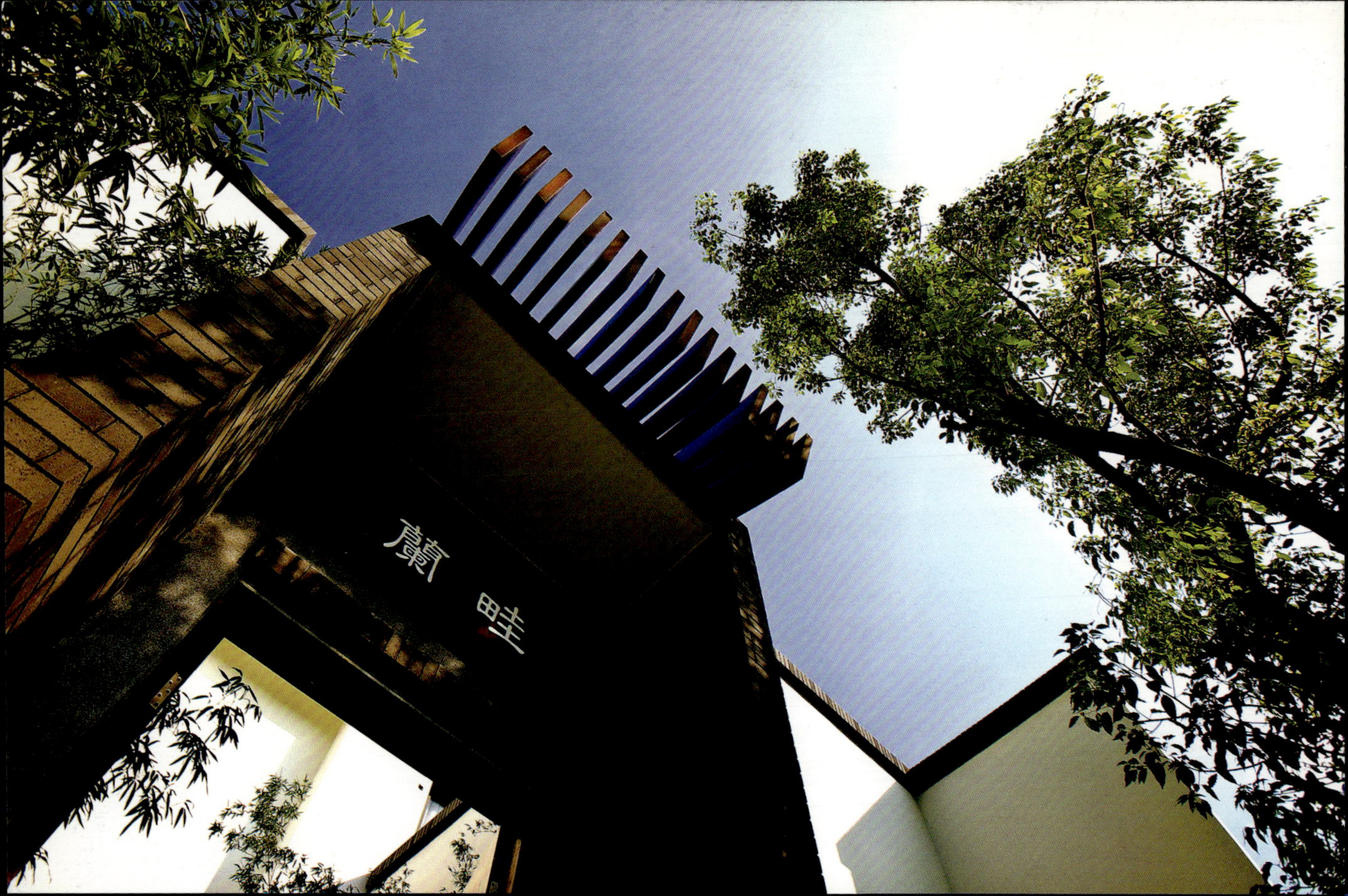

中山清华坊

开发商：中山市圣都房地产开发有限公司
项目地址：番禺区南村镇新城区兴业大道北侧镇政府旁
建筑规划设计：北京东方华太建筑设计工程有限责任公司
合作设计：广州集美组室内设计工程有限公司
规划用地面积：130 328 m^2
总建筑面积：51 137 m^2
住宅面积：44 113 m^2

2

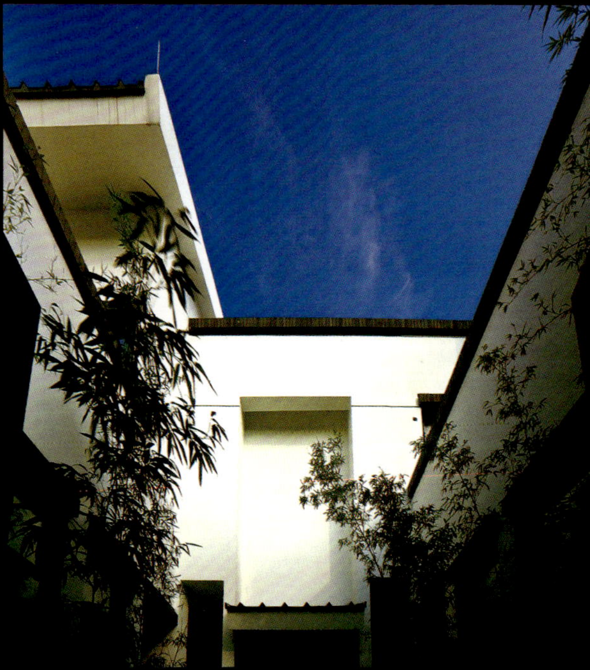

中山清华坊，是继成都、广州两个清华坊后，推出的第三代升级创新型产品，前面两个清华坊，着重强调中国传统民居和现代居住生活的平静对话。而中山清华坊，强调更多的是东西方文化的糅合。

中山清华坊坐落在著名的4A级旅游风景区——中山詹园的背后，一块依山傍水、风景秀丽的山谷之地。这里三面环山，林木繁盛，坐拥三个天然湖泊，自然环境优越。为了体现中国传统民居的“天人合一”，建造者们怀着一颗“无为”之心，遵循着大自然的原来的一草一木。建筑布局上，依托原有地势，各个院落组团错落有秩，使得整个“清华坊”像自然镶嵌于此的一个原始村落。

清华坊位于中山市老虎臀区内规划，分为A区、B区、C区地块，一期为A区的平缓地带，连排低层住宅部分为一期工程。北为山景，建筑物退缩用地红线6米以上；南面为规划道路，绿化退缩4米以上；西面为天然湖泊，东面相邻为厂房，退缩红线10米以上。地理环境优美，道路及市政设施较为完善，地块平整，利于建设开发。

本规划设计是一低层低密度的高尚住宅区，均为连排式，在每户的底层均设有独立的车库和私家花园，户主可以根据个人的喜好对私家花园进行自己的个性化设计，与整个小区的绿化景观和中心湖泊相呼应，使人感到犹如住在公园，充分体现了"以人为本"的设计理念。在单体设计中充分考虑住户的品位及要求，各个功能房间的布置及相互关系上，经过精心设计，确保住户在使用上达到一种居住的舒适性，并使其富于生活情趣，在建筑的后花园处理上，设计了一种高墙深院的传统民居中庭院的效果，在立面屋顶造型处理上，借鉴传统民居和园林亭阁的特点，与现代风格相结合。立面的处理采用了极具传统的、具有民居特色的屋顶，在立面上通过采用不同颜色的处理，使建筑融入整个大自然中，成为一个有机的整体。

在中心湖泊区为小区主入口，也是整个形成小区的中心景观，与山景遥相呼应，形成一个和谐整体的效果，结合自然的景观和各私家花园的小型园景形成层次丰富的、有生气的景观空间环境。在小区的环境道路两边，在适当的位置设置有特色的园林绿化小品和雕塑，增加小区的文化气氛，形成小区内部的景观轴线，具有很强的秩序感和丰富的个性空间，使住户更加亲近自然，与自然融为一体，并使邻里之间的交往更加亲切、富有情趣性。

4

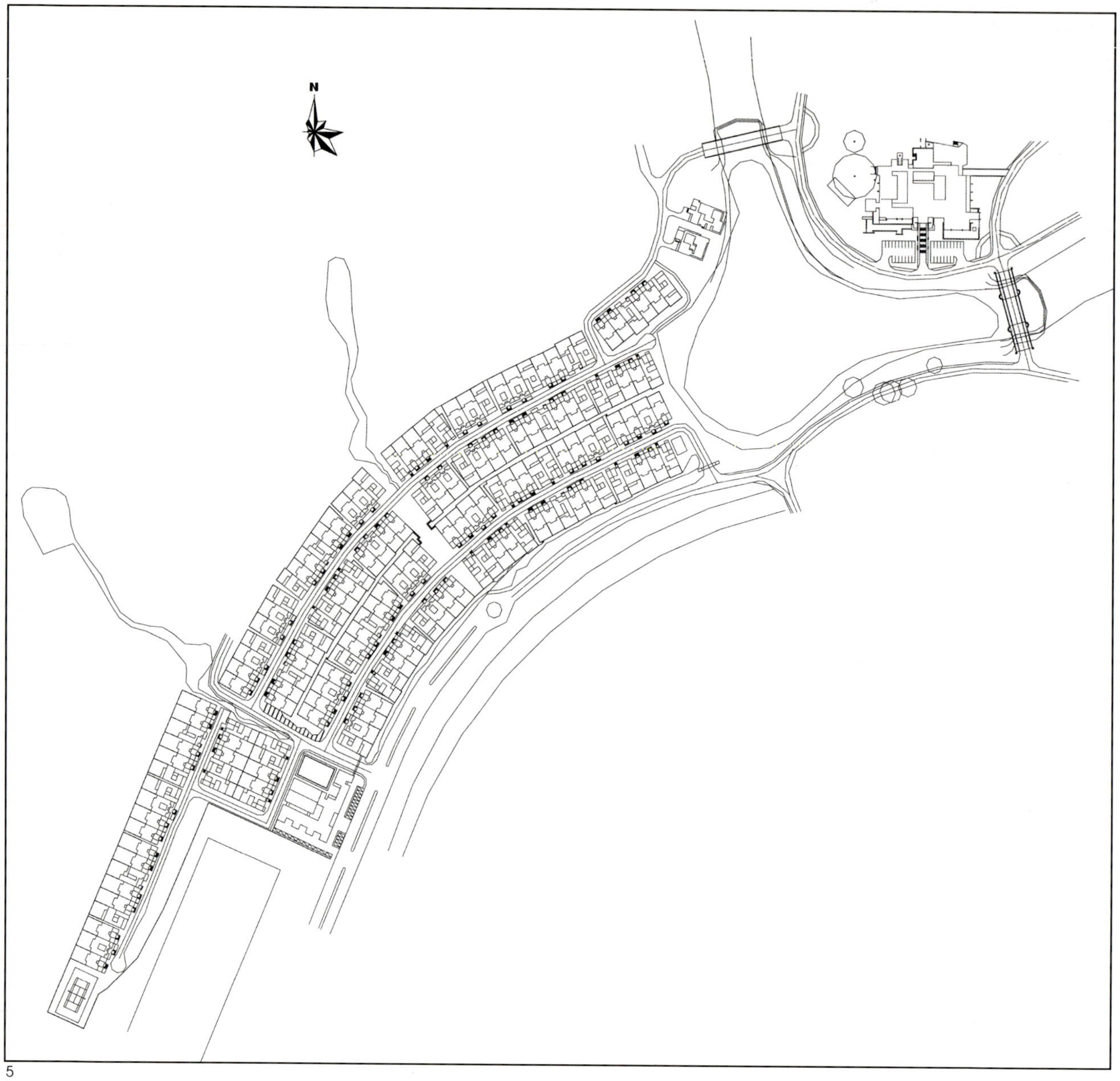

5

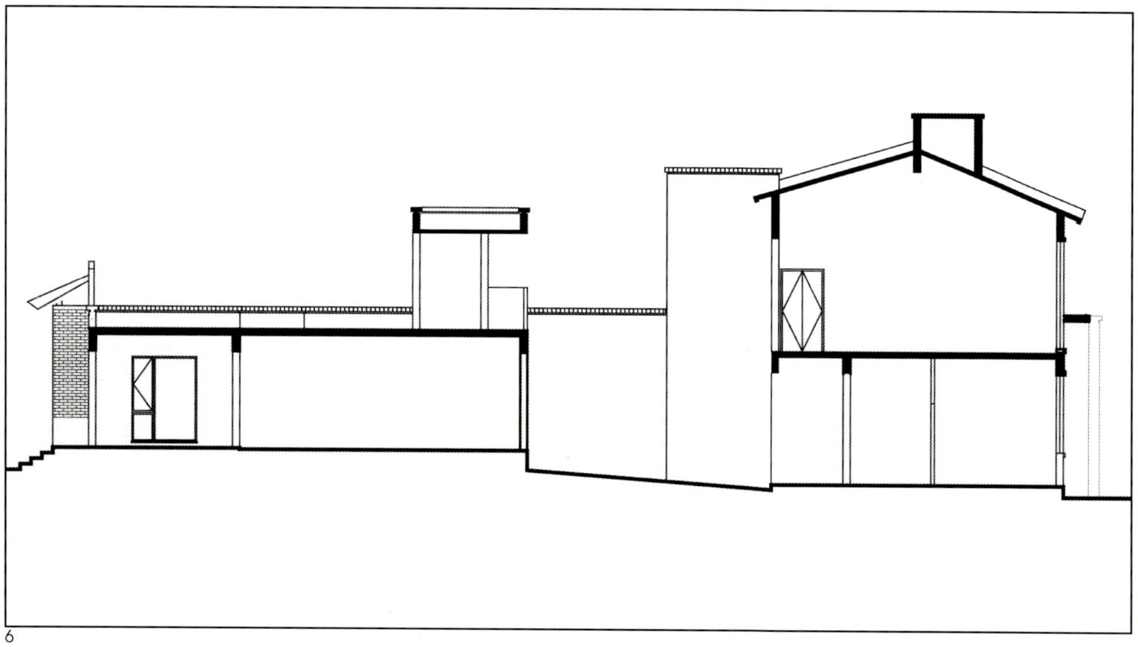

6

Developer: Sunnycity Land Development Group

Architectural plan and design: Beijing Sino-Sun Architects & Engineers Co., Ltd.

Cooperating designer: Newsdays Interior Design & Construction Co., Ltd.

Planned plot area: 130 328 m²

Gross building area: 51 137 m²

Residential area: 44 113 m²

Total residential units: 160

Total residential population: 560

Average residential number: 3.5 in each house

Construction density: 20.2%

Volume rate: 0.39

Afforesting rate: 50.9%

Motor–car space: 298

Zhongshan Qinghua Garden is a new product of the third updated generation promoted, following another two in Chengdu and Guangzhou. The former two pay attention to quiet balance between traditional Chinese folk house and modern living while Zhongshan Qinghua Garden emphasizes more in the integration of oriental and western culture.

Zhongshan Qinghua Garden is located in the back of Zhoangshan Zhan's Garden, a famous 4-A-grage tourist scenic spot lies against the mountain and along the water, it is also a picturesque valley scenery facing mountains in three directions, possessing dense forests and three natural lakes. So, to show the opinion of Union of Nature and Man into One of traditional Chinese folk house construction, architects have followed the object law of nature with an idea of not-doing, not to damage even a bush in this advantageous natural environment, but only to seek harmony with nature. In terms of the architectural layout, we build each group according to where they are and in staggered array, making the whole Qinghua Garden inserting it nature like a primitive village.

4. 门楼实景图
5. 总平面图
6. 立面图
7. 一层平面图
8. 二层平面图
9. 独立别墅实景图

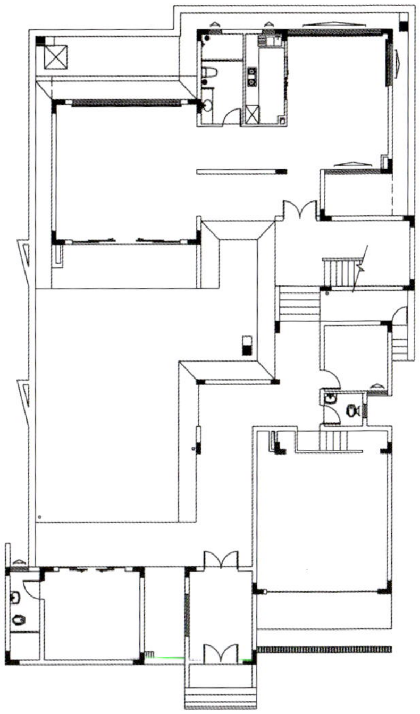

7

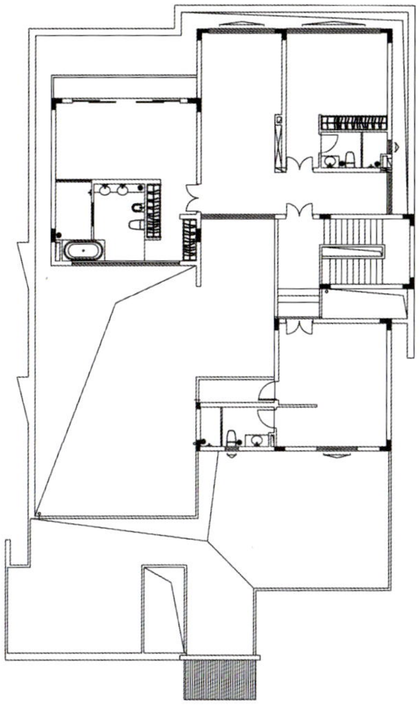

8

Located in inside Laohu District in Zhongshan City, the garden is divided as Block A, Block B and Block C. Part of the low-rise terrace houses are included in the first phase, also know as Block A, which takes the mountain feature in the north, and with a construction back red line of over 6 meters; planned road in the south with greening back red line of over 4 meters; natural lakes in the west and mill building in the east with a back red line of over 10 meters. Such beautiful geological environment and improved road, municipal utilities and flat section is beneficial for development.

This planning design is a noble low-rise and low –density residential quarter, all are terrace houses with individual garage and private garden in each bottom, and owners can design the private garden as whatever the want, echoing to the greening landscape and the central lake, such makes them feel as if living in a park, fully representing the people-first designing rationale. Owners' taste and demand is fully taken consideration into the individual design, and, the arrangement and mutual relationship of each functional room is elaborately design to ensure that they are comfort and interesting for living. Back gardens turn on an effect of traditional folk houses with high walls and large courtyards. In terms of elevation roof shape are combined with modern style referring to features of traditional folk house and garden pavilion. By process of different colors, the construction is merged into nature as an organic whole. The central lake is the entry as well as central feature of the residential quarter, echoing in distance with the mountain, forming a harmonious community. The natural landscape and mini-sized garden feature together form active landscape space which is rich in layers. In both sides of the road in the residential quarter there are special small greening ornaments and sculptures in proper spots which add to cultural atmosphere and become a feature axis inside the residential quarter. All this makes owners more close to nature, together with nature and for friendship and interest in the neighborhood.

10

10. 连排别墅立面
11. 巷口实景图
12. 聚落结构

13

14

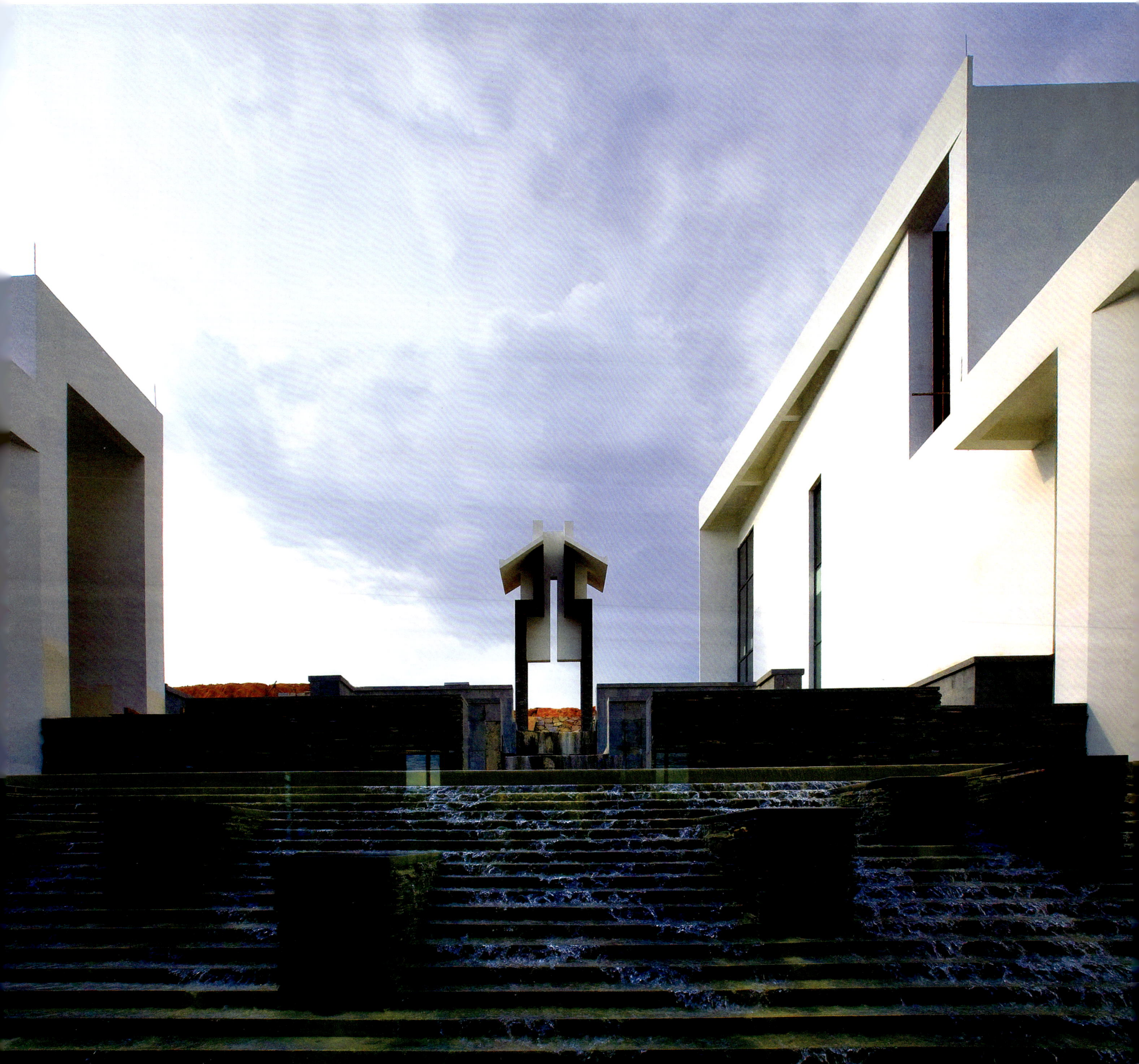

15

16

17

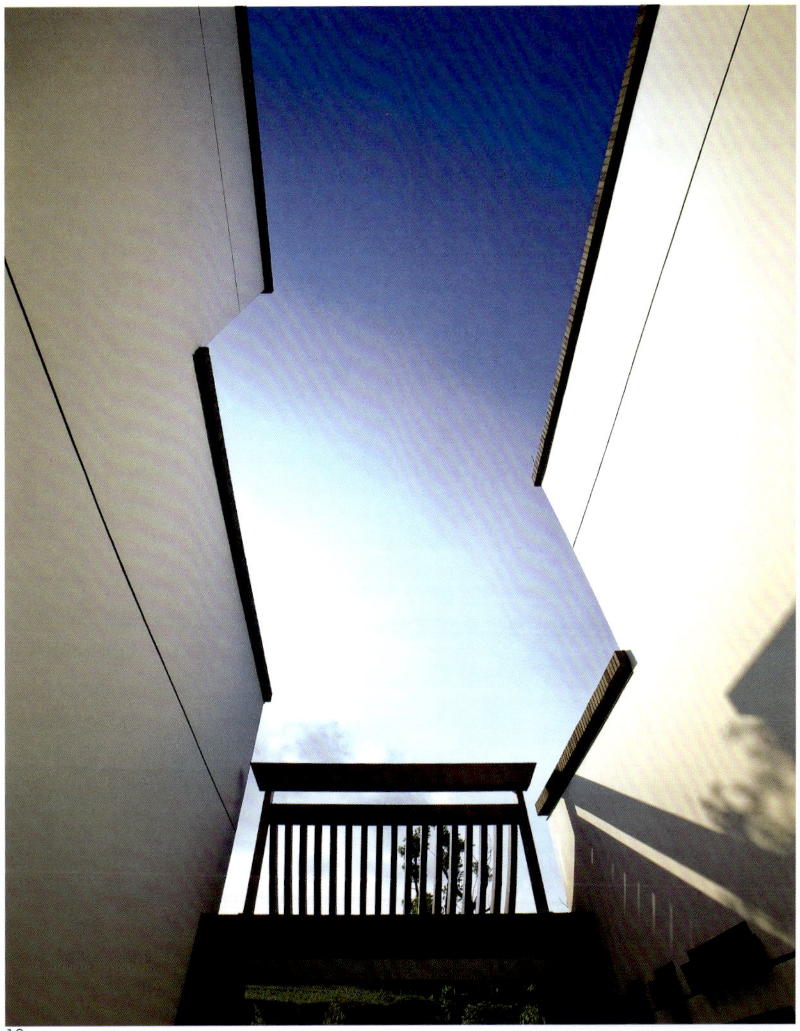
18

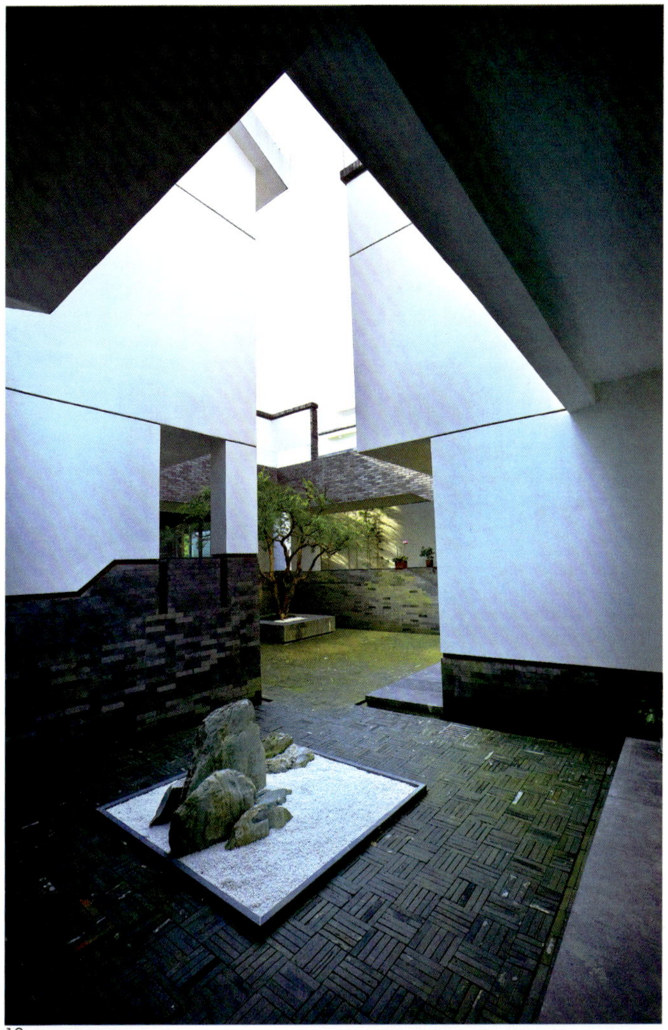
19

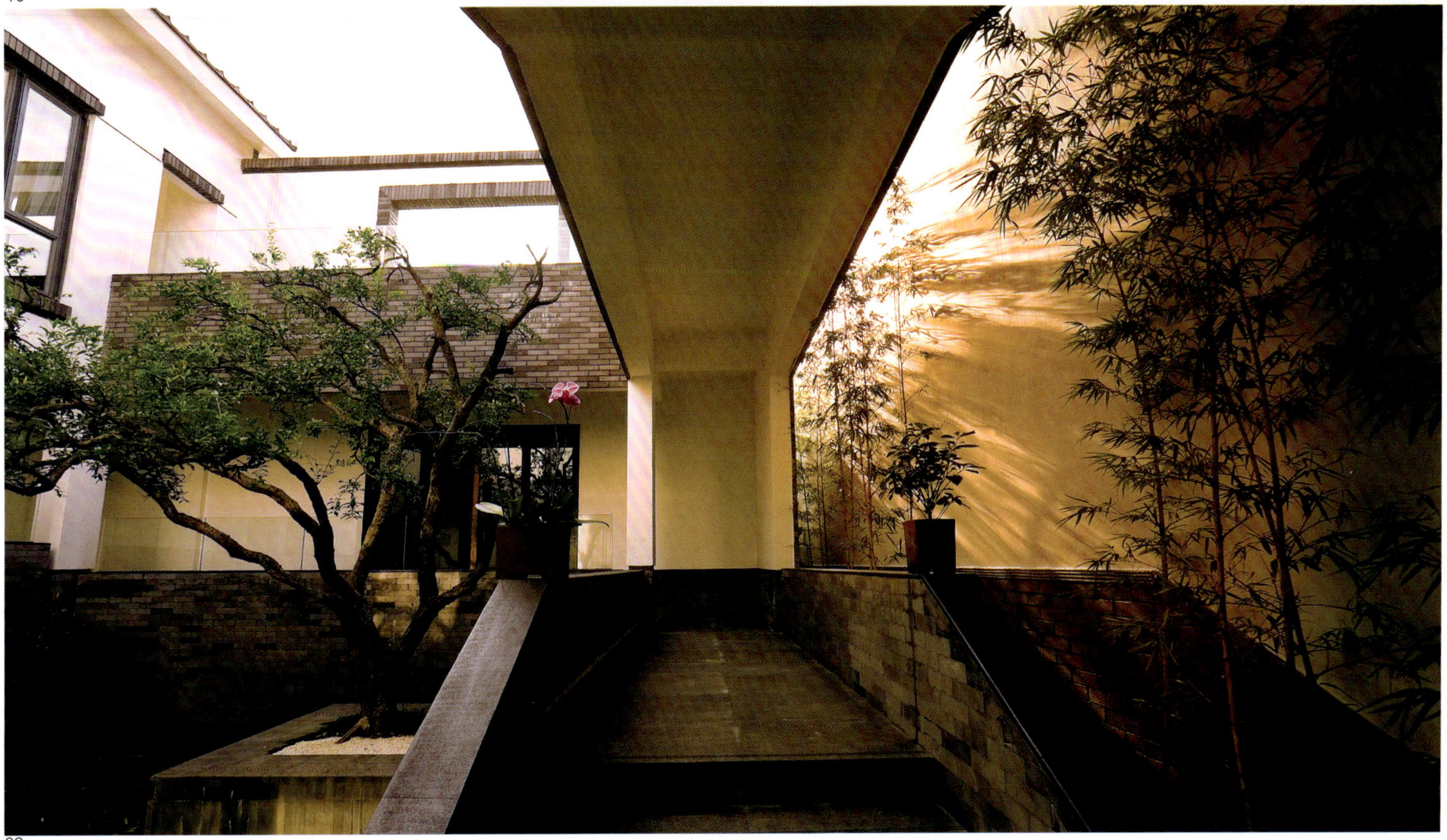
20

1

1/2. A型别墅效果图

3. D型别墅效果图

4. E型别墅效果图

广州圣普拉多

开 发 商：广州银业君瑞房地产开发公司

项目地址：广州花都区芙蓉嶂度假区内银湖湾

占 地 面 积：200 000 m²

建 筑 面 积：140 000 m²

容 积 率：0.43

绿 化 率：30%

总 户 数：320

2

项目区位

广州圣普拉多位于花都狮岭芙蓉嶂度假区内，项目占地20.97公顷，地处广东省省级生态旅游度假区，距花都中心城区15公里，距广州中心城区40公里，广州白云国际机场约17公里，离花都港约15公里，通过度假区前面的山前大道，往东可到106国道、京珠高速公路及街北高速公路；往西可到广花公路、广清高速公路及107国道，与广州及珠江三角洲各地联系极为便利。地块濒临芙蓉嶂水库，南、西、北三面均有开阔的水面，水质优良，周边山体植被茂密，有良好的生态人居环境。

总体规划

整个别墅区结合地块三面环水，沿等高线错落布置，建筑完

全与周边的地形条件融为一体。小区分为五个独立的组团，每个组团只设一个出入口，组团内部道路为环状，组团私密性强，便于管理，有利于给住户营造组团内部的私密性与安全性气氛。而且别墅的排列以最大限度利用湖景资源和小区开阔景观为原则，其中高档别墅尽量临水布置。竖向设计上因地制宜，合理利用地形，在竖向设计中以就地平衡土方为目标，进行场地设计，并设计台地式的建筑布局，使更多的住户足不出户就能欣赏到美丽的湖景。

交通体系

区内机动车交通采用"小区一组团"二级道路系统，通过环路式联系，交通效率高，组团私密性强。小区车行出入口位于

基地临城市道路的南北两端。南端的车行道出入口是小区的主要景观出入口，该处采用跨水景观桥的形式引人入胜的入口景观空间。刚由芙蓉度假村大门进入渡假的人们均会被这个由水、山茂密的树林组成的小区入口空间吸引。

采用有效的人、车分流，尽可能地避开小区内机动车交通，并结合小区内滨水公共空间，中场广场，林荫散步道及组团庭院并结合地开高差形成景观，变化丰富强调外部优质景观（例如湖水，树林，山体）与步行系统，绿化体系及公建设施的相互渗透和有机结合。

景观体系

景观主轴：由东面的酒店会所开始，依次布置了商业步行街，中心广场，叠水林荫广场和湖心小岛，一直向开阔的湖面延伸，形成收放有序的景观序列。

景观次轴：由一系列的景观廊道组成，其间布置各种小广场，雕塑，叠水溪流，凉亭，小桥等小品，起到了分隔各个别墅组团，丰富景观层次的作用。

湖边开放空间：沿湖边设置了亲水木栈道和各类休闲设施，如座椅，健身设施等，使业主能近距离享受优美的自然景观。

组团景观：通过为各个别墅组团布置各种特色植物造景，赋予其不同内涵和意境，提升小区的品味和质素。

建筑设计

圣普拉多的建筑风格源于西班牙建筑的古朴，典雅，风格给人感觉亲切，不张扬。建筑立面的精彩来自于经典的比例以及丰富的细节。建筑立面的各种比例都是经过反复推敲和论证，对屋顶的坡度，檐口的装饰，开窗的比例等外观细节都做出了精心的设计。建筑立面中对西班牙建筑的建筑元素的运用，将芙蓉嶂的自然山水环境和人文因素在建筑中反应，使圣普拉多成为富于创新的艺术作品。

3

5

6

7

01 主入口 MAINENTRENCE
02 入口中心水景 ENTRENCE MIDDLE WATER
03 岗亭 GRANDE HOUSE
04 双亭 PAVILIONS
05 景墙 LANTERN WALL
06 喷泉跌水 AERETED JET
07 景观系列跌水 FEATURE WATER
08 木栈道 WOODEN DECK
09 特色花钵 FEATURE FROWER FOT
10 木平台 WOOD DECK
11 特色构架 FEATURE STRUCTURE
12 溪流 BOURN
13 雕塑小品 SCULPTURE
14 景观水体 FEAT. WATER
15 休憩场地 RESTING PLACE
16 特色台地广场 FEAT. ROTUNDA
17 观景亭 FEAT. PAVILION
18 景观花架 FEAT. TRELLIS
19 景观步道 FEAT. DECK
20 滨水平台 VIEW WATER DECK
21 特色喷泉 FEAT. AERETED
22 特色水景 FEAT. WATER
23 柱廊 PORTICO
24 观景平台 VIEW DECK
25 景观桥 FEAT. BRIDGE
26 次入口 ENTRANCE
27 隐性车道 KECESSIVE ROAD
28 遮阳伞 UMBRELLAS
29 码头 DOCK

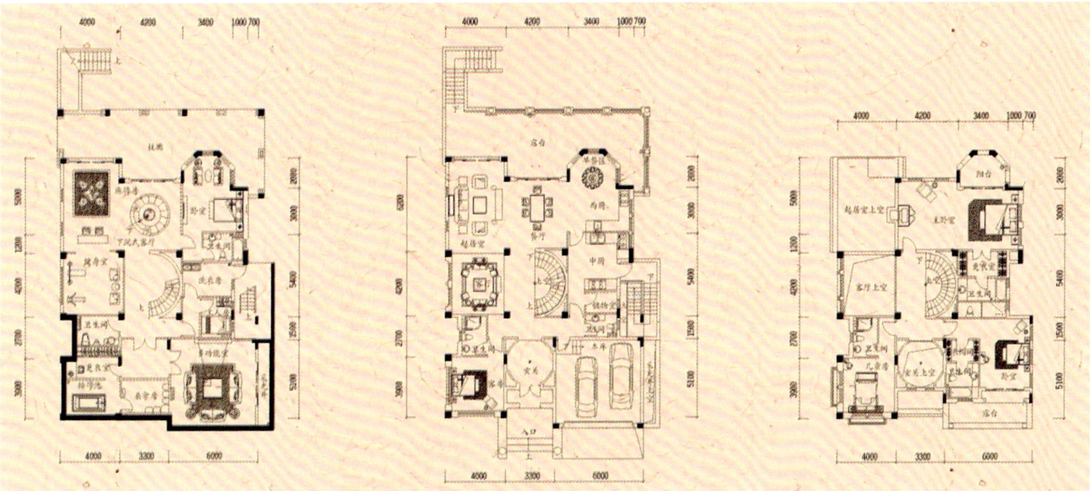

8

9

5. B型别墅效果图

6. B型别墅平面图

7. 总平面图

8. C型别墅平面图

9. 入口实景图

Project location: Yinhu Bay, Furongzhang Vacation Area, Huadu District, Guangzhou

Floor area: 200 000 m^2

Building areal: 140 000 m^2

Plot ratio: 0.43

Afforesting rateL: 30%

Total units: 320

Project location

Guangzhou Sent Prado is located inside the Furongzhang Vacation Area in Lion Hill, Huadu District, covering an area of 20.97 hectare. It is in the province-class ecological tourist and holiday resort, 15 kilometers away from Huadu's central area, 40 kilometers away from Guangzhou's central area, about 17 kilometers away from Baiyun airport and about 15 kilometers away from the Huadu Harbor. It gets a convenient connection to Guangzhou and Pearl River Delta. By crossing through the mountain main road in front of the vacation area, you can get to State Highway 106, Beijing-Zhuhai Expressway and Jiebei Expressway in the east, Guangzhou-Huadu Road, Guang-Qing Expressway and State Highway 107. The section is beside the Furongzhang Reservoir, facing broad water views in the direction of south, west and north, so it has good quality water source with dense forest around and fine ecological living environment.

10

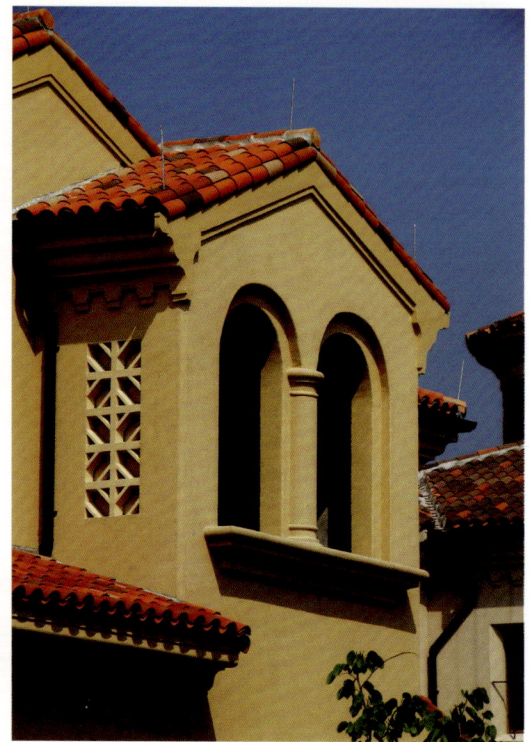
11

13

Overall plan

The whole villa area is arrayed along the contour lines, combined with its location facing waters in three directions, so the construction is totally integrated with its surrounding terrain. The residential quarter is divided into 5 individual blocks, and one entrance and exit is set for each block. Inside the block the road is in ring shapes, good for residents' privacy and safety and easy for management. Besides, the array of the villas is based on the principle of the maximum use of the lake feature resources and the broad landscape of the residential quarter, among those high-class villas are arranged alongside the water as much as possible. In terms of vertical design, the terrain is reasonably made use of, aiming to balance the land and to design the field with terrace layout. By doing so, more residents can enjoy beautiful lake features without stepping outside their houses.

Transportation system

In the residential quarter, motor vehicles are subject to second-grade road system of "residential quarter- blocks" to get connected with the ring roads. Entrance and exit for cars is the north and south ends of the base near the city roads. The south one is the main landscape entrance and exit of the residential quarter which introduces magnificent landscape space with over-water brides. People who enter the gate of the Furongzhang Vacation Village will all be attracted by this residential quarter full of green water and dense trees.

System of "pedestrians and vehicles being apart" is effectively adopted to avoid motor traffic inside the residential quarter. It is also combined with public water space, the middle plaza, the avenues and the block courtyards as well as the elevation difference to form its landscapes. Changing excellent external landscapes (such as the lake, the trees, the hills) and walking system, the greening system and the public facilities are organic combination.

Landscape system

Main landscape axis: starting from the hotels and chambers in the east, business walking streets, central plazas, water and tree plazas and lake-center-islands are arranged orderly, extending to the broad lake.

Sub landscape axis: it consist of a series of feature corridors with all kinds of small plaza and small ornaments arranged at intervals such as sculptures, streams, bowers, and footbridges and so on. They segregate each villa blocks, making them in abundant landscape layers.

Open space in the lakeside: water trestles are set in the lakeside with all kinds of leisure facilities such as chairs, fitting facilities, etc. so the

14

owners can enjoy natural landscape in close distance.

Block landscapes: by decorating each villa block with special plants and giving them different cultures and poetic imagery, the residential is promoted in its taste and quality.

Architectural design

Architectural style of Sent Prado is originated from the classic and elegant Spanish construction, making people feel friendly but not exaggerated. Bright spot of the construction elevation comes from the classic proportion and its abundant details after many times of test. Slope of the pitches, decoration for the eaves and proportion of the windows are all elaborately designed. Finally, the use of the Spanish construction elements in the architectural elevation and action of natural environment and cultural factors of Furongzhang has made Sent Proad an original word of art.

15/16/17/18/19. 别墅实景图

15

16

17

18

19

20

20. 宅间绿化与廊架
21. 丰富的立面造型
22. 底层骑楼休闲区

21

22

1. 别墅立面
2. 别墅局部
3/4. 别墅外观

长春万科兰乔圣菲

开 发 商：长春万科房地产开发有限公司
项目地址：长春市净月大街5399号
建筑设计：加拿大AEL建筑景观设计有限公司
占地面积：130 000 m²
绿化率：45%
容积率：0.65

万科·兰乔圣菲，得名于美国最富有的小镇Rancho Santa Fe，Rancho Santa Fe位于南加州镇圣迭戈北部，拥有500余年贵族传统，是2002年评出的美国最富有的小镇。小镇以其美丽的风光、一年320天普照的阳光吸引了包括比尔·盖茨，文莱·苏丹这样大名鼎鼎的世界首富比邻而居。2006年，万科地产秉持20余年对中国人居的执著与用心，在长春的净月潭打造万科·兰乔圣菲，为这个城市的名流巨贾提供顶级的高尚生活。

由万科地产开发的长春万科兰乔圣菲位于净月潭国家森林公园正门对面，项目占地面积13万平方米。该项目在上海兰乔圣菲

规划风格的引领下，结合北方气候环境和生活习惯的特殊性，注重采光、通风、保暖等细节的因地制宜，从而打造一座洋溢着加州风情的北方式社区。

长春万科兰乔公寓，是集结万科地产行迹二十余座城市开发经验的臻品之作，是长春万科首度呈献的全小户型住宅。万科兰乔公寓共规划出10栋建筑，面积区间在60至95平方米之间，它成功汲取西班牙艺术精髓，特别结合长春特有的地域人文特性，以精巧、灵动的艺术气质见长，突显出现代化、国际化的和谐人居理念，极具宝贵的收藏价值。

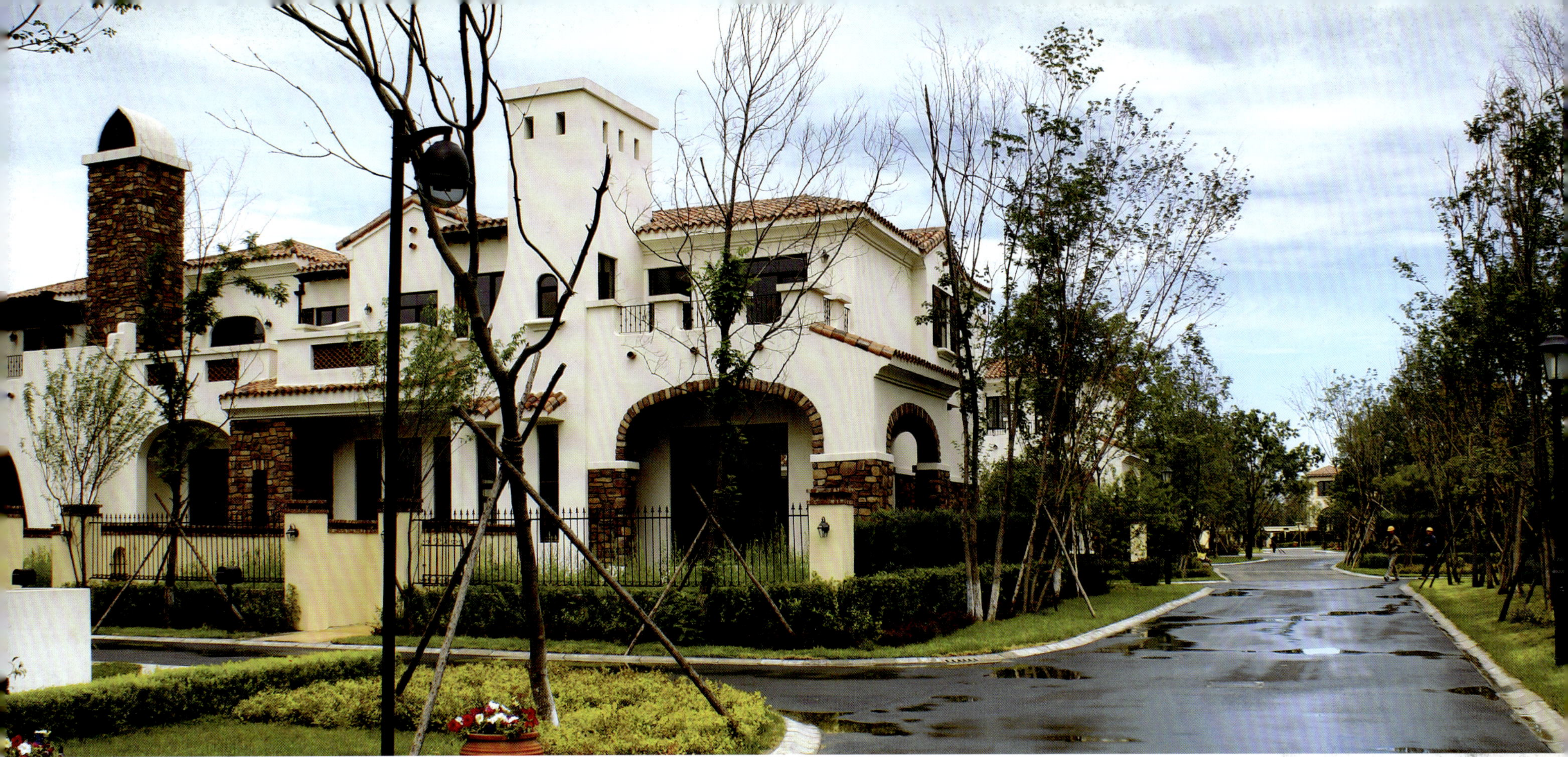
3

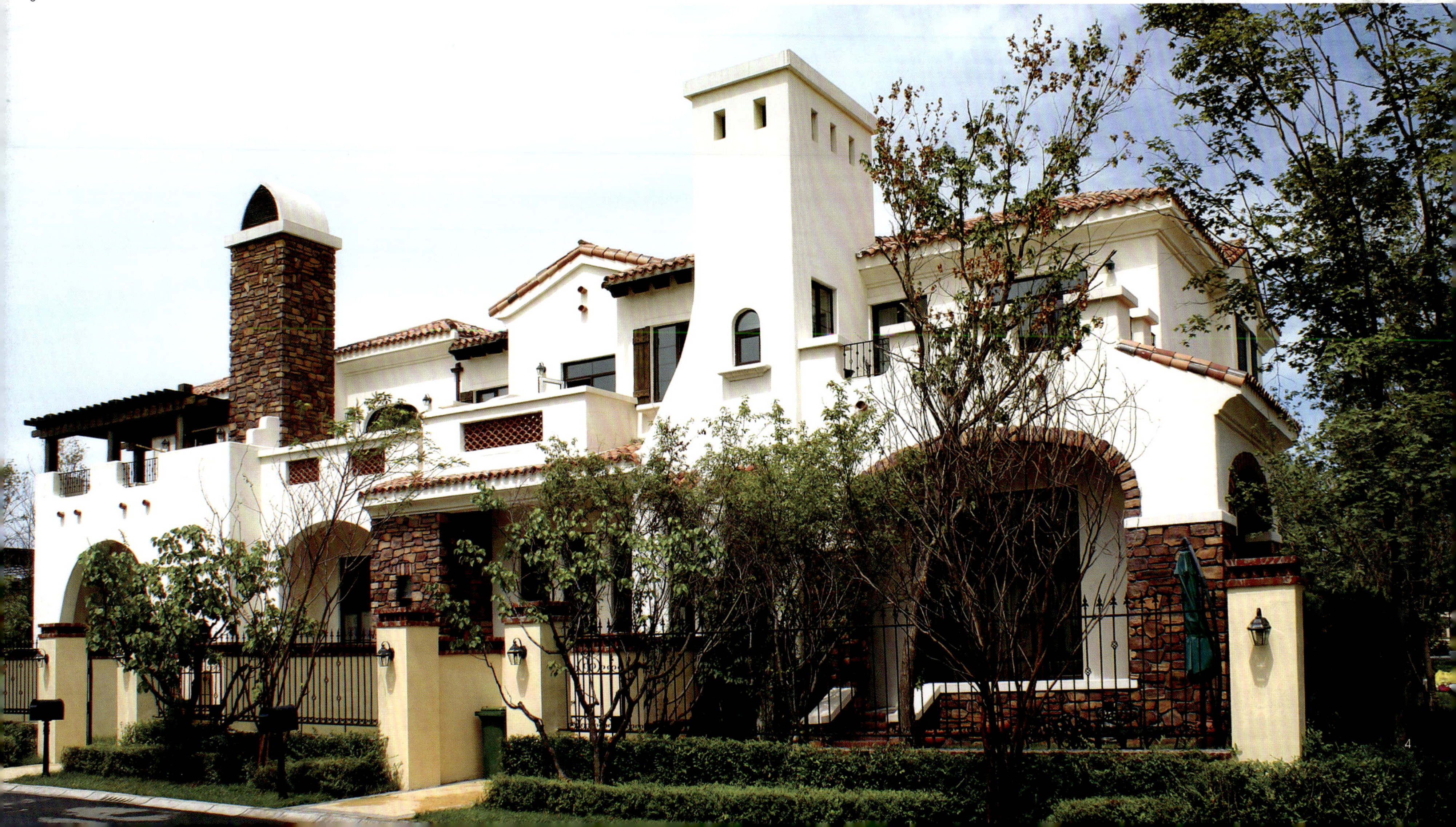
4

Project address: 5399, Jingyue Street, Changchun

Developer: Changchun Vanke Real Estate Development Co., Ltd.

Architectural design: Canada AEL Architectural Landscape Design Co., Ltd.

Floor area: 130 000 m²

Afforesting rate: 45%

Plot ratio: 0.65

Vanke•Rancho Santa Fe is named after a small town in the north San Diego in southern California which is called Rancho Santa Fe, selected in 2002 as the richest town in America.. With its beautiful scenery and sunshine covered for 320 days a year, the small town has attracted world-famous richest men such as Bill Gates and The Sultan of Brunei to be here as its neighbors. In 2006, by keeping its obsession and care for China residents along these 20 years, Vanke Real Estate has built Vanke•Rancho Santa Fe in Jingyuetan in Changchun to provided for celebrities and rich people of the city with noble life.

Changchun Vanke Rancho Santa Fe, developed by Vanke Real Estate, is located right opposite the Jingyuetan National Forest Park, covering an area of 130000m². Led by the planning style of Shanghai Rancho Santa Fe, the land of the project is made full used of with considerations of special northern climate environment and lifestyle, so details like lighting, ventilation and warm are paid attention to, a northern community of California style is therefore built.

Changchun Vanke Rancho Apartment, a best work gathering Vanke Real Estate's developing experience covered over 20 cities, is a total medium housing type residence presented Changchun Vanke for the first time. In Vanke Rancho Apartment there are 10 construction planned in total, with areas of 60 to 95 square meters. The apartment is of most precious treasure value because of its combination of essence taken from Spanish arts and particular area and culture feature of Changchun, showing its ingenious article temperament and harmonious living concept of modernization and internationalization.

5/6/7. 别墅外观

8. 别墅组团外景

5

6

7

8

9

9/10. 兰乔公寓外观

10

1

沈阳奥林匹克花园三期

开　发　商：沈阳奥林匹克置业投资有限公司
项目地址：沈阳市东陵区双园路269号
建筑设计：加拿大AEL建筑景观设计有限公司
占地面积：约280 000 m²
总建筑面积：约180 000 m²
容　积　率：0.79
绿　化　率：43%

2

漫步地中海是奥林匹克花园三期的产品，堪称是奥园的巅峰之作。

社区规模

占地面积约28万平方米、总建筑面积约18万平方米，是棋盘山区域规模最大的别墅项目。规划了约1000多户的产品，将来人口会达到3000人。

社区特色

社区最大的特点就是整体规划是建立在有坡度起伏的山地上的，地势最大高差约20多米！绿化、水景和建筑融为一体呈现出海湾式的社区形象，这样不仅使得每栋建筑都能够最大程度享受到日照，也完全打破了传统房地产项目呆板的规划格局，视觉效果非常好。

社区另外一个重要的特点就是景观资源优势非常强！社区南侧紧邻沈阳母亲河浑河，西侧有80米宽的绿化带环绕，这种得天独厚的景观资源是一般别墅项目无法拥有的。整个社区的生态环境先天条件就很优越，再加上40%以上的超高绿化率，整个社区就是一个天然的大氧吧。

建筑布局规划

社区地形纵向可以分东部、中部和西部三个区。以中央景观带为主轴向两侧呈扇形伸展。东部的坡地地貌特征很明显，景观视线非常好，在这里采用自由散落的组团形式规划了独栋产品。西侧和北侧设计了低密度花园洋房产品。整个社区中央则设计了连排别墅产品。

西班牙建筑风格特色

建筑外立面颜色是浅色调，设计用浅黄色来表达西班牙建筑阳光、充满活力的个性，明快而不张扬，非常自然踏实，不会像英式、德式建筑采用深色外立面给人的感觉那么沉重，而且西班牙风格也是目前国内最为流行的别墅风格。

双庭院设计，一个是入户庭院，一个是回廊围成的家庭庭院。入户庭院是设计传统意义上的庭院，主要强调的是会客功能，比如家里有客人来拜访的话，可以在这里烧烤、聊天，体现对客户的尊敬也能展示您的尊贵身份；而家庭庭院是户外到室内的过渡空间，家里人可以在这里乘凉、打牌，其乐融融。

3

4

5

在细节方面，采用红色陶制筒瓦坡屋顶、文化石装饰的外墙、圆弧形的檐口、大量的拱门、铁艺和陶艺的装饰等，也是西班牙建筑的明显标志，非常注重细节表达。

这种风格的建筑施工强调的是给人一种手工、生态的感觉，这个从铁艺的围栏、陶制的瓦片、门窗和外墙弧形的施工工艺都能看出来。而大多数房地产项目都是工业化施工，建筑外立面都是很呆板的直角，很难把细节做得这么出色。

西班牙建筑风格还有另外一个明显特色：创新联排独栋化，联排产品中相邻的两户无论是户型还是外立面都有很大差异，使得整栋建筑看上去有独栋的感觉，品质感很好，这点也是其他风格别墅难以达到的效果。

景观规划

三期的景观规划很有层次，设计景观规划的初衷是打造一个不只是用来欣赏，还可以让所有的居民、所有的男女老少都能够参与进来的有情有景的社区。

社区规划了两条景观带：一条是社区中轴的主要景观带，一条是从商业街主要景观节点处向东西两个方向延展的次要景观带。主要景观带上规划了小区的中央景观区，还将设计两三个主要的景观节点和大面积的公共绿地，这个主要景观带上水景设计

6

比较丰富，包括一处上千平方米的景观湖和会所附近的一条水景带。这条主景观带是提供给居民集中交流的场所，社区居民可以集中在这里散步、聊天，结识更多的朋友。而一般的别墅社区都缺少这样的集中景观设计，缺少活力。

社区里还设计了十多处主要景观节点和次要景观节点，不同的组团也会有不同的景观设计。景观节点处会考虑不同年龄、不同性别居民的生活要求，设置木椅、廊架、石阶、雕塑喷泉水系、矮墙、小桥等元素，使得所有的居住者都能享受到和谐的户外生活。比如小孩子可能就喜欢在开敞平坦的环境里游戏，比较安全；老人就喜欢能够坐在椅子上一起聊聊天；年轻人可能就愿意在小桥流水的地方散步等等。结合景观绿化，设计还布置了慢跑道、山地自行车道等健身设施，人们可以在社区里一边享受自然的景色一边锻炼身体，养成健康的生活习惯。

绿化

社区景观绿化考虑了北方的气候特点，在不同的组团会选择不同的灌木、乔木、色叶色干植物、常绿植物、草坪、宿根草花、应季草花等不同的植被搭配，达到遮荫、抗污染、减少噪音和尘土的功能，而且形成四季皆有景、景景不同、移步异景的视觉效果。

7

8

9

6/7. 独栋别墅效果

8. 合院效果

9. 商业区ABC区立面

10. 商业区D区立面

Project location: 269 Shuangyuan Road, Donglin District, Shenyang

Developer: Shenyang Olympic Property Investment Co., Ltd.

Architectural design: Canada AEL Architectural Landscape Design Co., Ltd.

Floor space: About 280 000 m²

Gross Building area: About 180 000 m²

Plot ratio: 0.79

Afforesting rate: 43%

A Walk on the Mediterranean is a pinnacle product of Olympic Garden Phase 3.

Scale of the community:

It is the largest villa project in area of Qipan Mountain, covering an area of 280,000 square meters and its gross building area is 180,000 square meters. The product is designed to contain about 1,000 houses with a prospect population of 3,000.

10

11

12

13

11. 会所和连排实景图
12. 会所南立面效果图
13. 会所
14. 多层效果图
15. 会所和多层实景图

14

15

Features of the community:

The best feature of the community is that it is generally constructed on an undulated slope on the hill with a maximum relief distance of about 20 meters! Green lands, water features and the architectures are integrated as a whole, turning on a bay community image, not only to enable every building to enjoy maximum sunshine but also to break through the inflexible traditional real estate's layout totally and make good visual effect.

Another important feature is that the community gets more advantageous landscape resources than other villa projects do. The community is right next to the mother river of Shenyang, the Hehui River, in the south, and is surrounded by an 80-meter-long greening belt in the west. The whole community is like a large natural oxygen bar with its excellent ecological environment and an extremely high afforesting rate of over 40%.

Architectural layout:

The terrain of the community can be lengthways divided into three parts, the eastern one, the middle one and the western one. Taking the central landscape area as a main axis, it spreads fanwise towards two sides. The eastern part has an evident slope feature which provides landscape view and single houses are planned here in groups in a freely dispersed block form. Low-density western garden houses are in the west and north side, and townhouses are in the center of the whole community.

16

17

Spanish architectural style and features:

The outer elevation of the architectures is in light color tones. Pale yellow is used in the design to express sunny character full vitality of Spanish architectures. It's bright, natural and practical but not flaunting or heavy as what the dark elevation have oppressed you in German architectures. It is also the most domestically popular villa style at present.

Design of double courtyards is adopted, one is entry courtyard and another is family courtyard enclosed by circular corridors. The entry courtyards are traditionally designed, focusing on visiting functions, for example, when there is a guest visiting, you can invite him to barbeque or chat here to show your respect for him while showing your elegance. On the other hand, family courtyards is a transitional space from outside into inside, family members can enjoy the shade, play cards in a harmonious atmosphere.

In terms of the details, distinctive marks of Spanish architectures such as red earthenware round tile roofs, exterior walls decorated by culture stones, eaves with arches, numerous arches, ironic crafts and pottery ornament and so on are adopted, paying attention to the detail expression.

An emphasis is laid on by this style and such construction to show a manual and ecological feeling, and that can been found from the ironic fences, pottery tiles, doors and windows, and constructing crafts of the arch-shaped exterior walls. You can hardly find details as excellent as these because nowadays most real estate projects are industrially constructed, making elevations of the construction in a rigid rectangle.

18

19

20

16. 连排实景图
17. 连排全景
18. 立面与阴影
19/20. 不同的立面

In the style there is another evident feature - the innovation of the townhouses' being single, because no matter judged from their house types or seen from the elevation, two neighboring residents in the townhouses are very different from each other, making the whole construction look individual and in good quality, and that's what villas of other styles cannot achieved.

Landscape planning:

The landscape design of phase 3 is stratified primarily to build a lively community not only for watching but also for the participation of all the residents.

There are two landscape belts in the community. One is the main landscape belt of the middle axis, and another is the sub landscape belt extending towards west and east directions from the main landscape nodes of the commercial street. The main landscape belt is planned with the central landscape area of the residential quarter as well as two or three main landscape nodes and large areas of public green lands. It is abundant in water features including one that is over thousand meters' long near the lake and the club. Here is where residents can communicate with each other, take a

walk and make more friends, and such vitality is just what ordinary villa community lacks.

In the community there are also over ten main and sub landscape nodes, and different blocks is differently designed. They are built with wooden chair, verandas, stone benches, sculpture springs, low walls and footbridges, and so on , so as to meet demand of residents at all ages, male and female. All of them can enjoy harmonious life in the open air, for example, children may like to play in wide and flat environment which is safe, the elderly will enjoy sitting on the chairs chatting and young people walk around the places like footbridges and along the waters. Also with the greening landscape, jogging tracks, mountain bicycle lanes help you to form healthy habit, strengthening your physical health while enjoy natural scenery.

Greening:

North climate feature has been taken into consideration of the community's landscape greening, so different plants are matched with different blocks to provide shade, avoid pollution and less noise.

21

22

21. 立面局部
22/23. 别墅效果图
24. 立面造型的变化
25. 细部

23

24

25

上海·绿城玫瑰园

开　发　商：上海绿城森林高尔夫别墅开发有限公司

项目地址：上海西南腹地的闵行区

占地面积：800 000 m²

总建筑面积：约220 000 m²

容　积　率：0.175

1. 光影
2. 入口
3/4. 带泳池的别墅设计

3

4

Total land–using area: 800 000 m²
Gross floor area: about 220 000 m²
Construction volume rate: 0.175

Shanghai Rosary is in No.1555, Qizhong Road, which is the core of Qizhong Forest Sports Center, Maqiao Town, Minxing District, Shanghai City.

Its design inspiration of layout and architecture came from the twenties and thirties' shanghai old western-style house, whose main conceive is contructing a block image which is made up of tall trees, high walls and bushes; piecing out water system, lakes ,rivers and plants form a country image. A mixture of pretty calm and elegant compromise architectural symbol, courtyards in different styles keep in harmony with the architecture. Details such as artistic creations have strong decorative flavor, floor, swimming pool, cast iron balustrade. Truly aligning with world class villa community, these elements are in a harmony composing a perfect impression of Rrosary.

According to the design of layout of imago about Shanghai old western-style house, styles concerned with Shanghai .Green Town rosary are French, English, and Italian and so on. Taking the end of the twenties' Western high-class community as based on the beginning of twenties' old western-style house, this case discover those design ideas that long vital principle community will follow. Such kind of villa community in Shanghai becomes a Culture of area mapping, a decision or creation that made between the comparison of Oriental and Western. Shanghai Green town rosary tries his best to become the first synthetically quality in high-tech villa product, become the marking pin of the third generation villa in Green Town.

5/6/7. 与环境的结合

6

7

8/9. 立面构成

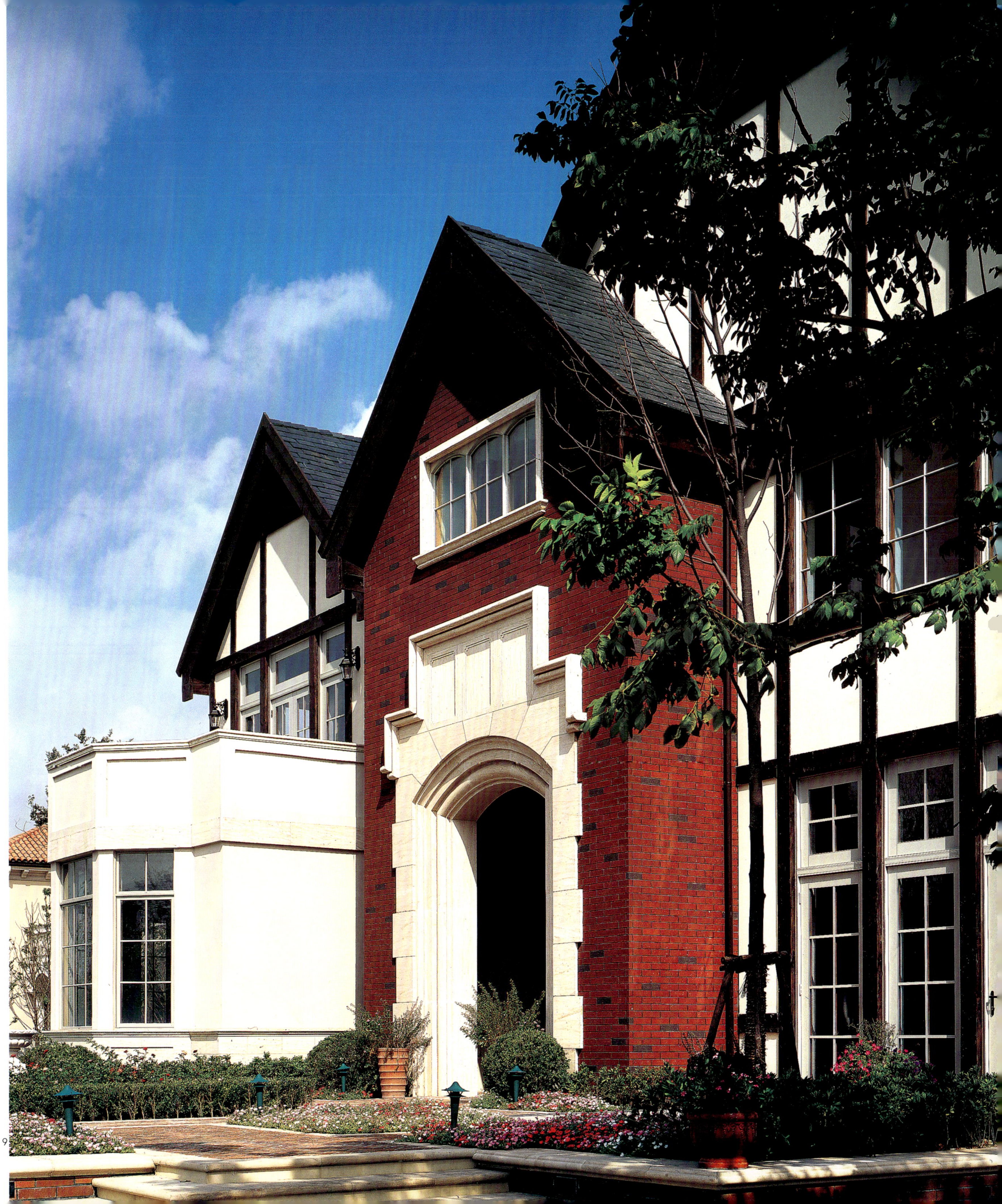

10

11

10/11/12/13. 细部与造型

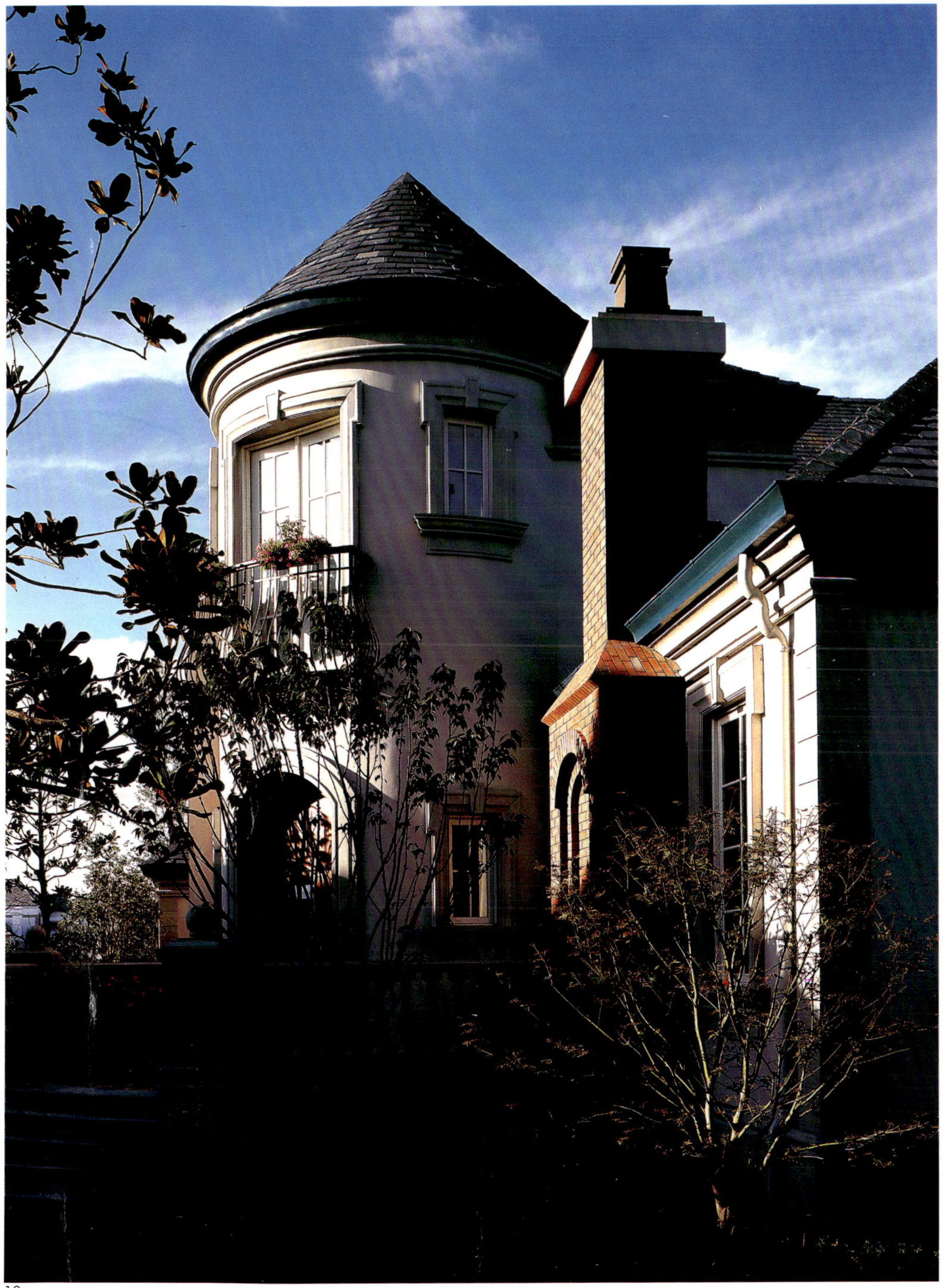
12

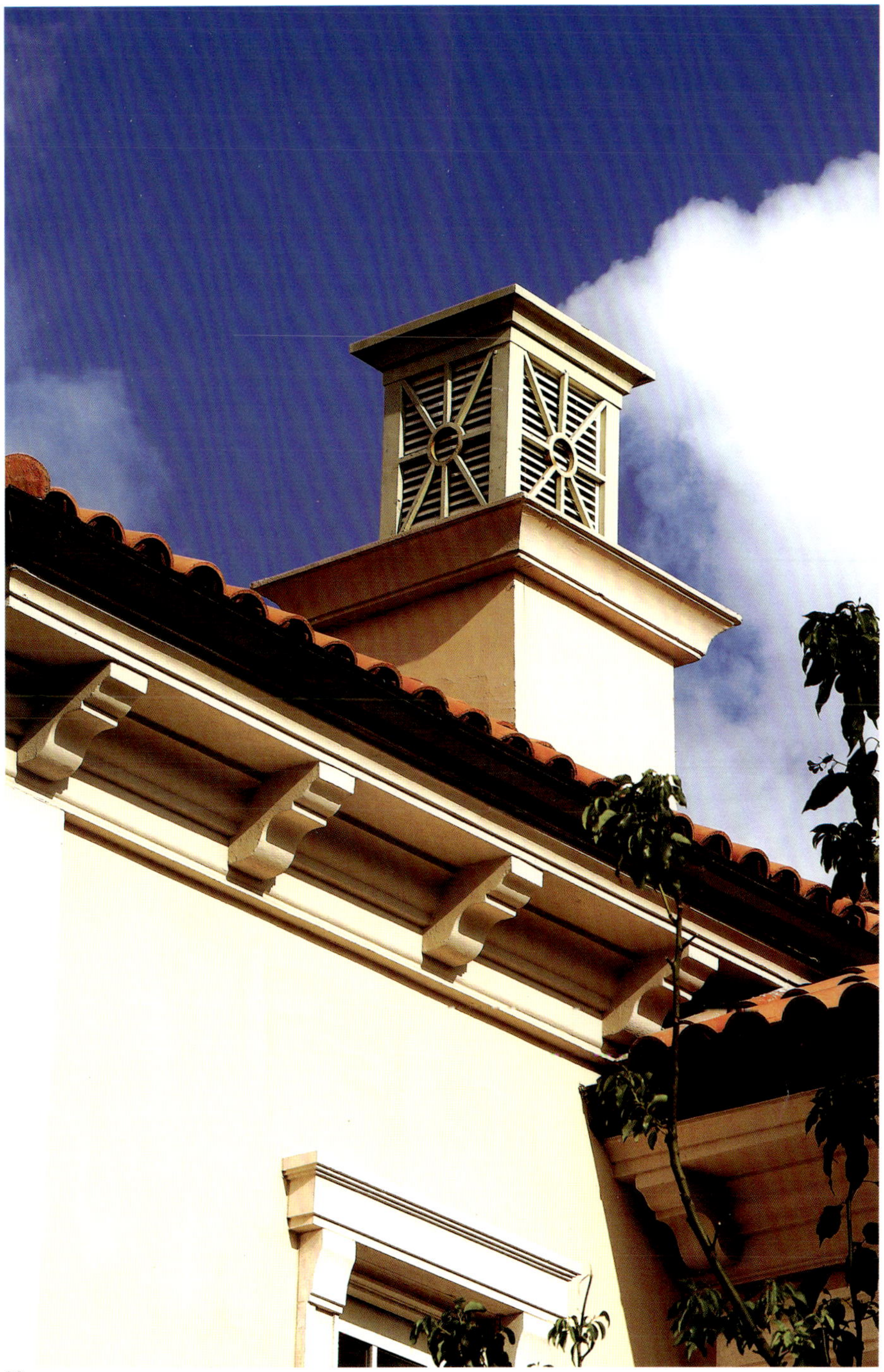
13

1

1. 别墅实景图
2. 会所
3. 鸟瞰图
4. 别墅内街
5. 围墙围合的通道考虑私密性
6. 庭院
7. 别墅透视

2

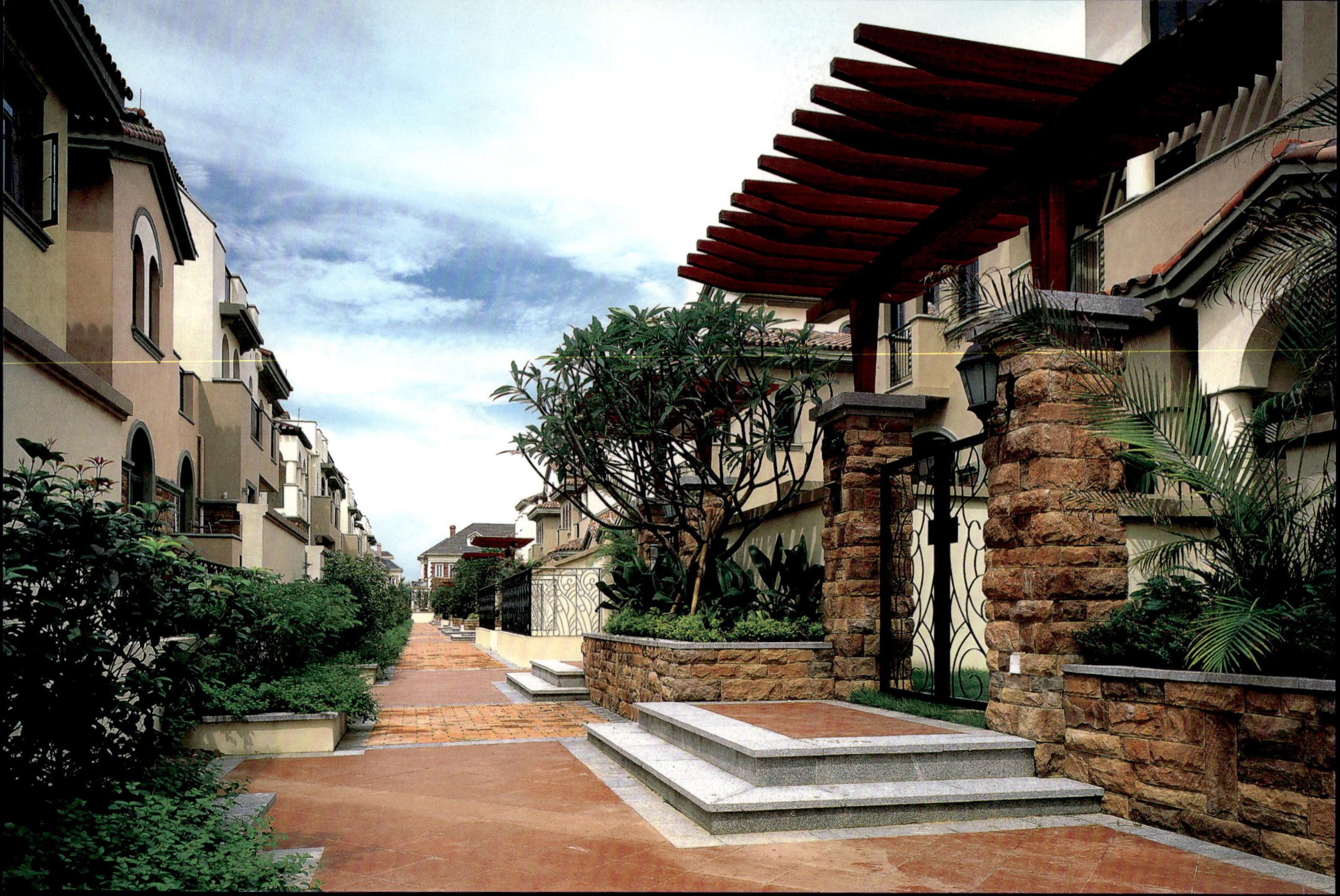

佛山万科城

开　　发　　商：佛山市万科投资有限公司
项　目　地　址：佛山市禅城区沿江路55号
建　筑　设　计：许李严建筑师有限公司
总 占 地 面 积：约380 000 m^2
住宅总建筑面积：约800 000 m^2
容　　积　　率：2.1
绿　　化　　率：35%
总　　户　　数：约7000户

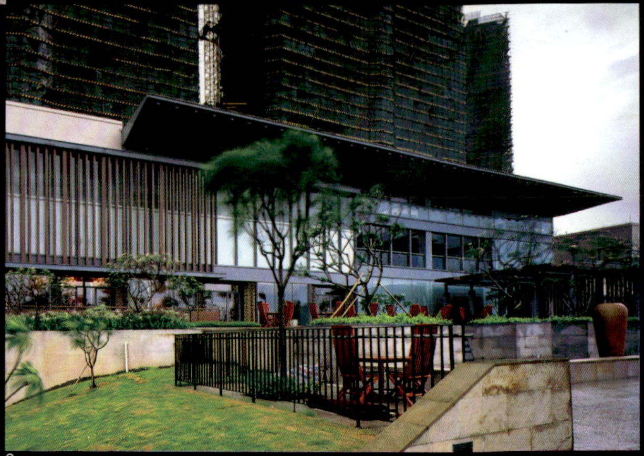

水是孕育一切生命的源泉。逐水而居，亲水、乐水，是人类始终不渝的追求。古有“智者乐山，仁者乐水”的生活理想，今有“回归自然，栖于水岸”的居住梦想。水，不仅有着调节气候，净化环境的物理功效，更有着修身养性，赋予人智慧的文化功能，有水的地方就有无限的生活意境。但凡江海湖畔，必是名流汇聚、名宅所据之地，举世闻名的世界住宅区，大多因其自然水景资源而弥足珍贵。

美国西雅图的华盛顿湖畔，日本的东京湾，法国的塞纳河畔，澳洲的悉尼玫瑰湾，香港的浅水湾，深圳香蜜湖等等都是城市贵族向往的居住地。水岸，是一座城市最珍贵的资产。

佛山禅城西南部片区，是广东佛山禅城经济开发区所在地，因其河网密布，水系发达，生态景观资源丰富，被政府规划为大佛山5+2中心组团的一座滨水新城。定位为以岭南水乡景观为特色，以休闲和居住功能为主的中心组团生态新区。

生态新区规划为“三个核心，两条长廊，双条轴线，五个功能区”。即生态绿心、景观核心、商业中心与东平河沿岸、佛山一环南北向两天生态绿化廊道，季华路和禅港路两条城市干道为轴线的商贸轴线和生活轴线，以及沿江居住片区、中央休闲片区和商贸片区等五个功能区。区域内还将建公益性市政公园，形成以水系、绿地和景观道路为骨架的大型生态滨河公园，有效优化居住环境。

万科在禅桂中心组团西南方向，西樵山下、北江、吉利河畔、顺德水道三江汇聚之处，成功竟得一块山水宝地，并开始以其23年水岸社区经验之大成，倾力打造比肩全球的江岸社区万科城。有专业人士指出，在佛山“5＋2”蓝图下，此地块将是佛山极罕有的上风上水之地，占据着佛山中心组团的山水骄傲，未来居住价值不可限量。

佛山万科城，以罕有80万平方米建筑规模，缔造一个纯粹高尚生活的城邦，开启大佛山居住时代。据了解，为了给居住者营造一个国际化、高品质的生活空间，营造建筑与景观、人与自然和谐共生的生活氛围。万科城不惜耗巨资为业主建造一座4000平方米的休闲会所，配置以健身房、旗牌室、餐饮、咖啡厅、茶座等高档休闲、运动、娱乐设施，让业主充分体验健康、快乐的生活气息。同时，项目还规划有600米长的商业长廊、750平方米室外天际泳池等配套，丰富了多姿多彩的社区生活。

在景观设计方面，除了社区内部的大面积西班牙风情休闲园林之外，万科城还在约1600米黄金江岸上修建了一个约5万平方米的原生态休闲滨江公园及一个约3万平方米江畔广场，置身于此，近可看吉利涌的风情，远可观壮丽的西樵山风光，尽享大城生活的从容与优越。

3

4

Developer: Foshan investment Vanke Co., Ltd.

Project Address: Chancheng area along the Yangtze River Road 55

Architectural Design: Rocco Architects Ltd.

Total area: about 380 000 m²

The total residential floor area: approximately 800 000 m²

Floor area ratio: 2.1

Green: 35%

The total number: about 7000

Water is the resource of everything. It is always the pursuit of humanity to live by water, close to water and enjoy water. In an ancient time, an ideal life is that "A wise man loves mountains, a kind man loves the sea". However, nowsdays, there is a dream of living that "Return to nature and live on the waterfront". Not only do water have physical effect on adjusting the climate and cleaning up the environment, but also it has a cultural function of cultivating onself and giving people wisdom. There is a place with infinite artistic life where water is in existance. It is the bank of rivers and seas that must be a convergent place of rivers and the place where the famous dwellings in. The world-famous residential areas are valuable because of the resource of natural waterscape.

Lake Washington in Seattle, US, Tokyo Bay in Japan, Banks of the Seine in France, Sydney Rose Bay in Australia, Repulse Bay in Hongkong, Shenzhen Xiangmihu and so on, are all longing for the noble residence in city. Waterfront is a most valuable asset of a city.

Chancheng Area In the southwest of Foshan is the location of the Guangdong Foshan Chancheng Economic Development Zone. Because of the clouds ricer, the developmented water system and the resourceful ecological landscape, the Chancheng Area has been planning to be a New City waterfront in the center of big Foshan 5+2 group. The New City waterfront is positioned as a new ecological area of group center which makes a feature of the Lingnan water landscape and is mainly to leisure and living.

New ecological area was planning for "three core, two promenade, two-axis article and five functional areas". That is, eco-green heart, the core of the landscape, the business center and Tung Ping River, the two promenade of the north and south linking in a two-way eco-green corridor in Foshan, the axis of trade and the axis of life which regards the Jihua Road and the Changang Road as the axis, and the five functional area, such as the inhabited area alonging the river, the central leisure area, business area and so on. A municipal public park which will be formed to be a large scale ecological park with a framework of the water system, green space and landscape road, is going to be built in this area to optimize the living environment effectiely.

Vanke is in the south-west of the Changui group center. It is successful to have this valuable land because it is in the roof of Xiqiao Mountain,and the convergent place of the North River, Lake Geely amd Shunde Flume. It was

5

began with its 23 years experience in waterfront community to build a global community of Jiangnan Vanke Town for its best. Professionals have pointed out that in the blueprint "5+2" of Foshan, it would be a valuable land with a good geomantic omen--the pride of the Foshan group center. The value is unlimited in the future.

Foshan Vanke Town with a construction's scale of rare size of 800,000 square meters, creates a purely noble life of the city, opening the living times. It is said that in order to create an international and high-quality living space, and the atmosphere of constructions and landscapes and human and nature living in harmony, Vnake Town spared no expense to build a leisure clubs with 4,000 square meters for the owner, equipping with the gym, the flag card rooms, restaurants, coffee houses, teahouses and other higher-end leisure, sports and entertainment facilities, allowing the owners to fully experience the healthy and happy life. Meanwhile, the project is also been planning 600-meter-long commercial corridor, 750 square meters outdoor swimming pools and so on to enrich the colorful community life.

In the desige of the landscape, Vanke Town built a Original Ecological Leisure Riverside Park with about 50,000 aquare meters in the golden riverbank with about 1600 meters and a Riverbank Square with about 30,000, besides a large area of Spain casual style garden in the community. Here, you can see Geely Bay's style on a nearer view and the magnificent scenery of Xiqiao Mountain on a farther view, enjoying the calm and superiority of the city life.

6

7

50

8

9

10

11

1

金碧湖畔

开 发 商：北京百顺达房地产开发有限公司
项目地址：北京顺义区顺平路与潮白河交界处
规划设计：新加坡SCDA建筑师事务所
占地面积：332 530 m²
建筑面积：165 359 m²
绿 化 率：45.2%
容 积 率：0.37

1/2. 别墅夜景图
3. 别墅内庭院

2

　　"金碧湖畔"总占地33.25公顷，是北京百顺达房地产开发有限公司2005年扛鼎之作，特别邀请了以"新亚洲"理念著称的新加坡SCDA建筑设计师事务所主导设计，将承载着浓厚的亚洲文化，以现代简约的手法，上演一部"新亚洲主义"豪宅乐章。

　　"新亚洲主义"是指在世界文化交汇、融合的趋势下，设计师们对当代亚洲建筑设计的一种再认识，是以浓厚的具有亚洲地域特色的传统文化为根基，融入现代西方文化，在功能上进行改良，让亚洲优秀的传统文化得以很好发扬，同时更加适合现代人的生活方式。"新亚洲主义"为亚洲、更为世界建筑设计带来了一种风情，一种气息，显现出一种新的东方人的态度、精神，是一种完全创新的居住理念。

　　空间特色："让别墅的形态舒展开来"，是"金碧湖畔"在空间展示上的最大特色之一。"一层半"结构，扩大了建筑占地面积，拥有完全独立的一层坡顶起居室，与家庭生活空间区隔开来；挑空、独立、宽敞、放松的一层社交空间，注重别墅居住的特性，更符合亚洲文化注重的礼仪需求，在现有别墅建筑结构中无疑是一创举。"U"型半围合的结构布局，是东方传统居住文化的一种现代表达，继承了中国传统四合院的居住文化精髓，以三面围合，一面向户外庭院开放的布局，既保留了东方人所注重的围合形式的归属感，也开放了居住的视野，延展了与自然空间的交流互动，使人的心理达至平衡，进入最舒适的状态。组团式布局，邻里间由庭院水景绿化、分段毛石院墙等相隔离，既保持了独栋别墅的私密感，同时强化了景观的整体性与共生性。在内部空间布局上，入室即是通透玄关，玄关落地大窗充分展示中庭风景，视线明亮、通透，完全契合亚洲文化讲求的"明堂净室"的理念。室内连廊设计的视野延伸至国际流行的东方禅味，强调"幽而不冷"，讲究线条的简约、质朴，光与影的交织，带动建筑的流线，步移景易，感受庭院景观的层次变化。

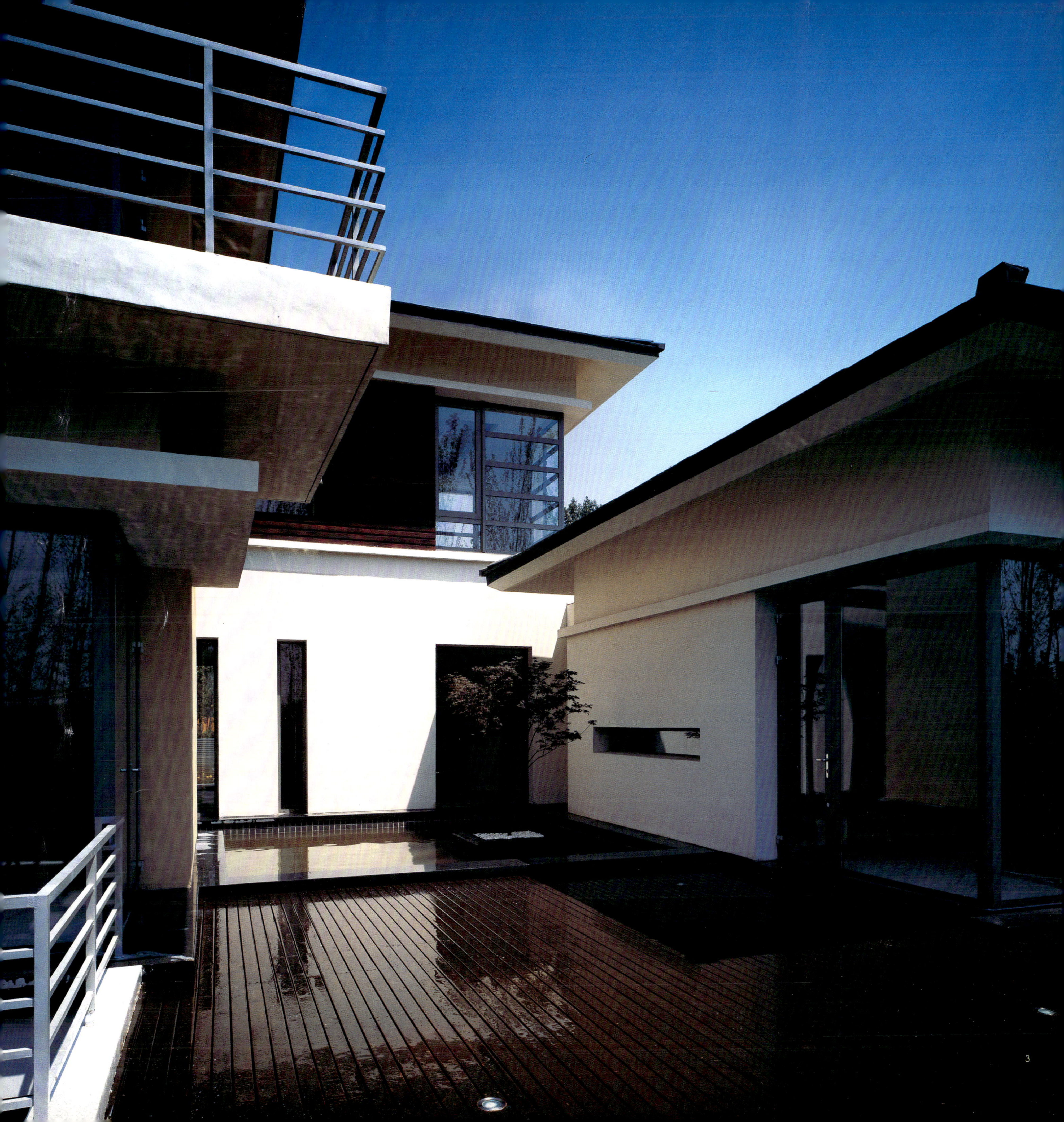

3

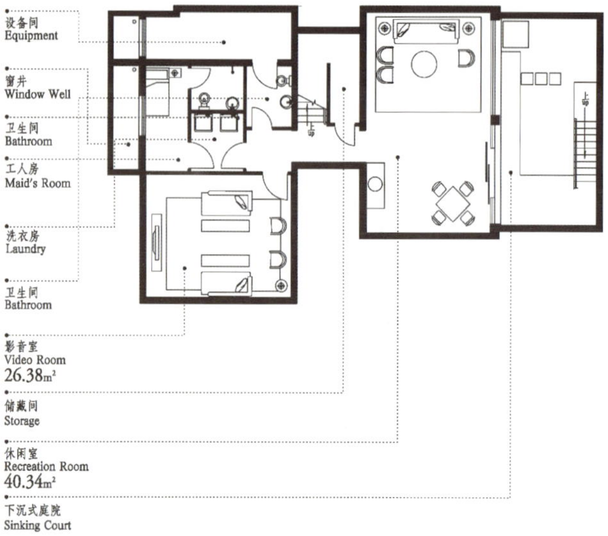

Plan 4 labels (地下一层平面图):

设备间 Equipment

窗井 Window Well

卫生间 Bathroom

工人房 Maid's Room

洗衣房 Laundry

卫生间 Bathroom

影音室 Video Room 26.38m²

储藏间 Storage

休闲室 Recreation Room 40.34m²

下沉式庭院 Sinking Court

Plan 5 labels (一层平面图):

西厨 Open Kitchen

中厨 Chinese Kitchen

餐厅 Dining Room 51.43m²

家庭室 Family Room 17.50m²

卫生间 Bathroom

客房 Guest Bedroom 15.53m²

卫生间 Bathroom

起居室 Living Room 31.25m²

车库 2-Car Garage 31.18m²

水池 Fountain

花池 Flower

下 down 上 up

内庭院 Inside Court

玄关 Foyer

别墅车库入口 up　　别墅入口 door

Plan 6 labels (二层平面图):

次卫 Guest Bathroom 5.40m²

主卫 Master Bathroom 20.53m²

衣帽间 Dressing Room 6.15m²

卧室 Bedroom 15.05m²

卧室 Bedroom 13.49m²

主卧室 Master Bedroom 29.66m²

阳台 Balcony

屋面 House Surface

down

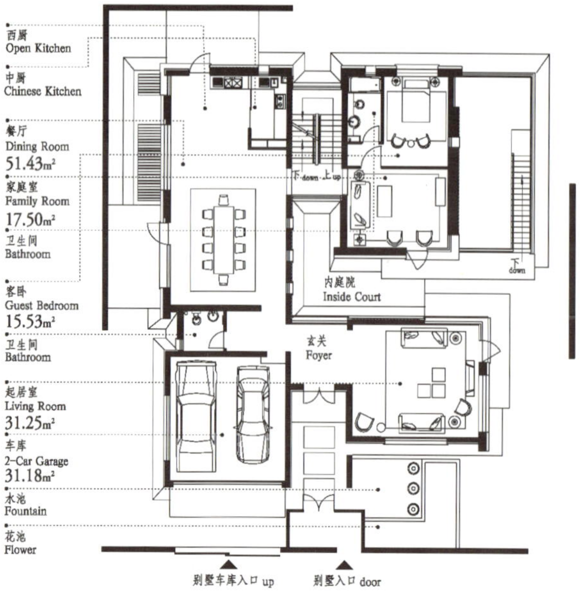

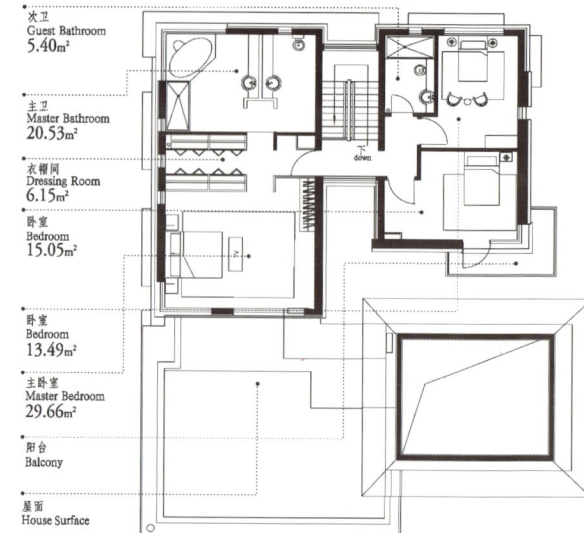

5

6

庭院特色：在东方，比如中国、韩国、日本等亚洲国家的传统建筑都具有与社会文化、周边自然环境相协调的布局和形式。讲究正面开阔、具有前庭、内院、后院的庭院空间，以求与自然共享空间。"金碧湖畔"沿袭了东方多重庭院的设计，开放式前庭，舒展明朗，与院子相连，有独立进出厨房的入口；水景庭院，与餐厅相连相通，使之成为观景餐厅，赏心悦目；围合式内庭院，形成别墅的风景核心地带，视线通达起居室、家庭室、餐厅、玄关，加强了家人之间的交流与沟通，使天伦之乐融于每一个角落；下沉式庭院，扩大采光面和活动区域。"金碧湖畔"的庭院设计不仅实现了功能需求，更注重东方对于意境的崇尚，体现"天人合一"的审美需求与精神满足。

园林景观：东方的园林自始受到道家的影响，强调与自然和谐，空间讲求平衡、对称感。"金碧湖畔"在园林景观环境的规划方面，注重与自然生态环境的和谐共生，因而人和建筑都成为自然的组成部分。双湖是"金碧湖畔"的中心景点，分布于南北两区，各具特色，以美丽的自然绿化和易于亲近而闻名。湖畔的细部强调公园式设计，与其相邻的有步行道、亲水台阶和延伸到水上的休息区，为人们提供交流、运动及休憩的空间。

4. 地下一层平面图

5. 一层平面图

6. 二层平面图

7. 别墅效果图

8. 地下一层平面图

9. 一层平面图

10. 二层平面图

7

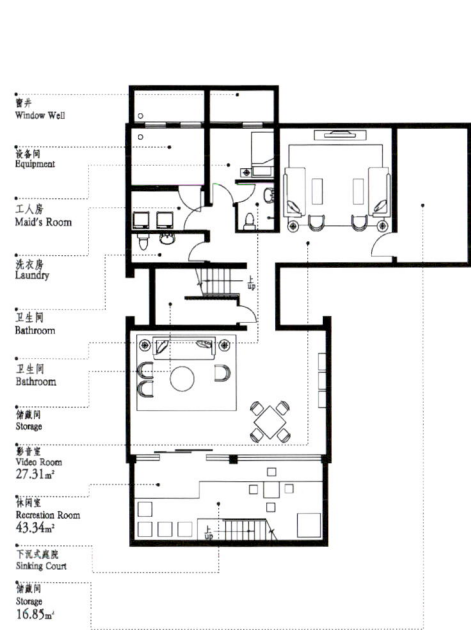

窗井
Window Well

设备间
Equipment

工人房
Maid's Room

洗衣房
Laundry

卫生间
Bathroom

卫生间
Bathroom

储藏间
Storage

影音室
Video Room
27.31m²

休闲室
Recreation Room
43.34m²

下沉式庭院
Sinking Court

储藏间
Storage
16.85m²

8

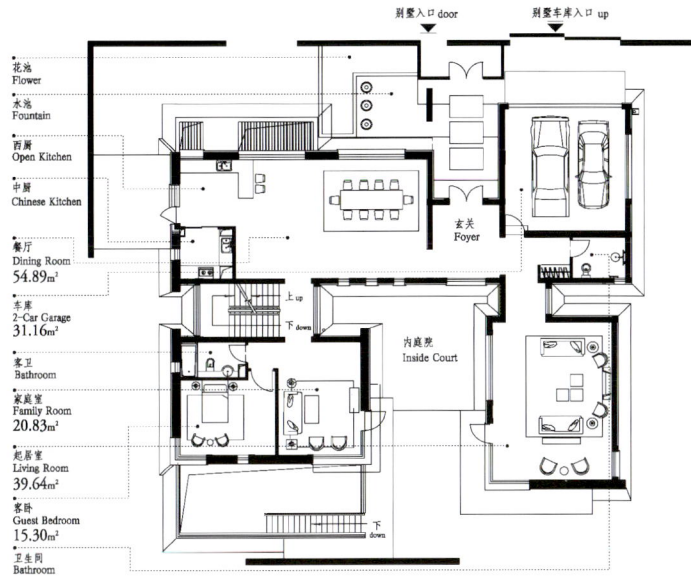

别墅入口 door　　别墅车库入口 up

花地
Flower

水池
Fountain

西厨
Open Kitchen

中厨
Chinese Kitchen

餐厅
Dining Room
54.89m²

车库
2-Car Garage
31.16m²

客卫
Bathroom

家庭室
Family Room
20.83m²

起居室
Living Room
39.64m²

客卧
Guest Bedroom
15.30m²

卫生间
Bathroom

玄关
Foyer

内庭院
Inside Court

9

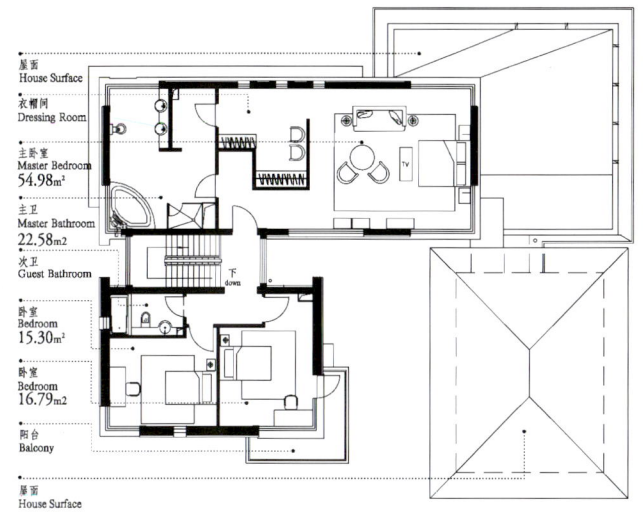

屋面
House Surface

衣帽间
Dressing Room

主卧室
Master Bedroom
54.98m²

主卫
Master Bathroom
22.58m2

次卫
Guest Bathroom

卧室
Bedroom
15.30m²

卧室
Bedroom
16.79m2

阳台
Balcony

墓面
House Surface

10

56

Developer: Beijing Baishunda Real Estate Development Co., Ltd

Project address: Junction of Shunping Road and Chaobai River, Shunyi District, Beijing

Planning design: Singapore SCDA Architects Firm

Floor area: 332, 530 square meters

Building area: 165,5359 square meters

Greening rate: 45.2%

Volume rate: 0.37

11. 别墅效果图1
12. 别墅效果图2
13. 别墅效果图3

11

12

13

the shape of "U" expresses traditional oriental living culture in a modern way, succeeding the sense of belonging paid attention to by oriental, opening living vision, extending interaction with nature. People can then feel best balanced psychological shape. Group pattern separates neighborhoods by courtyards, water features, greening area and rough walls, not only keeping privacy of detached houses, but also strengthen the harmony of the features. In terms of inner space layout, on entering the houses there are spacious porches, french windows of which brightly shows you the landscape of the quadrangles. It completely conform to the Asian cultural concept of "Bright halls and rooms". Sense of Zen meditation is reached by the design of extending indoor vestibule with the emphasis of peaceful quiet. Changes of layers in the courtyard features with brief lines, interlacing of light and shadow can also be felt as people walk around.

Courtyard feature: in the East, traditional architects in many oriental countries such as China, Korea and Japan, are full of layouts and forms in harmony with the social cultures and nature surroundings. It emphasizes a wide frontal that has a yard front court, an inner courtyard and a back yard so as to share space with the nature. Champagne Cove follows the multi-aspects of Oriental courtyard, the clear opening yard that stretch out connected with the yard have an independent entrance

14

Champagne Cove covers a total area of 33.25 hectare, best work in 2005 of Beijing Baishunda Real Estate Development Company Limited, is specially designed by Singapore SCDA Architects Firm which is famous for its New Asia concept, carrying strong Asian culture, a New Asianism luxurious musical movement played in a modern brief way.

New Asia refers to under the trend of world cultures mingled and fused designer's recognition toward Asian architectural design, which is taken Orient traditional culture of regional characteristics as foundation blended with modern western culture. Functionally improved, Asian splendid culture can be enhanced and it is more suitable for modern people's lifestyle. New Asia brings a kind of amorous feeling and flatus revealing a kind of attitude, spirit, and entire creative living concept for new Orients

Space feature: "Unfold the form of villa "is one of distinctive charms on display of space. The structure, "One and a half", enlarges constructive floor area .It own pitch living rooms that are complete independent on the first floor and separate with family living space. Features of sling slinging, independent, wide, and relaxing social space on the first floor and attention on the living in the villas at the request of the demand for Asian culture is no doubt a creation in modern villa architectural structure. Semi-enclosed layout in

15

16

17

through the kitchen; water courtyard through restaurants make the restaurants become a appreciate scenery that is good to hear , enclosure courtyards form a nucleus scenery in villa zone, while vision can through living room, family room and restaurant strengthening the communication between family members and spoiling the happiness in every corner. Sinking pattern courtyards enlarge the light and activity zone, not only meet the functional need, but also highlight orient lofty artistic conception and reveal the mental satisfactory and appreciation of the beauty demand for "integration of human and nature.

Garden landscape: the orient gardens have always been influenced by Taoist school, highlighting keeping harmony with nature. .the space insists on counterbalance and equity. As the planning of the garden landscape environment, for it emphasize keeping harmony with natural environment, human and architect become a part of nature. Double lakes are the central scene spot of Champagne Cove, distributed in north and south two regions with their own characters.   is famous for beautiful natural greening and familiarity. Details of the lake insist on design as a garden. Besides there are pedestrian streets, water-loving stiles and rest area along to water providing space for people to communication, sports and rest.

14. 别墅实景图
15/16/17/18/19. 别墅实景局部

18

19

1

1. 树丛中的别墅
2. 别墅局部
3. 别墅外立面造型

绿城·桃花源

开 发 商：杭州余杭绿城房地产开发有限公司
项 目 地 址：杭州市余杭区凤凰山南麓
建 筑 设 计：浙江大学建筑设计研究院
　　　　　　　浙江绿城建筑设计有限公司
　　　　　　　美国道林建筑与城市规划设计集团
占 地 面 积：1 800 000 m²

2

　　绿城·桃花源位于杭州市余杭区凤凰山南麓风景秀丽的丘陵地带，距杭州市区18公里。桃花源总占地达180万平方米，拥有真山真水的自然，是中国最低密度的别墅园区之一。整个园区分东、西、南三区开发，东区已全部售罄并交付，西区及南区正在建设中。

　　绿城·桃花源生态居住区以陶渊明《桃花源记》中所描绘的山野田园生活为创作蓝本，融合生态自然山水和田园人居生活为一体，创造深含人文理想的"桃花源意境"。桃花源东区与西区以全新空间布局体现了"豁然开朗、鸡犬相闻"的居住理念，南

区别墅则在此基础上向"芳草鲜美，落英缤纷、其乐融融"的自然理想进一步回归。园区内有众多山坡、河流、池塘、小溪，山美水丽、坡缓谷幽、秀林掩映、意境优美，是目前杭州乃至全国不可多得的经典别墅园区。

　　桃花源西区共规划建造形态各异的独立别墅570余栋。南区一期220余栋独立别墅于2007年交付，共分为山地别墅、园景别墅、水岸别墅、庭院别墅、小院别墅等五大类。2005年底推出的园景别墅建筑面积约420平方米，私家庭院面积平均约1000平方米，全面采用轻钢结构，并推行室内精装修和私家庭院精装修设

计，8套全装修样板房运用意式、法式、英式等多种别墅风格创造了身临其境的全装修精彩体验；水岸别墅地上建筑面积380平方米左右，地下建筑面积约200～250平方米，基地面积1000～1500平方米，多分布在潺潺溪流边。

　　绿城·桃花源相继荣获"住在杭州代表楼盘"、"全国人居经典综合大奖"、"影响中国的30大典范社区"，是绿城集团继九溪玫瑰园之后，别墅开发历程中的又一经典巨作，成为中国别墅史上的全新之作、中国第三代别墅的代表作品。

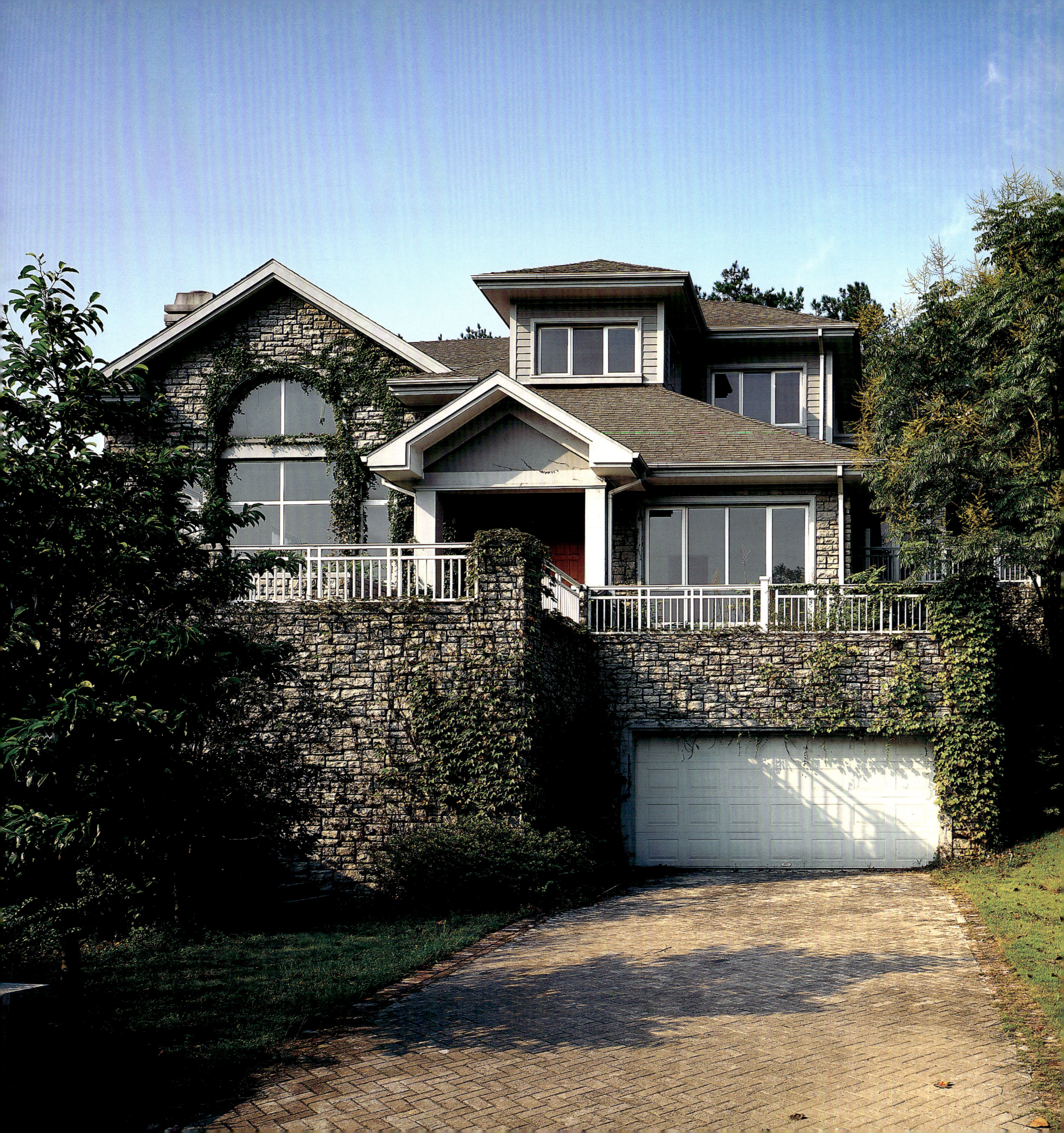

Developer: Hangzhou Yuhang Green Town Real Estate Development Co., Ltd.

Project address: South, Phoenix Hill, Yuhang District, Hangzhou City.

Construction design: Architectural Design and Research Institute of Zhejiang University

Zhejiang Greenton Architecture Design Co., Ltd

Floor area: 1800 000 m²

Green Town·Fountain Of The Peach Blossom Spring is located in a picturesque south foothill of Phoenix Hill, Yuhang District, Hangzhou City, 18 kilometers away from the downtown area, covering an total area of 180,000 square meters with beautiful nature. It is a villa group with a lowest density in china, developed separately in east, west and south area. The first one has been completed and delivered which the latter two are now under construction.

As a ecological residential quarter, Green Town·Fountain Of The Peach Blossom Spring is based on the country life described in A Tale Of The Peach Blossom Spring written by Tao Yuanming, integrating nature and country life into a whole, creating poetic imagery of cultural ideal. The residential concept of "the crowing of cocks and the barking of dogs were within hearing of each other was embodied in the whole new layout of the east and west area of the residential quarter. Then, villas of its south area are a further return to the natural ideal of "the fragrant grass was fresh and beautiful and peach petals fell in riotous profusion." And there are numerous slopes, rivers, ponds and streams, green hills and crystal clear water, winding slopes and quiet valleys, such magnificent environment is a rare classical villa group in Hangzhou even in China.

There are 570 detached houses of different outward design planned in west area of Fountain Of The Peach Blossom Spring. More than 220 detached houses of the first phase of the south area have been delivered in 2007, which are mountain villas, garden villas, waterfront villas, courtyard villas and yard villas. Garden villas with areas of 420 square meters introduced in the end of 2005 and private courtyard with an average area of about 1,000 square meters are all constructed with light steel, and interior decoration and private courtyard design are luxurious. Various kinds of styles such as Italian pattern, France pattern and Britain pattern are used in 8 completed sample houses which make you feel like being in any country of its style. Waterfront villas cover an area of 380 square meters, underground area of about 200 to 250 square meters, root area of 1,000 to 1,500 square meters, mostly spread along the ripply streams.

Green Town·Fountain Of The Peach Blossom Spring has been honored as "Life In Representative Residential Quarter In Hangzhou", "Comprehensive Reward Of Classical Human Habitat In China", one of "China's Thirty Example Influential Community", following Green Town Group's Jiuxi Rose Garden to be another classical hit in its developing history and become a whole new word of the history of China's villas, representative work of 3rd generation villa in China.

4. 总平面图

5. 别墅内庭院

4

听枫待霜

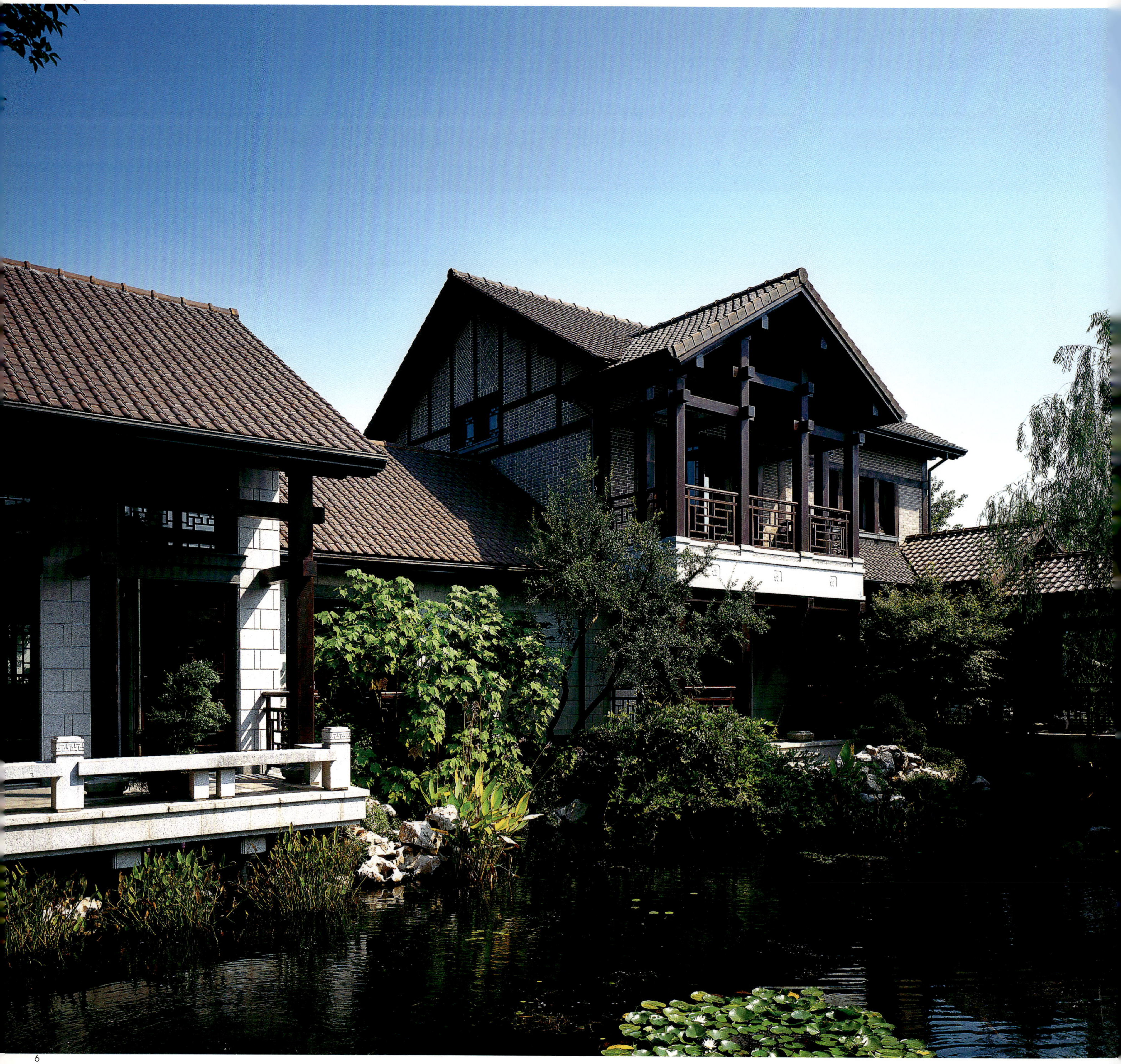

6

6. 临水而居
7. 回廊通幽
8. 典型的中式风格

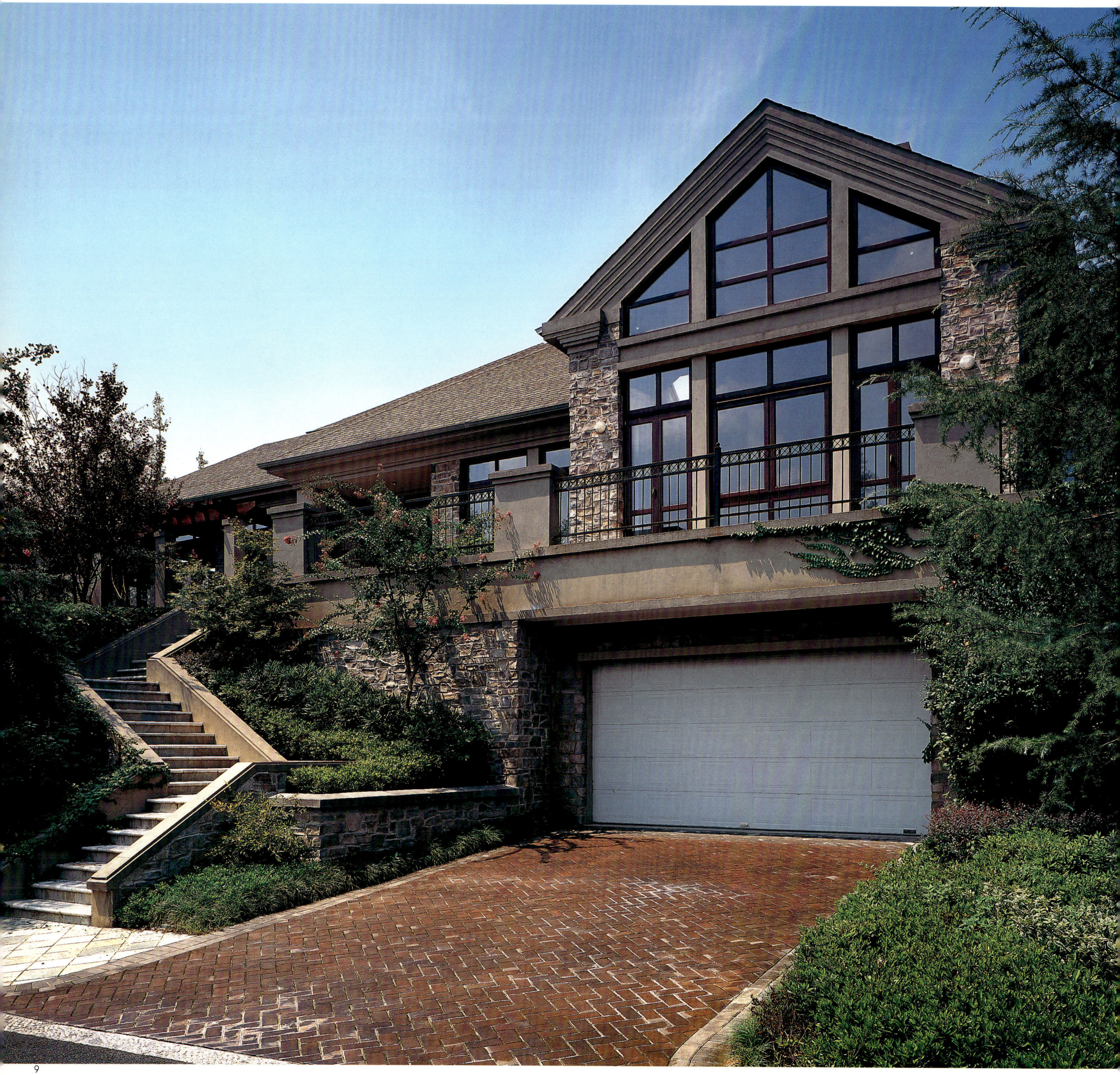
9

10

11

12

9. 依地势而上的梯级
10. 隔水远眺
11. "桃花源"意境
12. 山野田园生活写照

1

1. 南大门实景
2. 南大门手绘效果图
3. 别墅休闲区

碧水庄园三期

开　发　商：北京碧水庄园房地产开发有限公司
项 目 地 点：北京市昌平区京昌高速沙河第10出口
占 地 面 积：1 534 100 m²
总建筑面积：220 000 m²

2

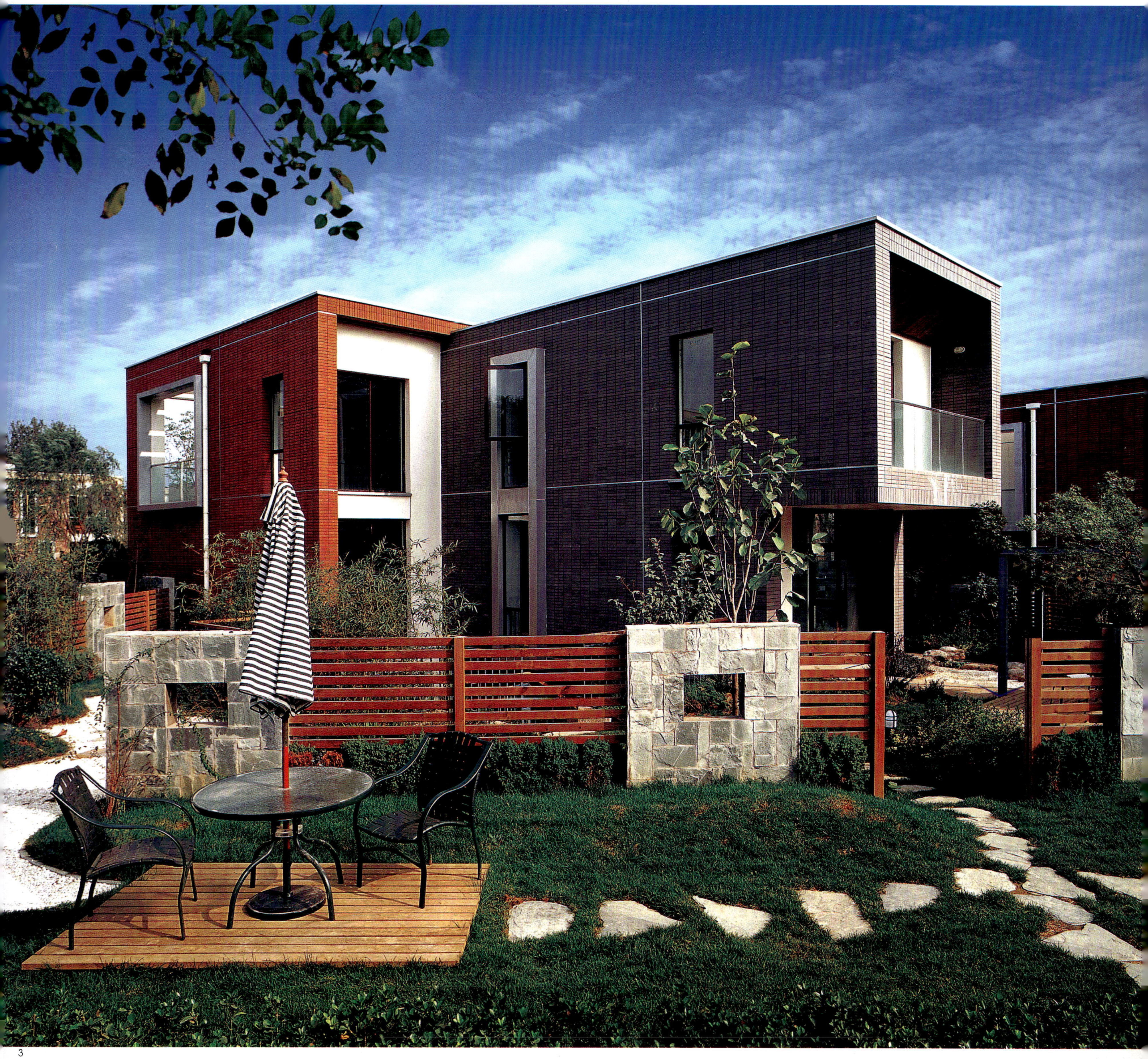

3

花架

聚会广场

休闲步道

观景绿化带

休闲一角

天然条石

日式枯山水

值班室

临时停车

水景

主入口广场

4

Floor area: 1 534 100 square meters
Building area: 220 000 square meters

Bishui Manor has a good location in Beijing, right in the North of "Longmai". In its West is the West Moutain which has wonderful and lively scenery in all seasons as well as strong cultural foundation. National Forest Park and Historical Relic Preservation Area are in its North. In the Southeast is the well-known Zhongguancun Science Park. Nansha River runs along it, right opposite the Shahe Reservoir distantly. Such geographical location with mountains and rivers is like a favor from nature, making it a noble villa project closest to the Beijing suburb.

The newly-promoted Manor inluding 3 phases is planned to build 343 villas which are divided into 15 kinds. And there are latest breakthrough in each floor areas, the maximum one is 1024 m^2 and the minimum one is 314 m^2.

Private lands vaires from 1,200 m^2 to 5,200 m^2 are enough to meet demands of personages from all circles, and the developer had spent large amount of capital in builing another 24,00 m^2 –large lake, this noble villa block therefore becomes the only one which owns a private lake feature of 28,800 m^2 .

We selected North-American-country-style villa modelling in the outward appearance of Manor. Natural slope top and extending plane perfectly explain what nature is like. Classical construction elements is adopted in all designs for a creation of new presenting form, such as splendid porches, magnificent wall hills, roofs in all shapes, roof-windows which add to bright spot and flexible view windows,etc. Tones of four seasons is a main part to the outward appearance, which is ornamented with natural or artificial bricks, interchangeable colors, sense of concovo-convex and abundant shapes. All these efforts shows upgrowing atmosphere and feature of nature, the whole manor seems like appearing indistinctly with the hills and water paths as backgrounds.

Bishui Manor promotes harmony between construction and nature. We combined it with traditional Chinese private park making rationale and brought more greenness into houses. Extraordinary large courtyard and their innerspace make the villas and the gardens more transparent. That's how gardens with villa inside and villas with gardens inside is represented. Atmosphere of peaceful quietness, plain lightness, active freshness everywhere, and sense of nature and health everywhere create comfortable feeling of living in heavenly place. Precious plants up to 100 kinds like magnolia denudata, lilac, oriental cheery and ginbo, etc is planted with little sculptures decorating them everywhere. Plants, pools, garden paths and lake stones are selected with tall trees and bushes at intervals. Such unique arrangement keeps a green and spacious space, layers of which is clearly represented. All scenic views will present you a poetic picture.

5

4. 东侧总平图
5. 别墅景观局部
6/7. 景观节点图
8. 别墅实景图

6

7

8

9

10

11

12

9/10/11. 立面设计
12. 别墅环境景观

13

14

13. 文化墙实景图
14. 小石路景观
15. 休息区景观

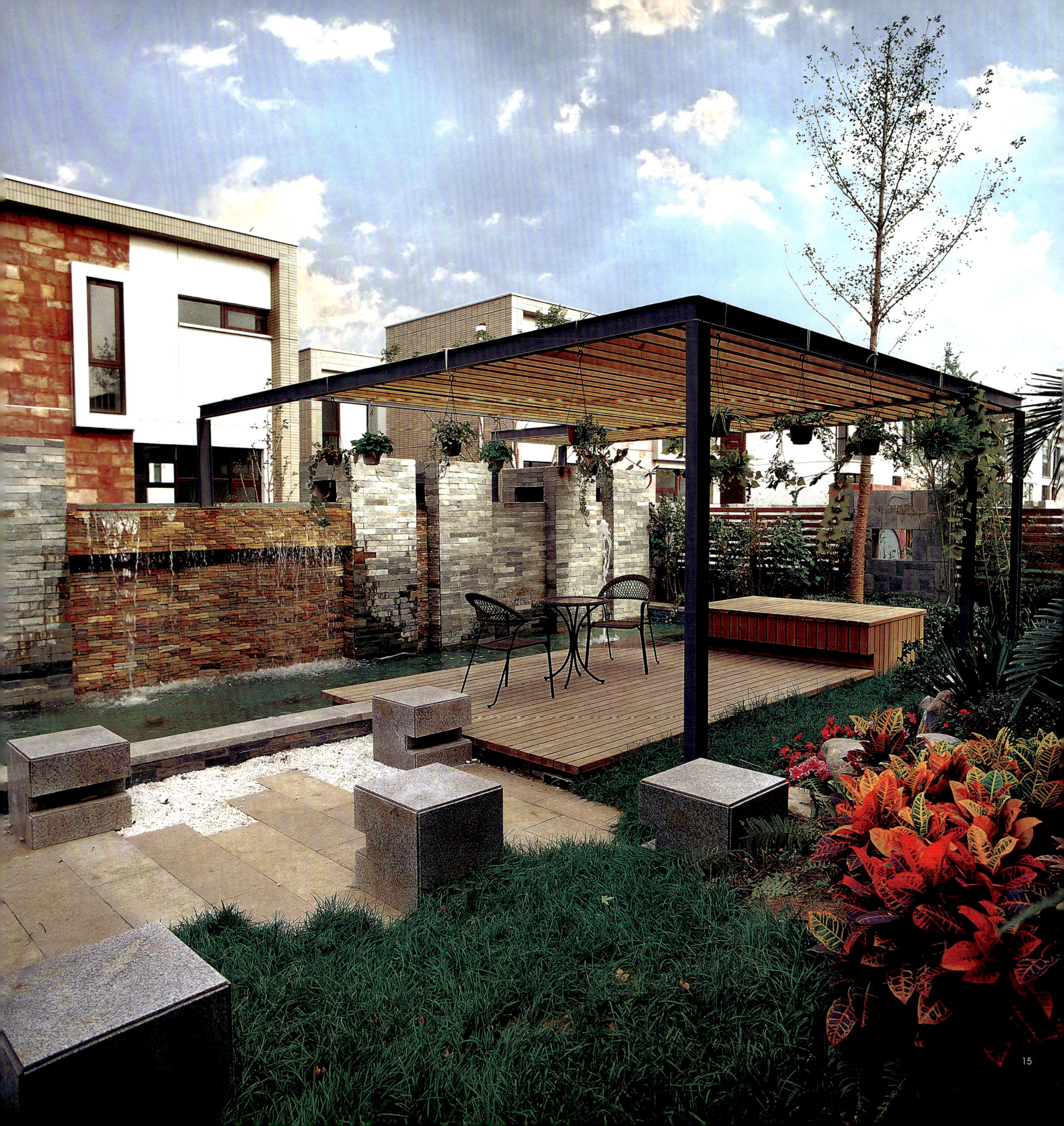

1

昆山世茂蝶湖湾

开　发　商：世茂蝶湖湾开发建设有限公司
项目地址：昆山市中心城区南部
建筑设计：香港王董建筑师事务有限公司
总占地面积：720 000 m²
总建筑面积：1 070 000 m²
容　积　率：1.5
建筑密度：25%
绿　化　率：65%

2

昆山世茂蝶湖湾位于昆山市中心城区南部，是连接国际化大都市上海和历史文化名城苏州的重要枢纽，地理位置优越，交通便利，是联合国发展计划中心、联合国生态安全科学院认证的"国际生态安全示范社区"。

蝶湖湾总占地面积近72万平方米，总建筑面积约107万平方米，容积率仅为1.5，建筑密度仅为25%，是名副其实的高绿化低密度高级住宅社区。社区由12幢沿湖面一字派开的高层公寓、数个分布于湖中的独立岛屿式联体别墅及独栋别墅、SOHO公寓、SHOPPING MALL、星级国际酒店、三大多功能主题会所、国际双语幼儿园和学校、商业广场及台湾风情街组成。

蝶湖·湾仔道作为其一期商业部分，总面积超4万平方米，是完全区别于普通社区商业配套的区域商业中心。蝶湖·湾仔道采用先行全球范围招商再进行部分商业销售这种全新的运作模式，与国际顶尖商业管理公司进行深度合作的商业项目。世茂蝶湖湾的社区充分利用湖景资源，蝶湖景观公寓均采用弧形单排设计线形排列，并配合独立岛屿式别墅的高低错落布置，令社区轮廓更加生动，有效避免视野阻隔，实现"家家面湖，户户有景"。268套分布于湖中的蝶湖Townhouse，以及珍藏于湖心岛的20套双拼别墅和16套独体别墅，坐拥社区内10万平米蝶状湖水，40万平米社区湖畔公园，将低密度高绿化亲水生活诠释得淋漓尽致。

世茂蝶湖湾社区以生态湖景为规划主题，容积率仅为1.5，而绿化率高达65%，整体湖面面积将近12万平方米，其蝴蝶状的水体布局，融城市生活与居住生活于一体。社区是由12幢沿湖面一字排开的高层公寓、6个分布湖中的独立岛屿式联体别墅及独栋别墅、3个不同规模和功能的主题会所（专业运动主题会所、老人儿童会所、名流会所）1座四星级酒店、SOHO、SHOPPING MALL、中小学校、幼儿园、商业广场及台北特色风情街组成的超大型全功能综合性高尚亲水社区，不同的功能分区使水岸生活从城市公共的商业空间过渡到居住社区的公共空间再到完全私密的生活空间，而绿岛的设计模式使社区空间与城市空间相对独立，避免了快速交通干道对社区内生活品质的干扰，社区内独有的人

车分流的"无烟城"人性化设计，真正达到机动车和步行者的流线完全分离，机动车进入社区后直接进入地下通道，在交通的便利性与住户的居家品质间，取得最完美的平衡点，让生活在此的业主尽享安全无忧的漫步，呼吸不受尾气污染的清新空气，形成住宅区内舒适、安全、安静的高品质现代人居生活环境。世茂蝶湖湾是昆山首个设立独立别墅私人游艇码头的社区，业主在工作之余可以携亲友荡舟湖上，享受阳光湖水的悠闲雅意，营造了一处别致、幽雅、生态、环保的世外桃源。

世茂蝶湖湾户型设计充分利用了生态湖景，高层公寓均采用弧型单排线形排列设计，面积从85平方米的两房至200平方米的四

房，可以满足不同客户的不同需求，配合岛式别墅的高低错落布置，令社区轮廓更显生动，并有效避免了视野阻隔，真正实现了"家家面湖，户户有景"，每户的客厅及主卧均朝南朝湖面，配以昆山罕有的大面积落地玻璃窗，将园林湖景引入室内，使业主享受真正的亲水生活。社区内繁华便捷的商业商务区，分为东西两大区域，东部是以大型商业、酒店、办公为核心的商务区，提供便捷高效的商务办公服务和优雅舒适的酒店服务，西部是以具有江南水乡特色及台北风情商业街为核心的休闲区，提供一站式的生活体验和亲水生态环境，24小时的商业活力区，不断延续城市中心区的活力。项目将分为两期开发，一期为中央运动主题会所轴线以西的部分于2008年全部建成，整个项目的开发期约为5年。

Architect design: Hong Kong Wong Tung &partner Ltd

Overall covering area: 720 000m^2

overall constructive area: 1 070 000m^{2}floor-space

floor area ratio: 1.5

building depotition: 25%

greening rate: 65%

Kunshan shimao diehu wan is located in the south of central area of Kunshan City, major junction of shanghai ,International metropolitan and Suzhou, famous historical cultural city. With such advantageous location and convenient traffic, it has been approved of by UN development and plan center and UN ecological security model as an International ecological safety model community.

Total area of Diehuwan is near 720,000m^2, and its total floor area is about 1,070,000m^2, rate of volume is only .5, and finally, its construction density is only 25%, all these make Diehuwan an high-class residential quarter of highly greening and low density. The community is made up of 12 blocks of high-storey apartment spreading along the lake as a line, and several undetached villas of independent-isles-pattern and detached villas, SOHO apartments, shopping mall, relics international hotels, three major theme multi-functional chambers, bilingual international kindergarten, commercial plaza and Taiwan flavorful bar street. Community consists of 12 blocks high-rise apartments

4

5

as a line along the lake surface, several separate island-style conducted villas lying in the lake, single-family villas, Soho-style apartments, shopping mall, international stars hotel, three multi-functional theme clubs, international bilingual kindergartens and schools, commercial plaza and Taiwan style steet.

Butterfly bay. Wan Chai Road, the first phase of of the commercial part a total number of more than 40,000 square meters. totally different from general communities which equipped with the communities of SHIMAO die hu wan make full use of Butterfly bay landscape apartments all adopt curve uniserial design linear series, with villa that have independent island pattern scatterhed randomly, making the rough sketch of communities more vivid, avoiding visual gap and making "each home face lake, every flat has scenery.", diehuwan Townhouse, 268 units, distributed in the lake, coupled with 20 villa independent enshrined in eyot . To sit on a seat which is covered with 100.000 square metre butterfly lake and 400.000 square metres residential waterside park inside the community. Interpret hydrophilic life that have low density and high greening incisively and vividly.

SHIMAO diehuwan community regard the ecological lake as the layout theme, floor area ratio is only 1.5, but the greening rate reach 65%. the overall lake surface approach 120.00sq.m.Its butterfly shape waters layout combine urban life and residence. community consist of 12 high-rise apartment stood in a solid line along the lake surface, 6 conjuncted and independent villas distributing in lake, 3 theme chambers in different scale and function(professional sport theme chamber, olds and children chamber, celebrities chamber), one four-star hotel

,SOHO, shopping mall , primary and middle school, nursery school, mall and super water-loving community of Taibei character that has gracious comprehensiveness and full function. Different functional zones make the water's edge life grade from urban common commercial region into common residence, which move on totally privacy living space. However ,the design pattern of green island make the living space and the urban region relativity independent, avoiding disturbing the living quality by the transport artery. "no smoking quarter" dividing man and car is a special humanization design, it reach definely separated flown line of men and car .while cars enter community, they can directly drive through underground. There is no doubt that it choose a perfect balance point between transport convenience living quality. you may take a wander here without any worries and breath getting rid of air pollution while living here. enjoy a comfortable and safe, silent human living environment. SHIMAO diehuwan is the first community that settled independent villa and private in HUNSHAN. Owner can go boating on lake together with friends enjoying leisure Creating a special , elegant, environmental-friendly Shangri-La.

die hu wan butterfly way design make full of ecological lake.hign-rise apartment all adopt curve uniserial design linear series. coverage consist of 85 two rooms from 200 four rooms can meet the need of different customers in harmony with island pattern villa scatterhed randomly. making the rough sketch of communities more vivid, avoiding visual gap and making "each home face lake, every flat has scenery.", master bedroom and parlor in each flat face south and lake, equipped with the Hunshan rare big window that fall to the ground, bring the landscape and lake indoors making owner enjoy the real water-loving living. The luxious and convenience Commercial area is divided into the eastern one and western. The project can be developed into two periods, one consider the center sports theme chamber as the spool thread. Eastward will be built overall in 2008.the develop period of the whole project is about 5 years.

6

7

8

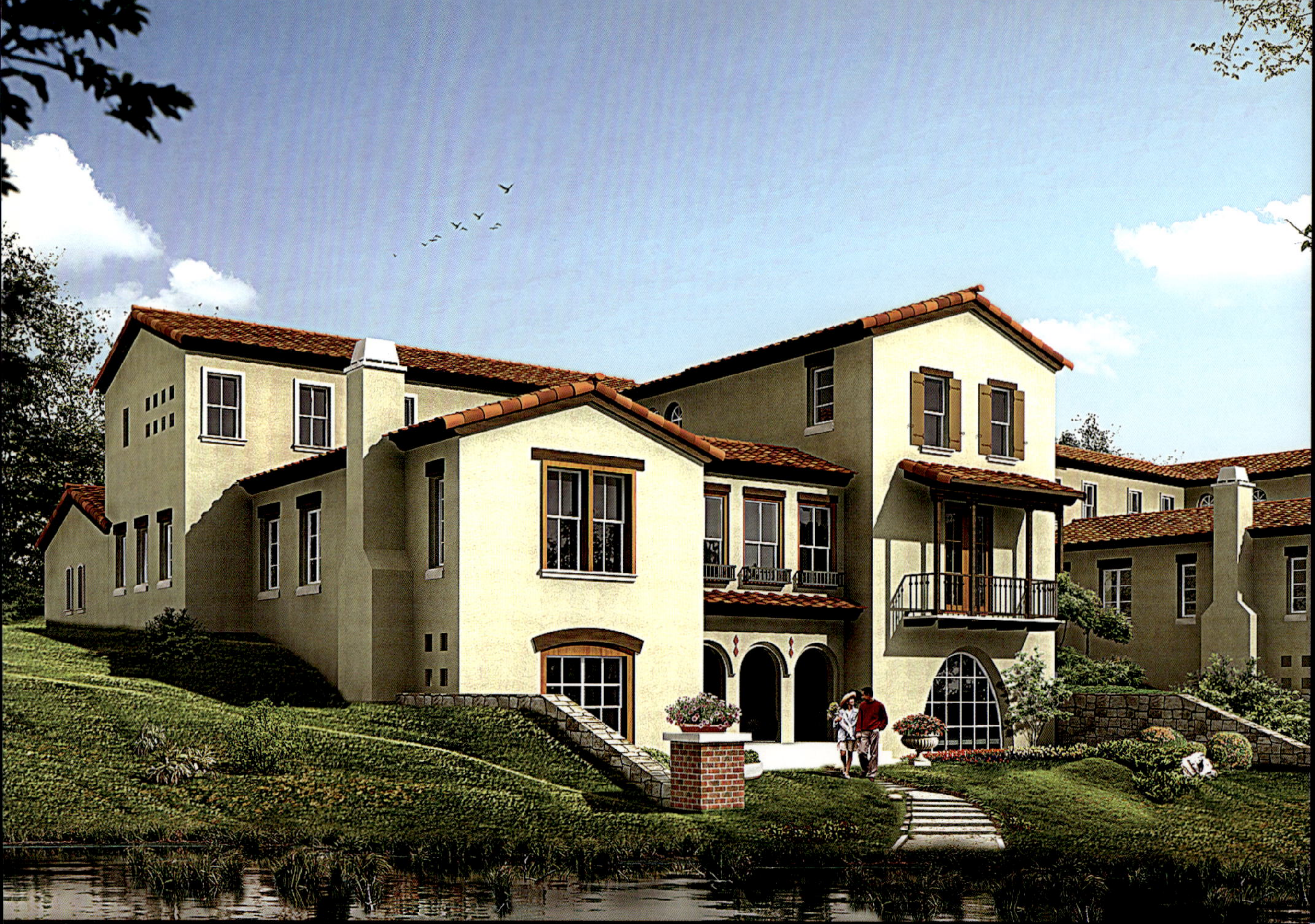

1

1. L1型别墅效果图
2. S1型别墅效果图
3. A组团庭院组合放大图
4. 总平面图

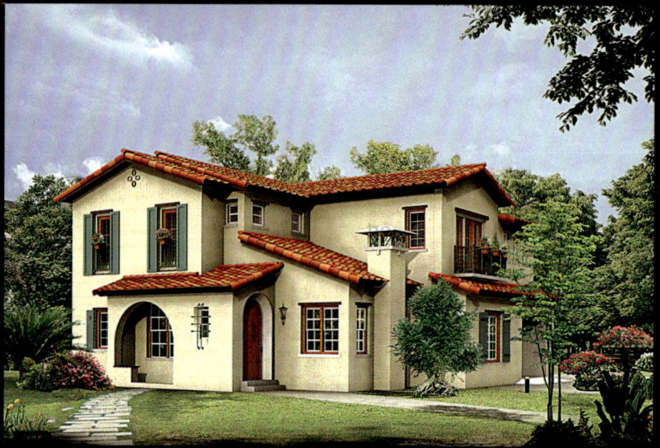

2

格拉斯小镇

开 发 商：北京海港房地产开发有限公司

景观设计：美国EDAW公司

美国BELT COLLINS公司

建筑设计：加拿大ARCHISPACE

美国JWDA

占地面积：2 240 000 m^2

建筑面积：约600 000 m^2

绿 化 率：60%

容 积 率：0.3

建筑形式：独栋别墅

3

Floor area: 2 240 000 m²

Building area: about 600 000 m²

Greening rate: 60%

Volume rate: 0.3

Glass Town is located in the East bank of central villa, along the Yunnyu River to southeast, inherited the central villas and accordig to topography features, from here, you can keep watch with embassies, and the capital airport. What's more, it approaches to core of city CBD. Walk along the Yunyu River to the South, Glass Town make a 6 kilometres exclusive riparian Boulevard ---- Glass Road for owners. The river which is 3 kilometres acrooss the community, the widest area even reach to 80 metres. Furthermore, 20,000 square metres orginal ecology of wetlands seeping the reflashing forest and whispered by the river.

Such 30,000 square metres buildings in the center of town provide totally upper living facilities for the construction community of Versailles Valley. Such as festival hall, inside and outside-commercial pedestrian street and libraries, cafes, and other cultural living facilities improved not only thoughtful ,but also garanteering the residents's international life style and special state. Luneng real estate combine with UK top-class international educational institutions---- Harrow School, with an entire journey educational pattern includes kingdergarden, primary, secondary and pre-university edeucation as to build a lead-to-world exclusive school. Harrow school as one of the most famous prvate schools, known for strict discipline, the people who study in it are quite rich. There're famous alumnus in Harrow Technology such as the poet Buron, British Prime Minister Winister Churchil and India leader Jawaharld Nehru and so on.

N

4

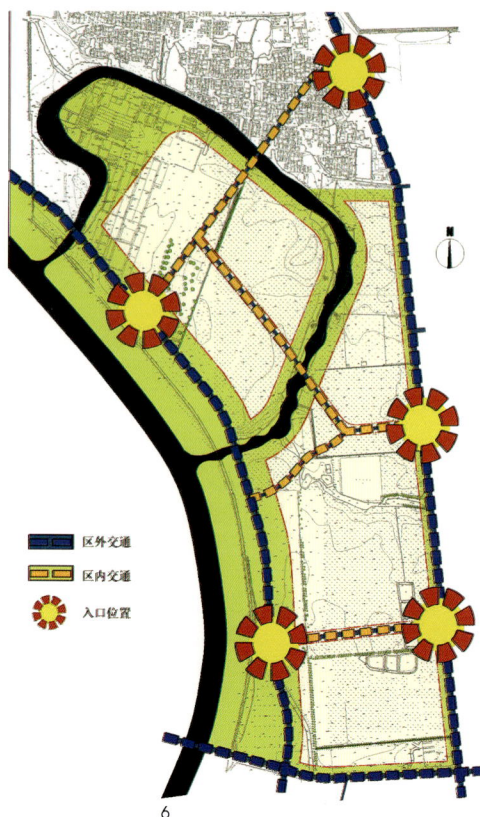

河道

缓冲绿化带

区外交通

区内交通

入口位置

文化生态带

节点

入口景观

绿色长廊

温榆河湾

主入口景观着重处理,树立起步区的社区形象,与区域内部景观中心形成呼应。

充分利用基址原有生态体系和优越的先天景观条件--80米宽的温榆河老河湾--与区域内部建筑形成相互借景的绿色长廊,成为联系公共休闲的绿色纽带。

区域内绿地分布均匀,强调居住均好性,使人可以真正亲近景观和绿化,突出别墅区的亲切感,使居住者充分享有宅前屋后的自然景观。

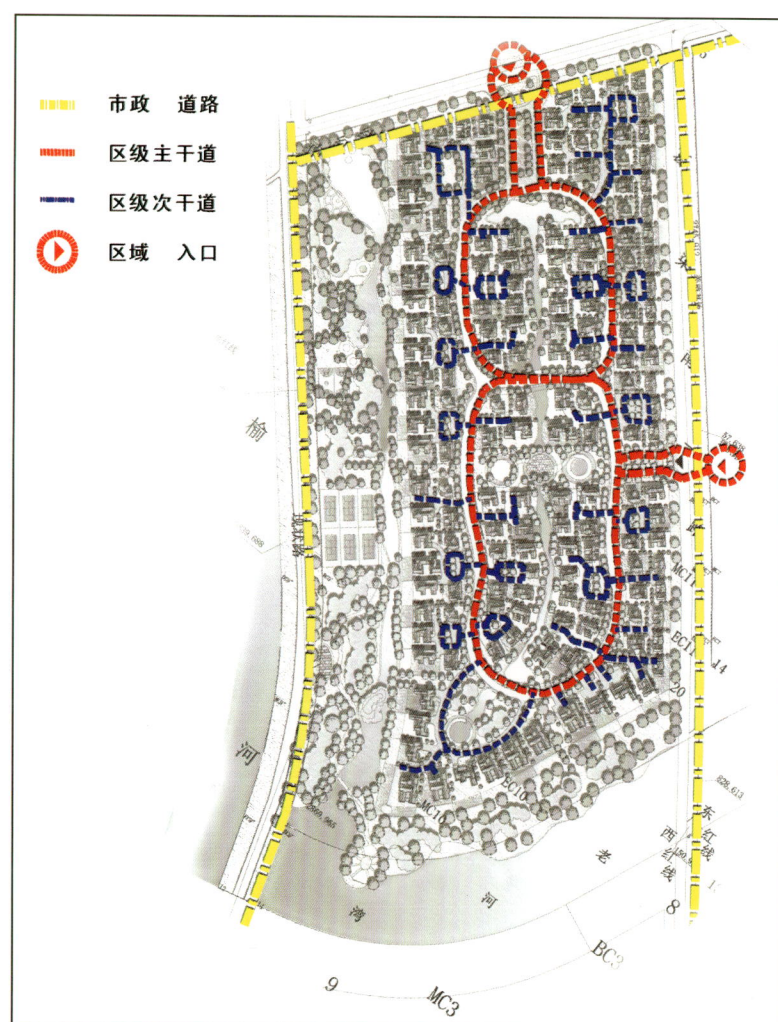

市政 道路

区级主干道

区级次干道

区域 入口

N

S系列小户型:250-300平方米别墅区域

M系列中户型:340-400平方米别墅区域

L系列大户型:450-500平方米别墅区域

S系列户型组成B型组团，作为最主要同时也是形式最丰富的户型分布在区域内部。

M系列户型结合S系列户型构成A型组团沿温榆河绿色景观长廊延展开来。

L系列户型作为面积最大豪华级别最高的一种住宅类型，占据了起步区南端临老河湾的一处最为有利的地块。

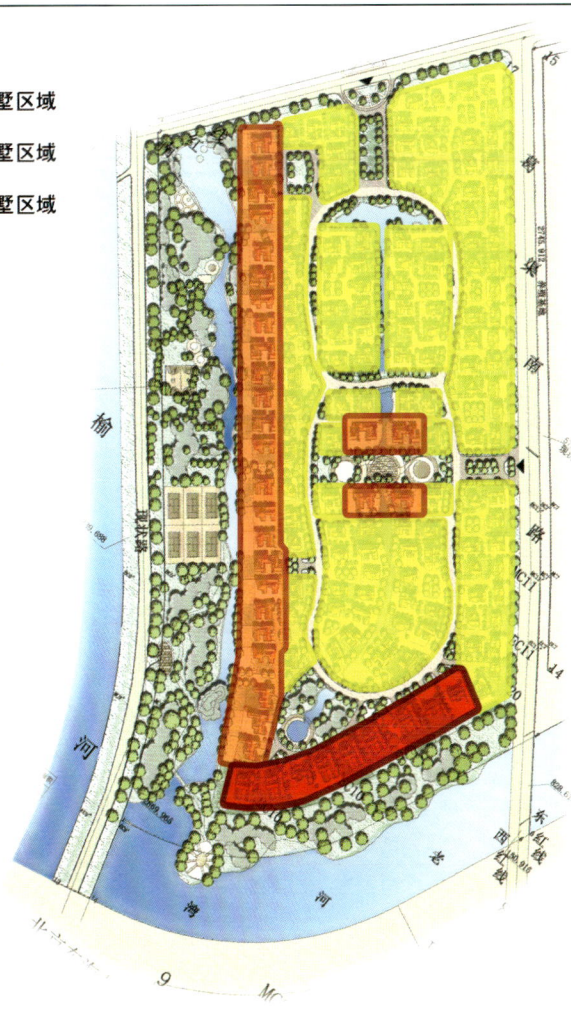

5. 绿化分析图

6. 交通分析图

7. 生态分析图

8. 起步区绿化分析图

9. 起步区交通分析图

10. 起步区规划总平面图及户型统计

11. 起步区户型分析图

11

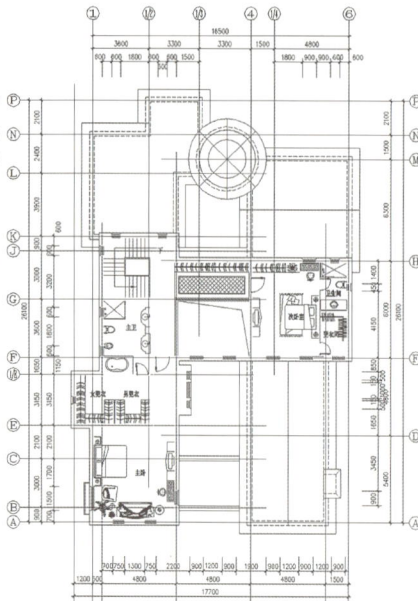

12
13
14

15

12. L1型别墅地下一层平面图
13. L1型别墅一层平面图
14. L1型别墅二层平面图
15/16. L1型别墅效果图

16

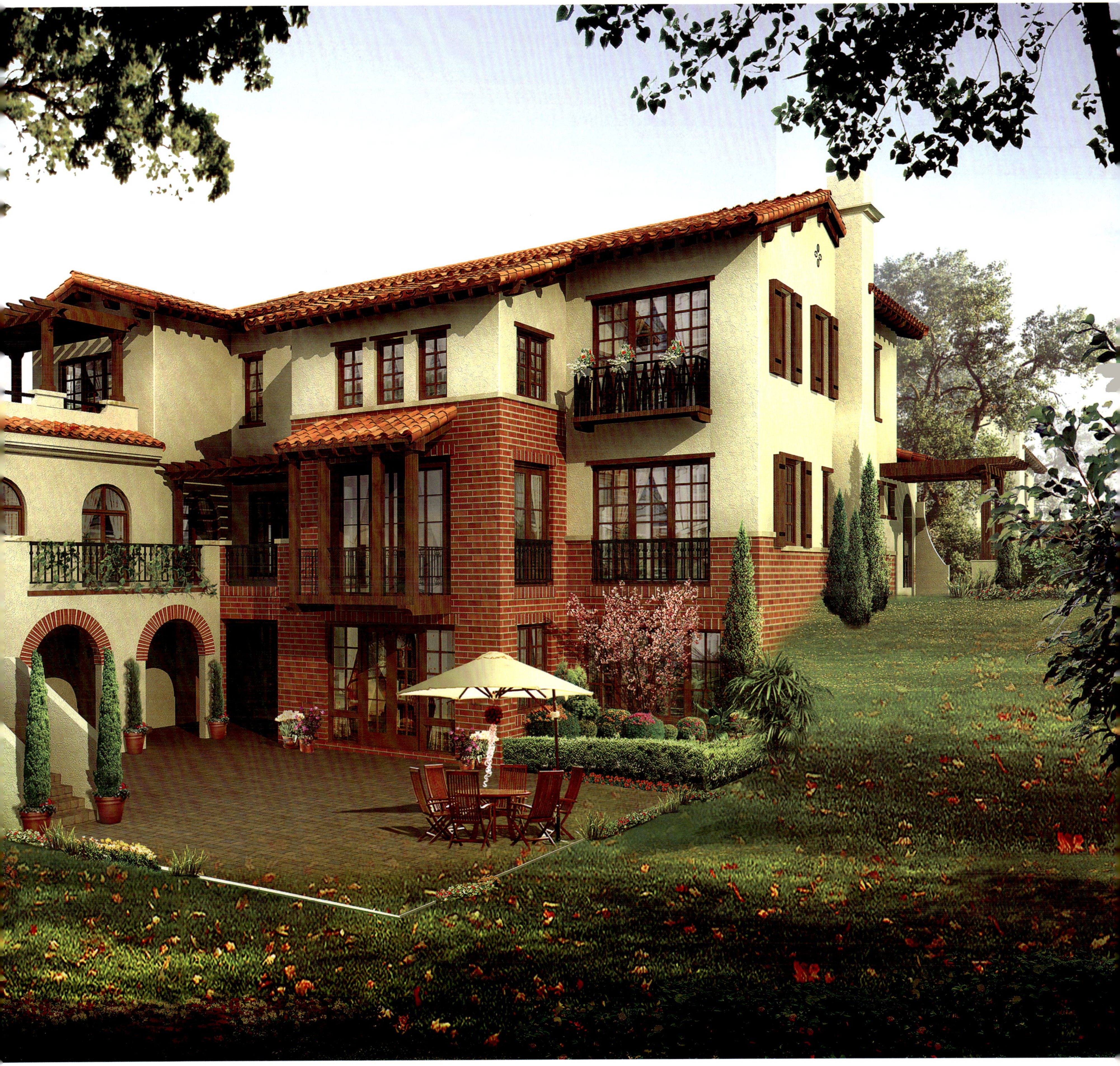

17

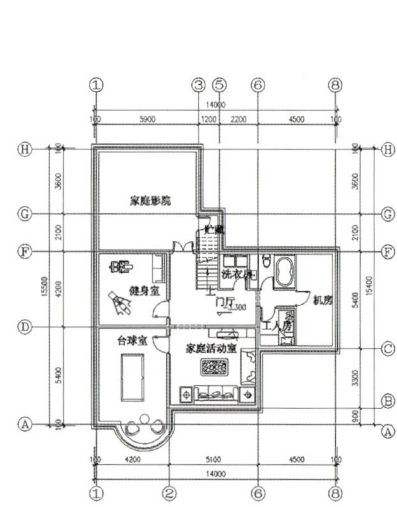

18

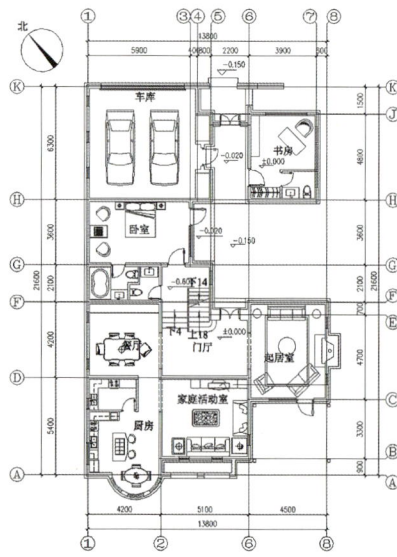

19

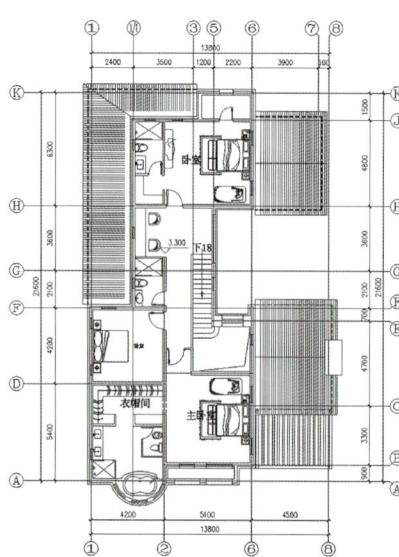

20

21

22

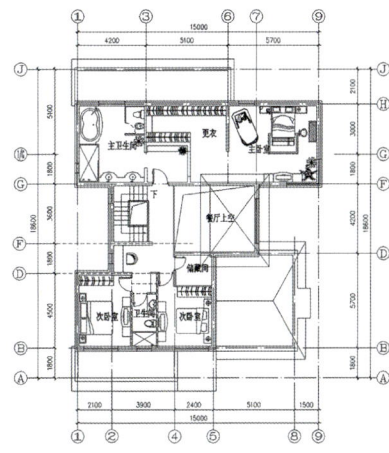

23   24   25

26

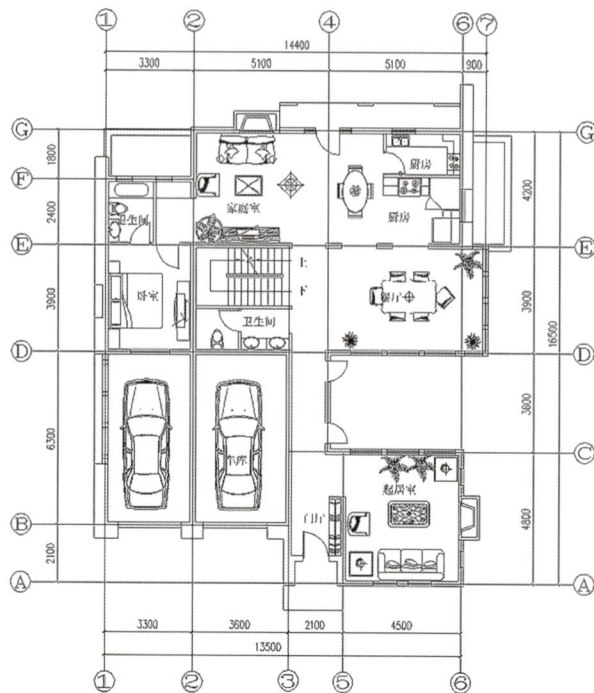

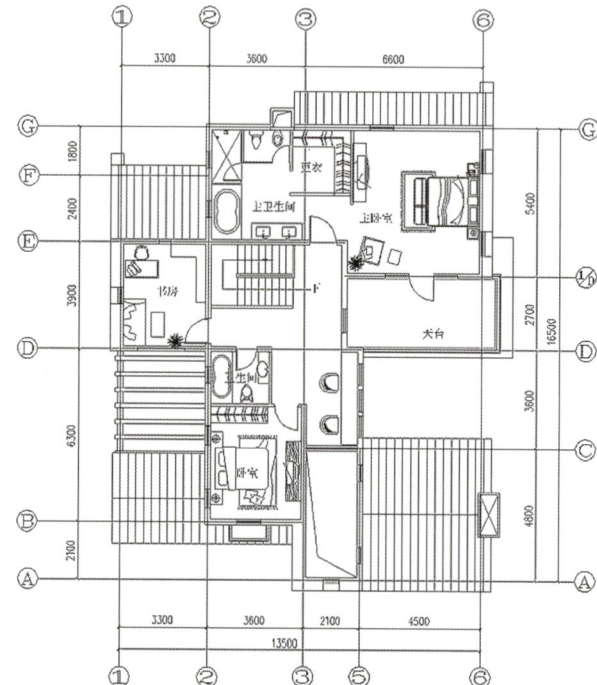

27

28

29

30

31

32

1

西贵堂

2

停车位：1:1
容积率：1.60
绿化率：40%

开　发　商：成都纵合世纪房地产投资有限责任公司
项 目 地 址：成都外光华片区芙蓉古城旁
景观设计单位：加拿大嘉柯景观设计公司
建筑设计单位：加拿大HDII国际设计公司
占 地 面 积：40 000m²
总建筑面积：60 000m²
总 户 数：280户

Floor area: 40 000 m²

Building area: 60 000 m²

Greening rate: 40%

Volume rate: 1.60

The number of dwellings: 280

Triassic villa in courtyard dwellings is in Guanghua Area, with only 280 cityhouses. Such innovating design of cityhouse perfectly combines the land use capability with life quality in villas. Among the courtyard groups they themselves echo each other but are respectively independent. We aims not to create environment that are profound, quiet, but not explicating. The 28-meter-long open chamber enlarges your view,the 6-meter-tall balcony provides you with place to plant, make your vision full of natural green color. The unique garden ladder keeps residents on the third floor avoid public so that they have direct access to the units. Humanization and Privatization is emphasized in the project, and as a extending function of the construction space, a sunken audio-visual room of 70 square meters will be presented along with the first or second floor which you decide to buy.

1. 别墅入口
2. 小景
3/4. 别墅采水区实景图

5

6

7

8

5/6. 别墅景观实景
7/8. 过道局部
9. 花园
10/12. 别墅玻璃挑檐
11/13. 休息区

低密度住宅

Low density
Residence

低密度住宅

1

规 划 设 计：SWA GROUP
罗麦庄马香港有限公司
梁黄顾建筑师（香港）事务所
深圳大学建筑设计研究院
美国凯斯设计有限公司
深圳市城脉建筑设计有限公司
建筑设计：深圳大学建筑设计研究院
深圳市城脉建筑设计有限公司
景观设计：香港易道规划设计有限公司
室内设计：美国凯斯设计有限公司
梁志天设计师有限公司
香港高文安设计公司

1/2. 现代风格立面
3. 亚热带景观园林

深圳星河·丹堤

开　发　商：深圳市星河房地产开发有限公司
项 目 地 址：深圳市福田区彩田路北、银湖西
总占地面积：200 322 m²
总建筑面积：360 580 m²
容　积　率：1.8
绿　化　率：38%
建筑密度：24%

2

星河·丹堤位于彩田路北，临近南坪快速、梅观高速。项目的南、北、东三面为13.47万平方米的银湖山郊野公园，西面为9.1万平方米的原生活水湖，是深圳唯一同时拥有自然山、湖、郊野公园三重资源的住宅。其F组团其北侧为别墅和民乐水库湖面，项目本身地势南高北低呈台地状，F组团与湖面高低地势约25米左右的高差，所以，即使是低楼层也都能欣赏到幽静的湖面景观。在地段和资源的基础上，星河方面一直将星河·丹堤定位于"深圳顶级别墅"。和传统的TOWNHOUSE相比，星河·丹堤在强调私密性的前提下，很注意空间的营造。项目的前后庭院、跃式客厅、两层架空地下室等设计，就为业主提供了高附加值的第一生活居所。此外，这一项目的合府、阔庭、厢庭、跃庭等设计，还是国家专利产品。星河·丹堤作为建设部科学技术项目计划智能建筑生态人居导入体系试点楼盘，凭借其得天独厚的原生湖山资源及优越的地理位置，受到地产界和各界人士的密切关注，是深圳美誉度最高的楼盘之一，产品品质也代表了深圳目前最高级别。"山、湖、林、草，是奠定星河丹堤品质的四大元素，星河丹堤极大地占有环境资源，让建筑最大限度地融入自然，让建筑空间与生态资源完美地自然结合。

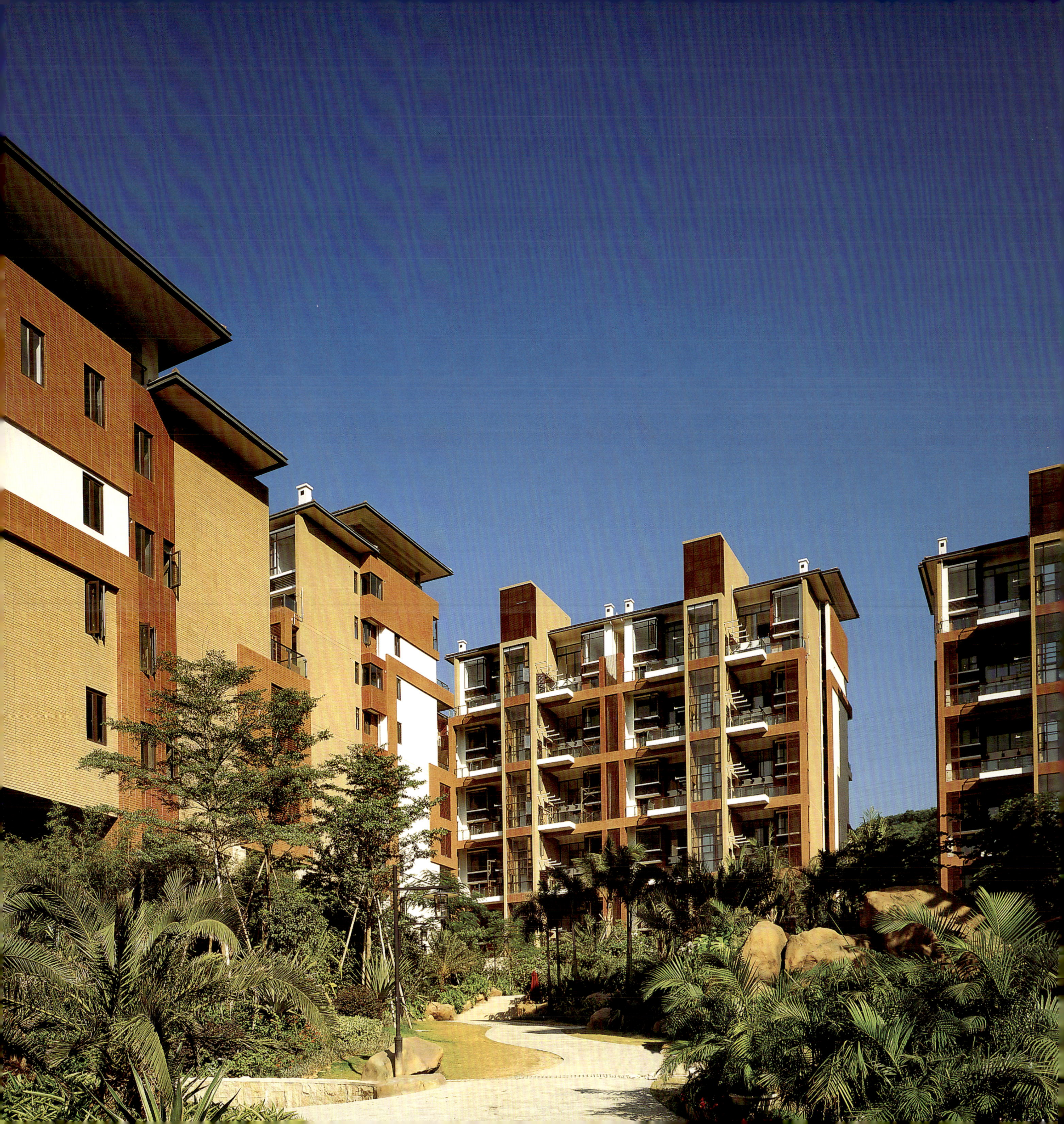

4

Developer: Xing, Shenzhen Real Estate Development Co., Ltd.

Project Address: Futian District, CaiTian (N), Yinhu (W), Shenzhen

Total area: 200,322 square meters

The total floor area: 360,580 square meters

Floor area ratio: 1.8

Green: 38%

Building density: 24%

Planning and Design: SWA GROUP

Luo Mai Ma Chong Hong Kong Ltd.

Huang Liang Gu Architects (Hong Kong) office

Shenzhen University Architectural Design & Research Institute

U.S. Case Design Co., Ltd.

Shenzhen City Architectural Design Co., Ltd. pulse

Architectural Design: Architectural Design and Research Institute of Shenzhen University

Shenzhen City Architectural Design Co., Ltd. pulse

Landscape Design: Easy Road Planning and Design Co., Ltd.

Interior Design: The United States Case Design Co., Ltd.

Steve Leung Designers Ltd

Hong Kong's high-security design text

SOURCE: City of Shenzhen pulse Architectural Design Co., Ltd.

5

项目紧邻的银湖山郊野公园，目前已由深圳市市委、市政府批准立项，星河地产更是捐资600万投入银湖山郊野公园建设中，预计这一公园将在三五年内完成建设。在景观改造上，本着尊重自然、保护为主、生态优先、合理利用的原则，规划中的银湖山郊野公园不仅占据城市核心的地理位置，而且又顺应地域的原生态，顺势营造了随性自然的环境。对此，深圳市城管局局长吴子俊曾向星河地产颁发了"以人文视野关注人居环境"的牌匾，并对星河地产热衷于城市生态建设给予高度的评价。更加难能可贵的是，银湖山郊野公园的初步规划方案还提出，要充分利用自然的绿色景观，发掘人的感悟能力，增强人与自然的交流，用眼、耳、鼻、舌、身体验"五境"，即银湖山的色、声、香、味、触不同感受，从而达到立体的自然感知，园内的自然植被主要是南亚热带季风常绿阔叶林、常绿灌木林，在一些沟谷地段还保存有较为完好的具有雨林性质的沟谷常绿阔叶林。

星河·丹堤2期160套叠层TOWNHOUSE，预计将在今年10月发售。项目E组团高层单位共732套，以3房以上的大户型为主，面积从88平方米到300平方米不等，户型包括二房至五房单位，还有复式单位，不少单位拥有270度景观视野。九仰·丹堤是星河·丹堤位于山脚下的最后一批别墅单位，产品类型为连排及独立别墅，面积介于210-400平方米之间，2008年9月入市。

6

7

Xinghe•Dante is located in Caitian Road (N) and close to Nanping express way and Meiguan highway. There is the Yinhushan Country Park with 134,700 square meters on the three sides of the project —South, North and East. To the west, it is a native fluviatile lake, covering 91,000 square meters. Xinghe•Dante is a unique housing with resources of native hill, lake and country park. To the north side of the group F, there are villa and surface of the Minle reservoir. The terrain of the project itself is high in the south and low in the north which forms terrace. The elevation difference between group F and the height of the lake is about 25 meters. All of these make it possible to enjoy the quiet lake landscape, even in the low floor. On the basis of location and resources, Xinghe always aims the Xinghe•Dante at the top-villa in Shenzhen. Compared with the traditional TOWNHOUSE, Xinghe•Dante pays great attention to making space, in the premise of emphasing privacy. The design of the project, such as the fore and back yard, the hop living room and the two piers overhead basement, provide the high added-value of the first living accommodation to owner. In addition, the design are the national patent products, such as co-house, wide court, wing-room and hop court, and so on. As pilot of the properties for sale in the import system of MOC's science and echnology projrct, which combines both intelligent building and Ecohabitat, it is of primary concern, depending on its advantaged primary lakes and hills' resources and excellent geographical location. It is one of the real estate which has the highest reputation, and the quality of product also represent the top-level in Shenzhen. The mountains, lakes, forests and grasslands- -are the four elements which lay the quality of Xinghe•Dante. It makes

9

the buildings merge into nature to nature to utmost extent, the architectural space combine with the ecological resources perfectly bu full use of the environment resources.

The Yinhushan Country Park, next to the project, has been approved to set up program at present by the Shenzhen Committee and the Shenzhen Municipal Government. In particular, the Xinghe property donated 6,000,000 to build the park, which is expected to be completed in three to five years. In the transformation of the landscape, it is based on the principle of respecting nature, putting protection, ecology priority and rational use. Not only does the Yinhushan Country Park in the project occupies the heart of the city's geographical location, but also it accommodates original ecology of the area, which takes advantage of opportunnity of creating a haphazard and natural environment. For this, the head of the Shenzhen KVR Wuzijun has presented the Xinghe property with a tablet "Focus on the human settlement with humanistic perspective", and speaks highly of its enthusiasm for urban ecological construction. What's more, in the rough plan, it mentions that there is gonging to take full use of the natural green landscape, discover human's sensibility and enhance exchanges

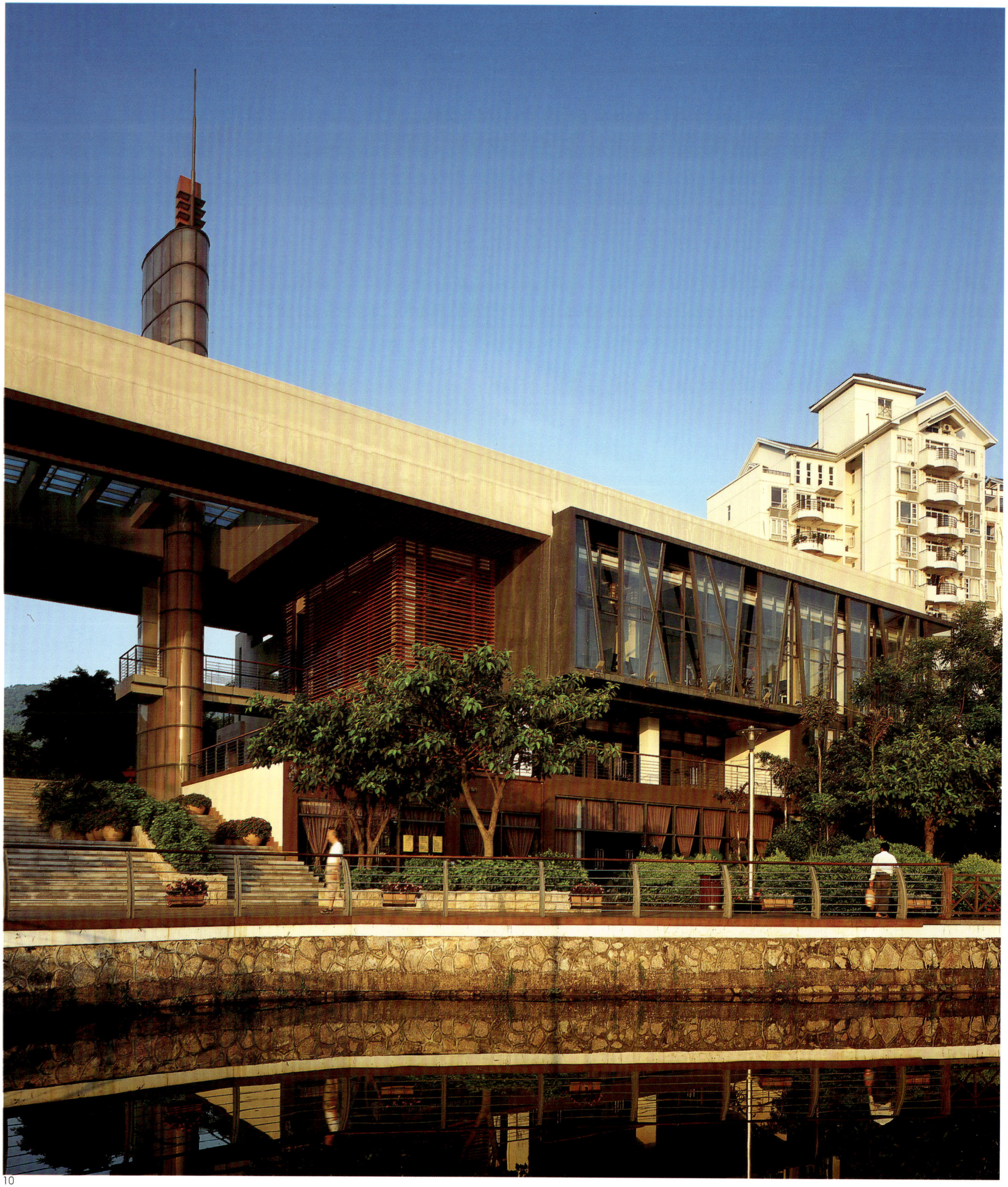

10

between man and nature. Using the five state—eyes, ears, nose, tongue and body, that is, the different feelings of the color, smell, taste and touch of Yinhushan, so as to reach a 3D-natural sense. The natural vegetations in the park are mainly the subtropical monsoon evergreen broad-leaved forest and evergreen sbrub. In some valley, there are still lots of valley evergreen broad-leaved forests in well-preserved with the nature of rainsforest.

Xing•Dante has 160 sets of laminated TOWNHOUSE in second phase, which is expected to be available for sale in October this year. Group E in the project delegations have 732 sets of housees in high-level in all. The main dwelling is a house with 3 rooms or more. The area of these houses are about from 88 square meters to 300 square meters. There are two to five rooms' unit, multiple units and many units which have 270 degrees Landscape vision. Dante is the last batch villa unit at the foot of hill. The types of product include the row villas and independent villas,with an area of between 210-400 square meters. Xing•Dante entered the market in September 2008.

11

10. 会所
11. 组团内街
12. 立面造型

12

13

14

13. 别墅实景
14. 勒石的使用
15. 翠竹的点缀
16. 立面光影

15

16

1

半岛一号

开　发　商：新城市地产开发有限公司
项目地址：惠阳淡水镇东华路
建筑设计公司：华森建筑与工程设计顾问有限公司

2

用地面积：248 400m²
一期总建筑面积127 502 m²
其中商业部分面积10 817 m²
会所1 033 m²
临时幼教中心711 m²
住宅部分面积114 941 m²
地下及室内架空车库面积12 902 m²
其中联排别墅 210~260 m²
叠加别墅 200~220 m²
宽景 77~140 m²
情景 120~150 多 m²
半岛文化体育公园约占地93万平方米
其中淡水河水域32万平方米，分3期开发
容 积 率：1.24
绿 化 率：40%

1. 别墅景观
2. 效果图局部
3. 总平面图
4. 一期鸟瞰图

半岛1号地块位于惠阳规划 8.8平方公里新区，淡水镇与秋长镇结合部，深汕高速公路以北，西邻棕榈岛27洞高尔夫球场及别墅区；东北临半岛1号规划二号路，南临沿淡水河，惠阳半岛文化体育公园，直通东华路，西北临北环路。

项目规划有情景洋房、宽景洋房、叠加别墅、联排别墅。建筑引入西班牙风格的规划设计理念，采用无釉赤陶桔色瓦屋顶，白色拉毛墙面，将阿拉伯风格与欧洲传统建筑形式融合起来；围合、半围合及行排式组团排布，人车分流；独栋别墅区中采用部分院落式的中式别墅。

一期和三期主要为休闲和运动场所，三期将建成以旅游发展、餐饮服务、商业娱乐和酒店住宿为主要功能的配套区域。目前一期已经基本完成，设有入口广场、服务建筑、停车场、儿童公园、健身区、滨水散步道、音乐喷泉广场、露天剧场、门球场、足球场一个、篮球场五个、网球场四个、露营烧烤区、木栈道垂钓区和休闲广场。三期初步规划有野战射击区、滨水广场、露天泳池、乡村俱乐部和高尔夫练习场等。

Greening rate: 40%

Volume rate: 1.24

Floor area: 248 400 m²

Plot of Peninsula Ⅰ is located in Huiyang's new planned district, combination of Danshui Town and Qiuchang Town. It is in the North of Shen-Shan Expressway, the West of the golf course and the villa area in 27 hole, Zhonglv Island. In its Northeast is the Second Road which is included in the Peninsula Ⅰ plan. Along with Danshui River in its South there's Huyang Peninsula Cultural and Spots Park, which goes directly through Donghua Road and faces Beihuan Road in the Northwest.

The project plan includes western-style houses with beautiful scene, town houses and row villas. Spanish style of designing rationale is introduced into the construction plan to adapt orange roof and white walls, fusing Arabic style and traditional European construction form together. Enclosure, semi-enclosure and rowly distribution is to separate pedestrians and vehicles. Individual villa area takes parts of the residential form from Chinese villa idea.

1st and 3rd phases are mainly for leisure and sports, and the latter one is to be completed as attached area with major function of travel development, catering services, commercial entertainment and hotel accommodation. The 1st phase is being completed with entry plaza, construction services, parking lot, children's park, gymnastic zone, waterfront promenade, musical fountain plaza, open-air theatre, gate-ball field, a football field, 5 basketball courts, 4 tennis courts, camping and barbeque area, zone of wooden platform for fishing and leisure plaza. The 3rd is primarily planned with field shooting, waterfront plaza, open-air swimming pool, country club and golf drilling courts,etc.

3

4

5. 一期总平面图
6. 人视效果
7. 中轴线景观分析图

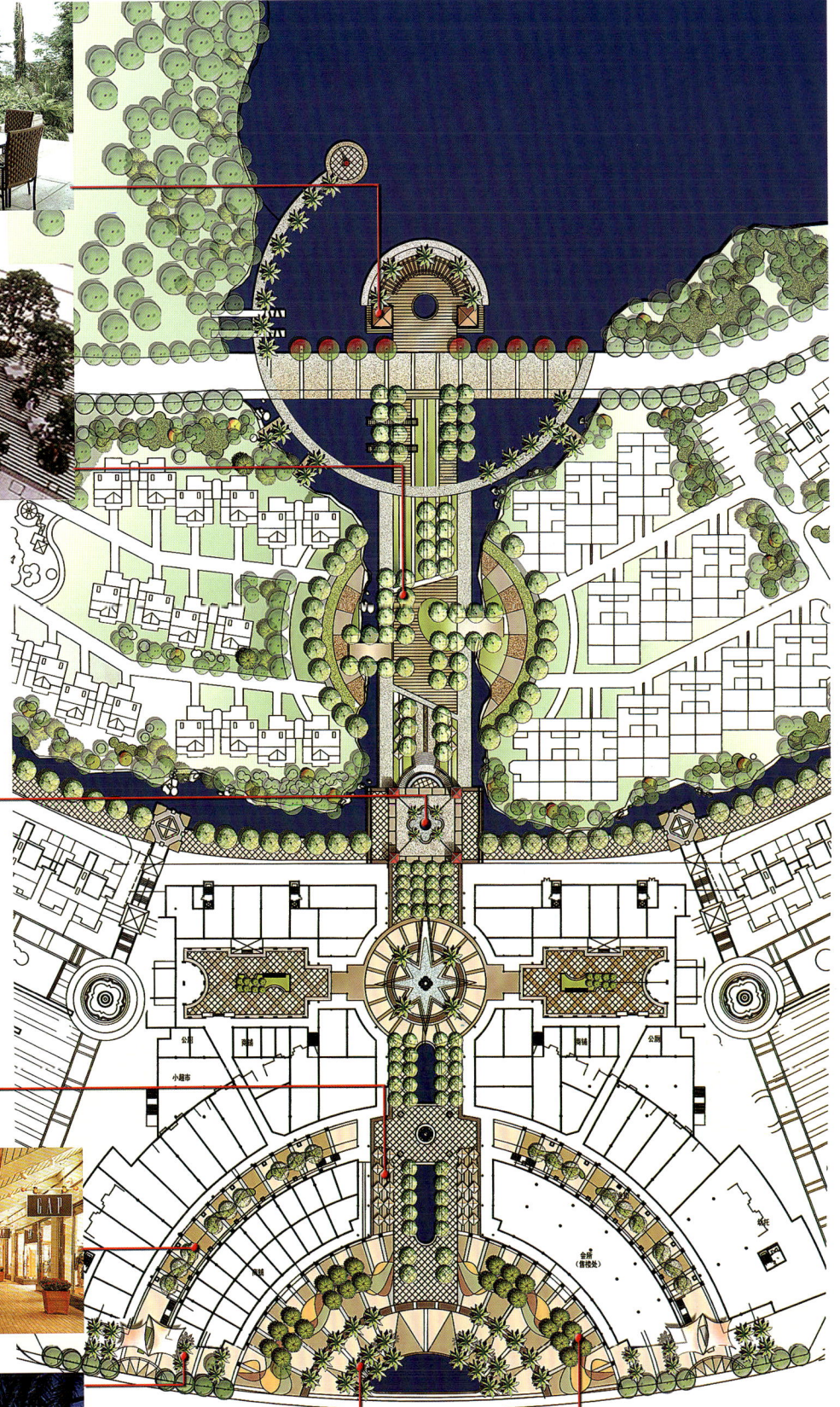

8. 街区商业入口透视
9. 立面集锦
10. 屋顶部分节点大样及示意图
11. 檐口大样
12. 别墅实景图

8

多层洋房透视1

多层洋房透视2

多层洋房透视3

多层洋房透视4

9

111

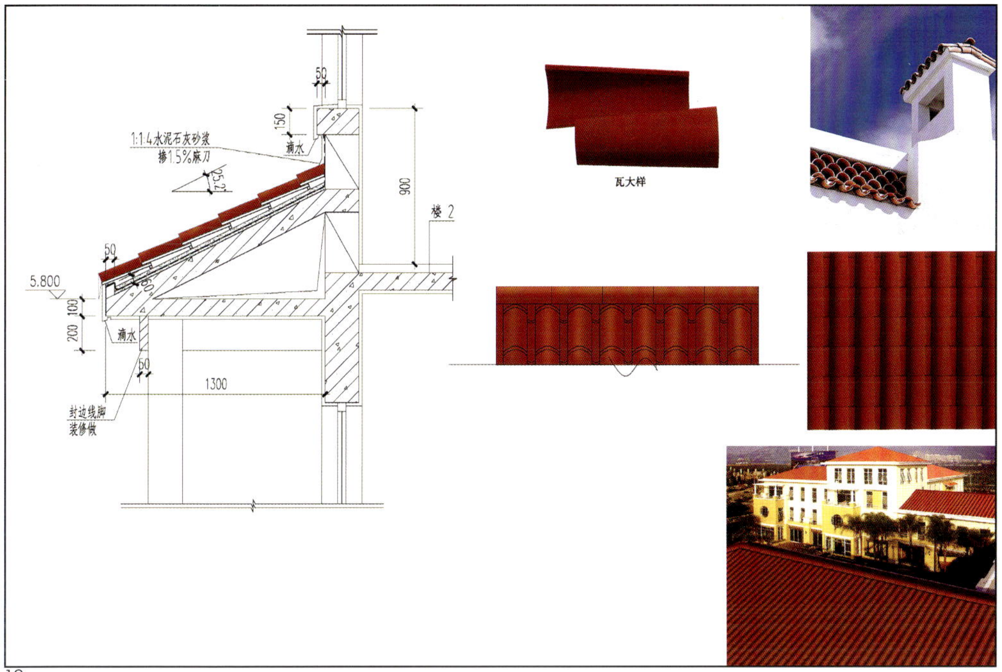
1:1:4水泥石灰砂浆
掺1.5%麻刀
滴水
楼 2
5.800
1300
封边线脚
装修做
瓦大样

10

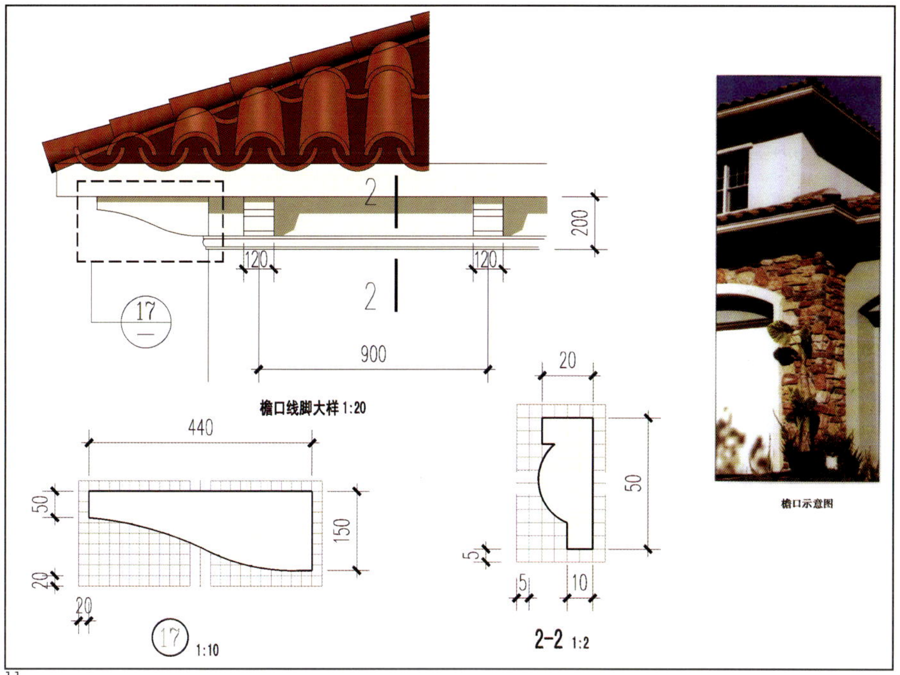
900
檐口线脚大样 1:20
440
17 1:10
2-2 1:2
檐口示意图

11

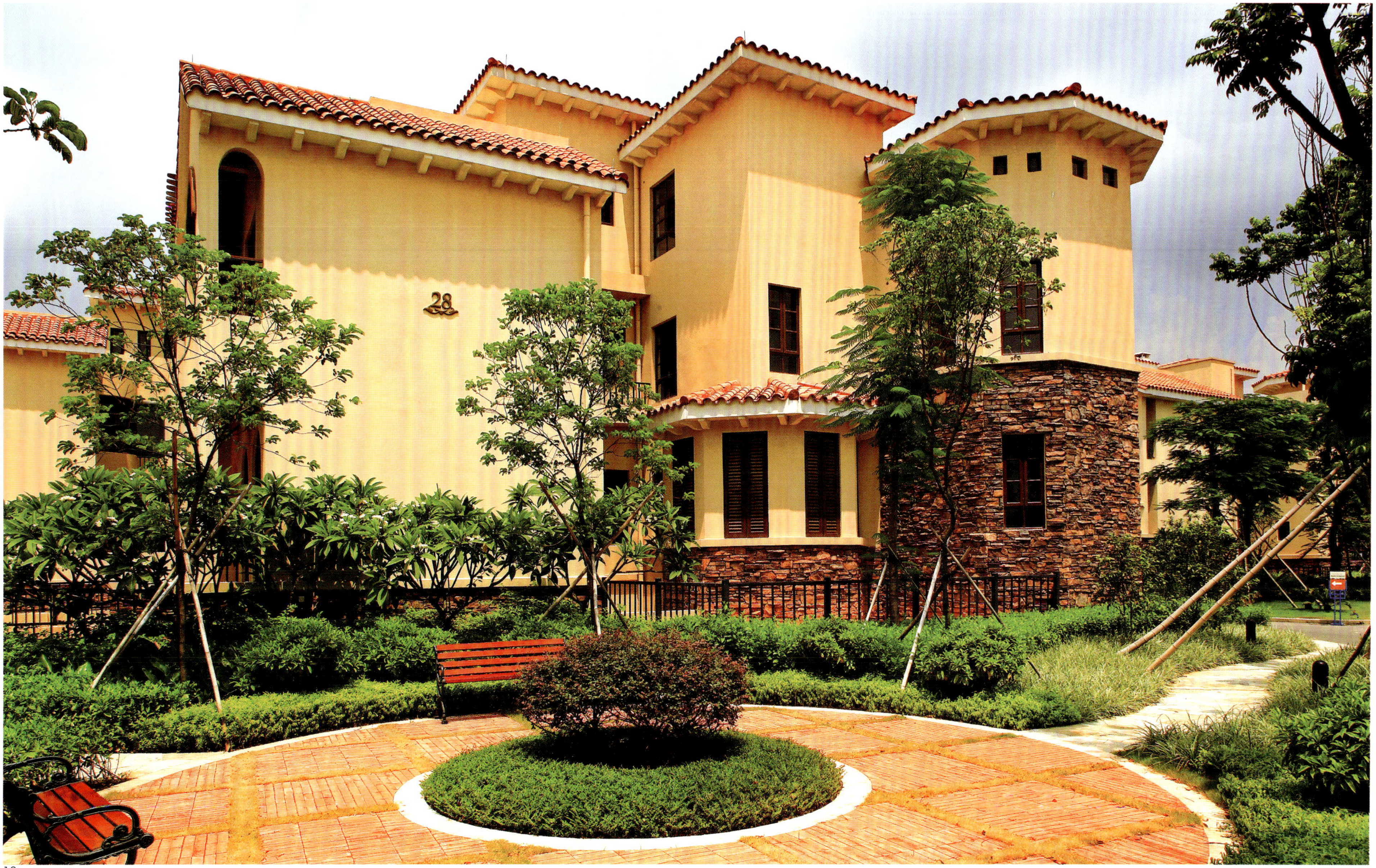
28

12

13

14

13. 主入口中轴景观透视

14. 广场地花

15. 阳台栏杆节点大样及示意图

16. 花架节点大样及示意图

17. 小区环境

阳台栏杆节点大样及示意图

18

19

20

21

22

23

22/23. 商业街区广场实景图
24. 商业街骑楼

24

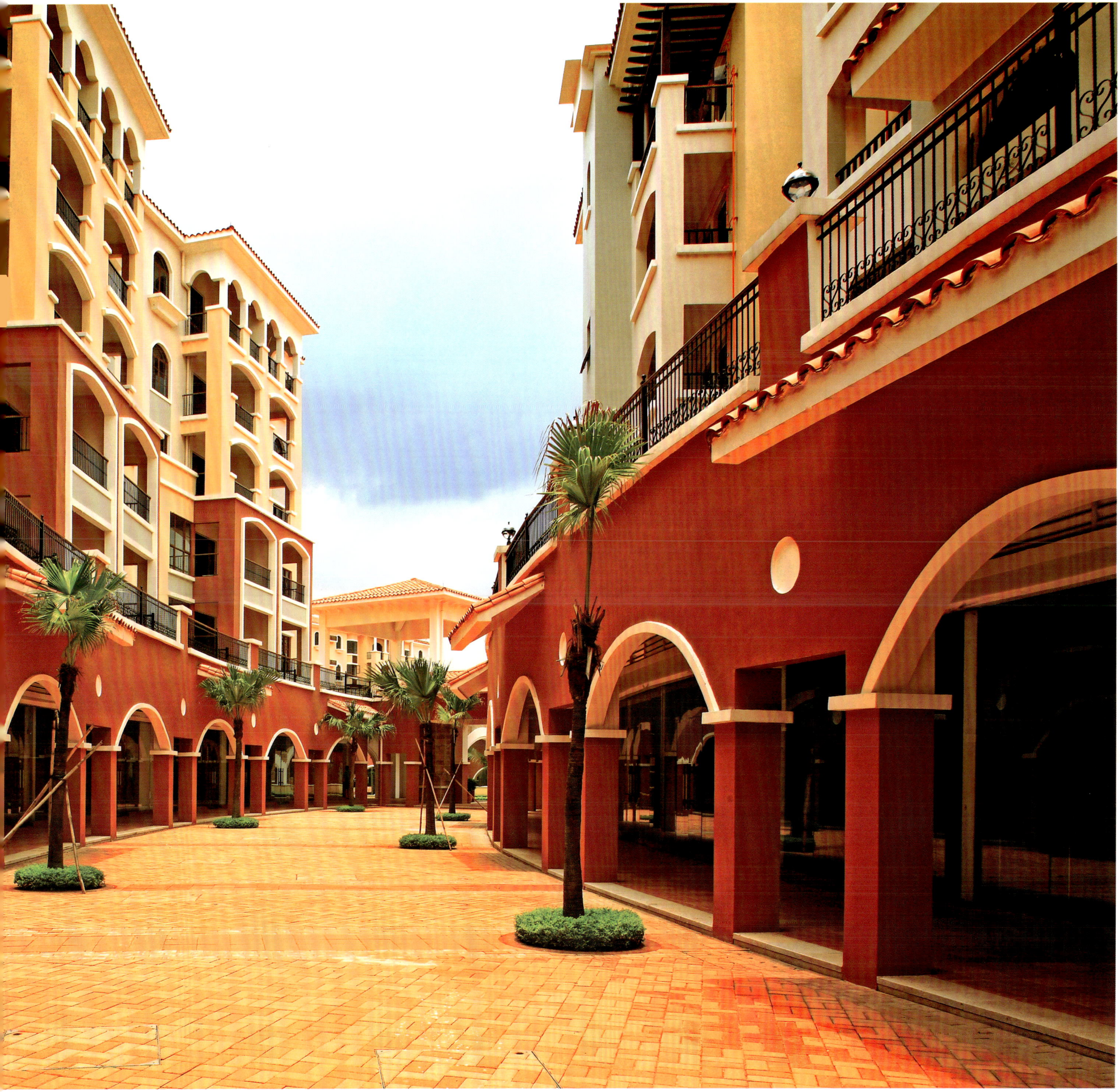

1

杭州大华·西溪风情住宅区

开 发 商：浙江大华房地产开发有限公司
项目地址：杭州西湖文一路延伸段，西溪风景区以西
建筑设计：华森建筑工程设计顾问有限公司
景观设计：泛亚国际
占地面积：800 000 m²
建筑面积：487 000 m²
容 积 率：0.6
绿 化 率：>50%
建筑密度：<23%
总 户 数：130

2

大华·西溪风情位于杭州城西，距市中心约10公里，正居珍稀的西溪湿地，约400米近距离贴近西溪国家湿地公园，规划将建成建筑面积达500000平方米的城市湿地水景排屋别墅社区。大华·西溪风情带着对现代生态住宅建设的思索，带着对中国古人"天人合一"精髓的感悟，在保护性开发的旗帜下将现代生活融入西溪岸边桃红柳绿的盎然，把现代环境景观设计曲径通幽的意韵巧缀于城市的繁华。在保留大华·西溪风情3500余米生态河岸线和内部约30000平方米天然水域的基础上，尊重自然，崇尚生态，尽可能在开发过程中保持西溪地区生态环境，延续原有的西溪湿地自然风貌，把大华·西溪风情打造成一个城市中的诗意家园。

临近西溪风景区，社区环境优美，拥有30000余平方米中央生态景观湖，70余处园林景观。"大华·西溪风情"大型低密度水景社区拥有有联排排屋、叠排排屋、双拼排屋和独立别墅等多种物业类型，绿化率高达50%以上。在规划设计上，涵盖了建筑、绿化、交通、文化等方面的设计思路，坚持社区规划的整体性和统一性，确保了居住的生态环境和人文环境。

大华·西溪风情以先进合理的景观设计理念，延续与发展原生态景观，用最经济合理的方式和最好的景观形态解决小区的防洪排涝问题。采用外高内低的场地地形，基地外围设置连续的生态土堤以达到50年一遇的防洪能力，减少填方降低内部场地高程。同时，将整个基地内部组织为一个独立的排水系统。小区规划设计成功的保留原生态景观资源形成一个完整的景观系统。整

个景观系统由占地3万平方米的中央生态湖，延续3.5公里长的河滨休闲绿化带及多条联系两者的景观河道共同形成一套完整、均好的景观网络。偌大的生态景观湖将绵延曲折的湖岸、婉转自由的流水、生机盎然的草木和饶有生活情趣的居住者融于一体，充分发掘人和水的互动关系，鼓励居住者近水、接触水，真正体现回归自然的现代生活理念。

大华·西溪风情采取组团式开发，目前已经开发三期，一期水岸茗苑共建有联排排屋364套；二期水榭茗邸共建有联排排屋、叠排排屋共347套；三期水榭香堤共建有双拼排屋、联排排屋和独立别墅312套，后续组团将陆续进行开发。

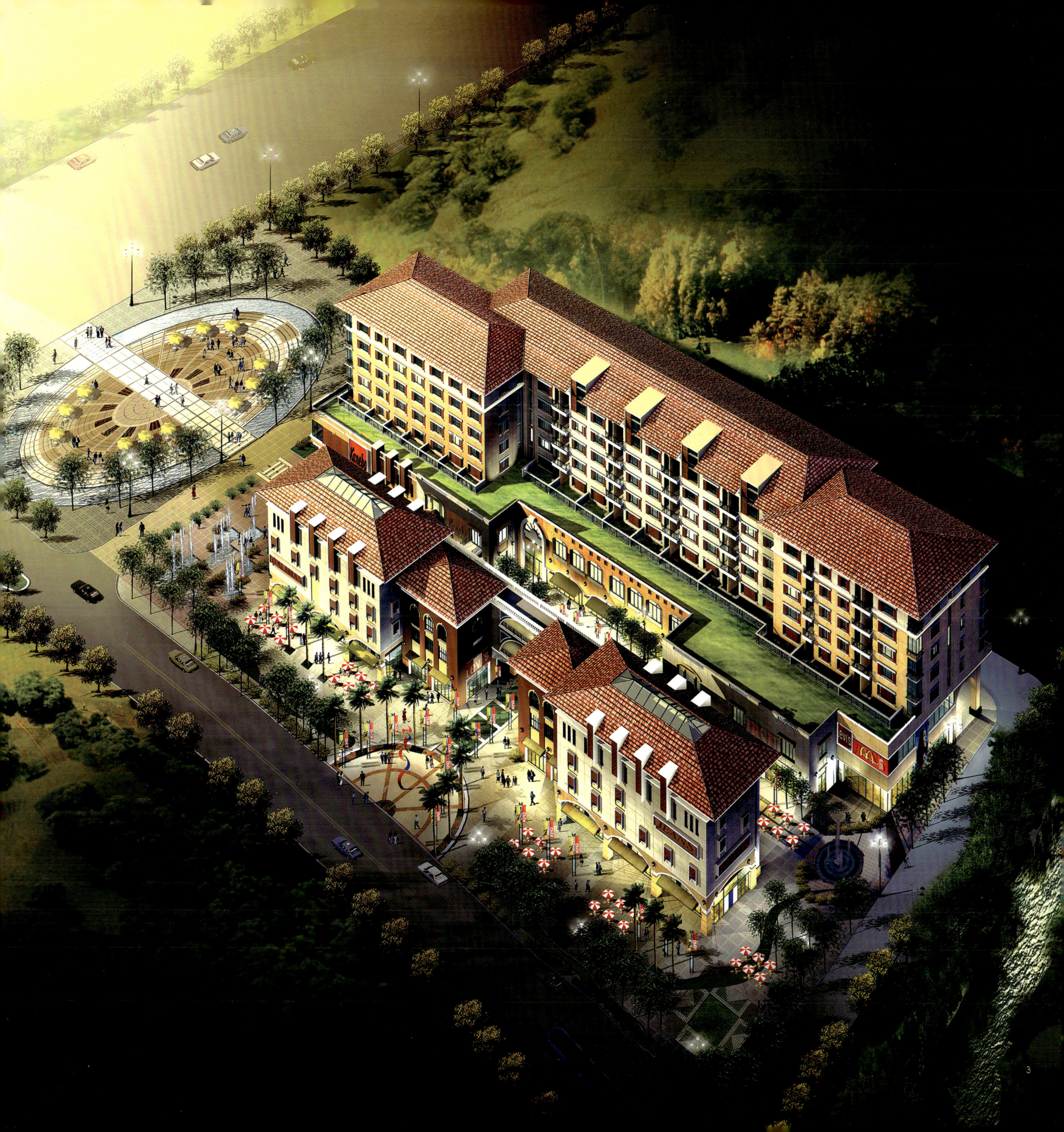

4

5

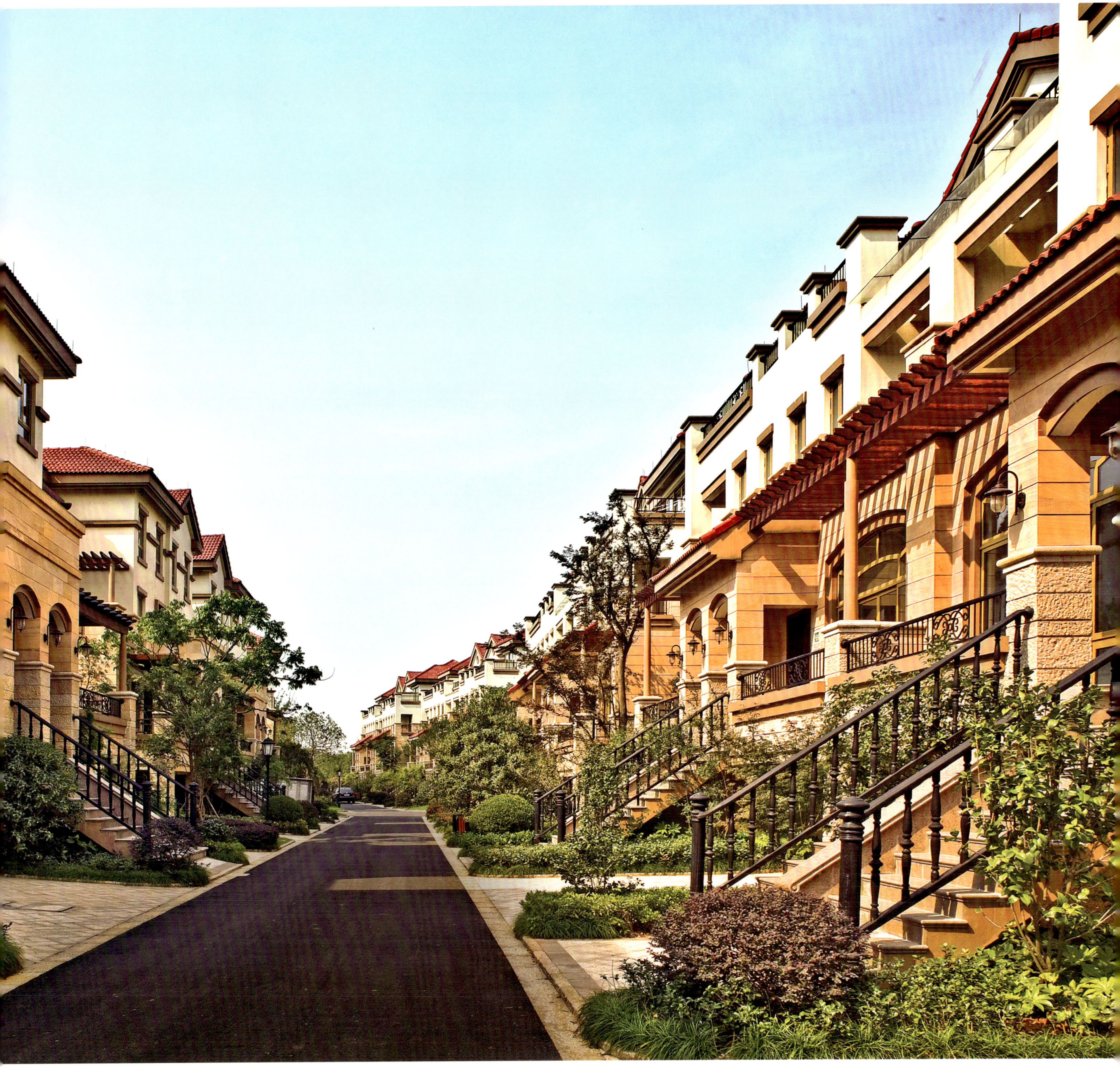

Developer: Zhejiang D.H. Real Estate Co.Ltd.

Project Address: Stretch of Xihu Wenyi Road, the West of Xixi Scenic Spot

Architectural Design: Watson Engineering Consultants Ltd.

Langscape Design: Pan Asia International

Area: 800,000 square meters

Building Area: 487,000 square meters

Volume Rate: 0.6

Green: >50%

Building Density: <23%

Total Number: 130

Dahua Xixi style is located in the West of Hangzhou, about 10km away from the city center. The area is occupied the rare xixi wetland, and it is close to the Xixi. National wetland park, just about 400 meters away. It is planned to build villa community of urban wetland with a construction area of 500,000 square meters. With the thinking of housing construction in the modern ecological way and the ancient Chinese "Harmony", it blend the modern life well with the exuberant scene on the Xixi shore under the banner of prtective development. The quiet of the modern environmental landscape design merge into the prosperity of city. On the basis of retaining 3500 meters of ecological bank lines in the Xixi customs and about 30,000 square meters ofnature internal water, it respects for nature and ecology. In the process of development, the ecolofical environment in the Xixi is maintained, as far as possible to keep the original natural state of wetland ang build the Xixi style into a romantic home in city.

The environment in community is very beautiful near the Xixi scenis area. There are 30,000 square meters of central ecological landsacpe , and about 70 gardening landscape. The large low-density waterscape community have row of townhouses, stacked row bungaows, townhouses, independent villas and other property types. And the green rate is as high as fifty percent. In the design, it covers the construction, afforestation, transport, culture and other aspects of design. The planning of community uphold the integrity and uniformily, to ensure the living ecological environment and human environment.

Under the advanced and reasonable concepts of lanscape design, Xixi style carry on the ecological landscape and develop it with the most ecological and reasonable way and the best shape of the landscape to solve the problem of antiflood and drainage in the district. With the outside the high-row, the external base set up a continuous and ecological embankment to achieve the high ability of antiflood which is unhappened in a period of five dacades, to reduce earth fills and elevations of space. Meanwhile, the entire innerbase is organized into a independent system of drainage. It is success to retain the original ecological landscape of the plannng, and the resources of the landscap form a complete system. The entire system of the landscape is composed of a central ecological lake which covers 30,00 square meters, a 3.5km long recreational green belt in the riverside and a number of watercourse which is contact with the common river, to form a complete and good network of landscape. The huge lake of ecological landscape made the stretching and tortuous lakeshore, the free water, vigorous plant and basically dweller blend well with each other. It explores entirely the interaction between water and human, and encourages the residents to approach and contact with water which really reflect the return to nature of modern living philosophy. Dahua Xixi style is developed by groups., so far 3 phases have been developed. There are 364 sets of row-houses in first phase court in the water's edge, 347sets of townhouses and stached-row-townhoses in the second court in the water's edge and 312 sets of Semi-detached townhouses, townhoses and detached houses. The following groups will do further development.

6

7

8

9

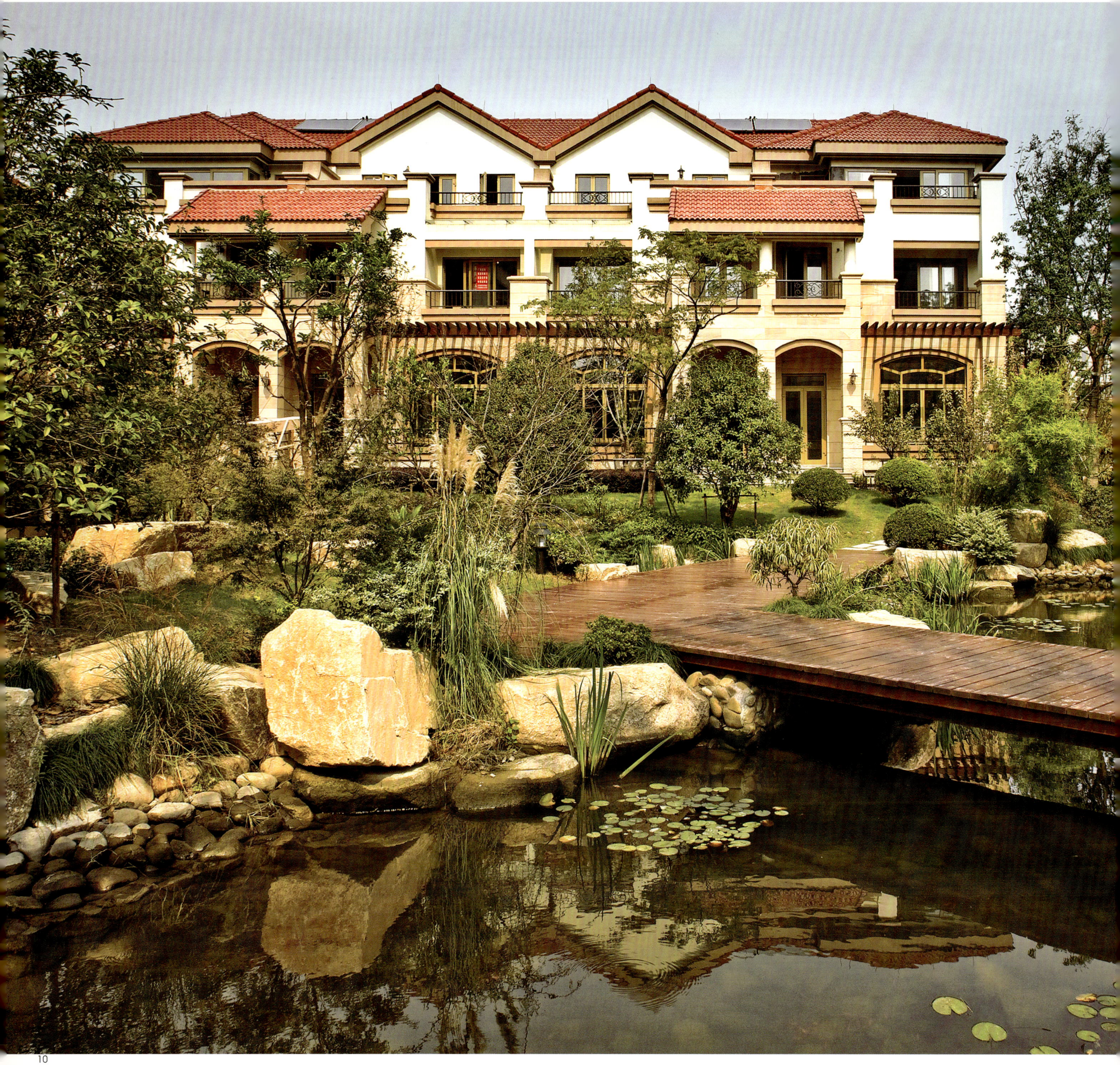

10

11

12

126

1

Developer: Chengdu Yuxing Real Estate Development Co., Ltd
Project Address: Chengdu Renminnan Road Line
Planning Design: Hanshi Internation
Area Covered: 99 694 m²
Total Building Coverage: 85 355 m²

1. 宅间关系
2. 住宅局部
3/4. 效果图

南郡七英里

开 发 商：成都裕鑫房地产开发建设公司
项 目 地 址：成都人民南路沿线
规 划 设 计：翰时国际
总占地面积：99 694 m²
总建筑面积：85 355 m²

2

　　本项目位于成都市南人民南路沿线，西邻成仁公路。项目用地面积为99693.8平方米，规划总建筑面积85354.8平方米。规划整体布局借鉴欧洲小镇住宅的型式，同时融入中国传统庭院理念，通过建筑体量围合，形成相对私密、尺度宜人的组团院落，营造既保持传统文化又独具西方舒适性的生活氛围。组团绿化向中心水系渗透，将每个组团包围在环状绿岛中央。项目户型设计极大的满足了使用的合理性及舒适性，每户均享有前后庭院及围合式内庭院，在户内形成露天空间，各种不同庭院空间的设计，继承川西文脉，同时又加以创新，以邻里空间的塑造为主题，创造都市内的庄苑生活。

This project is along Renmin South street and Chengren Street, in the south of Chengdu city. The site area of this community is approximately 99693.8 m² and the planned area is about 85354.8 m² in total. The planning of this community is according to the type of European town and adds in exoterica of Chinese traditional courtyard. It orientated on combining Chinese traditional culture and western comfortable life style through buildings' collocation and private courtyard group. The virescence pervades to center water system, which encircle each group in the center of orbicular green islands. The layout of the room is reasonable and comfortable, and each room has a front garden, a back garden and an inside one, which succeed to the Chuanxi culture and also innovate by the designer.

3

4

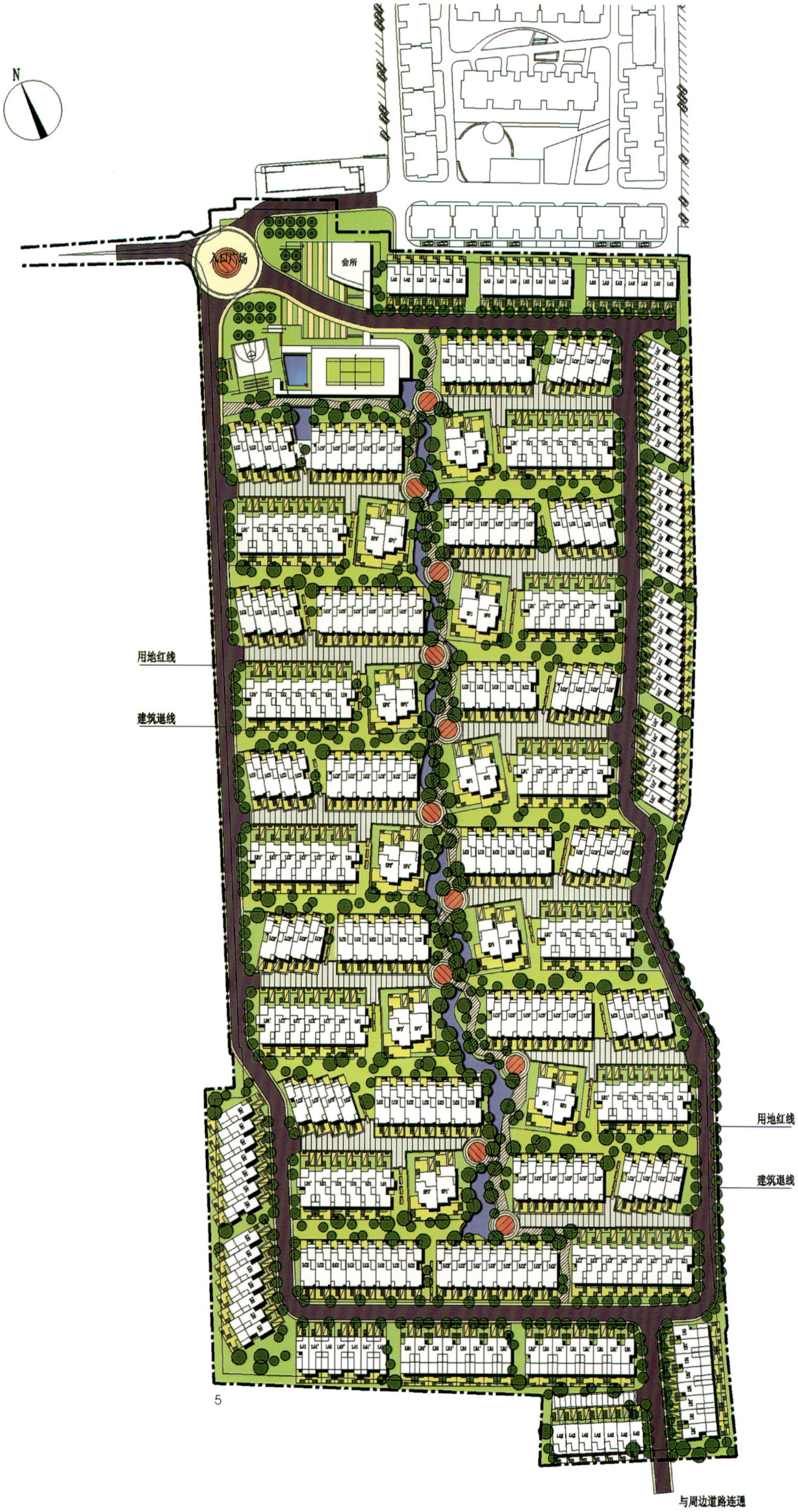
N
入口广场
会所
用地红线
建筑退线
用地红线
建筑退线
与周边道路连通

5

5. 总平面图
6/7. 住宅立面
8. 联排住宅二层平面图
9. 联排住宅立面图
10. 联排住宅一层平面图
11. 联排住宅实景图

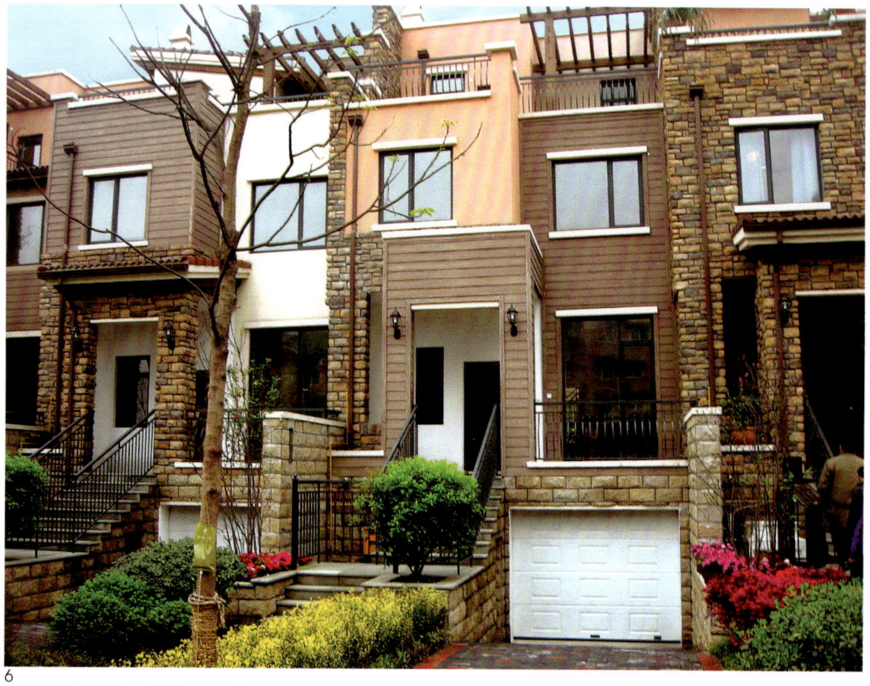

6

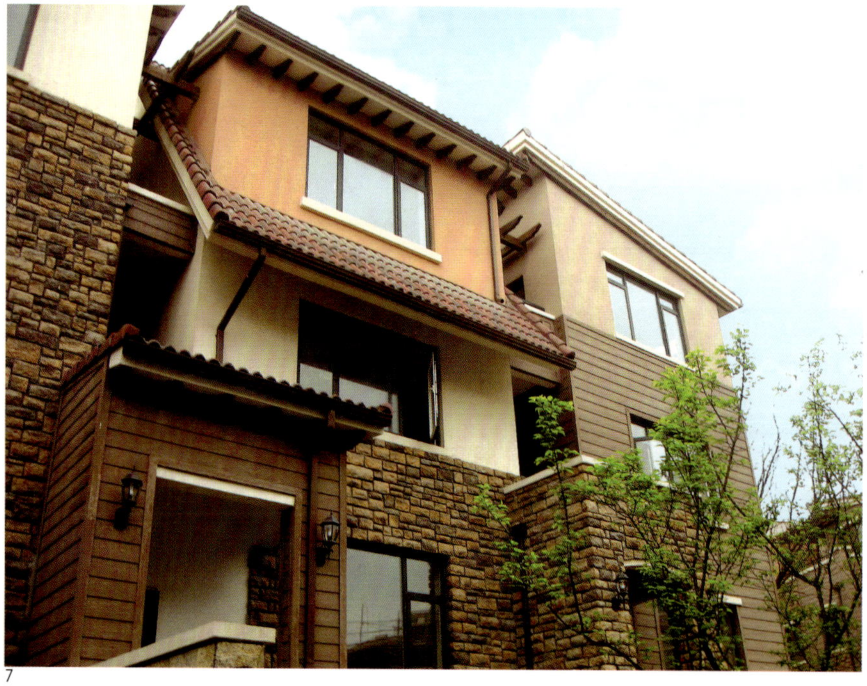

7

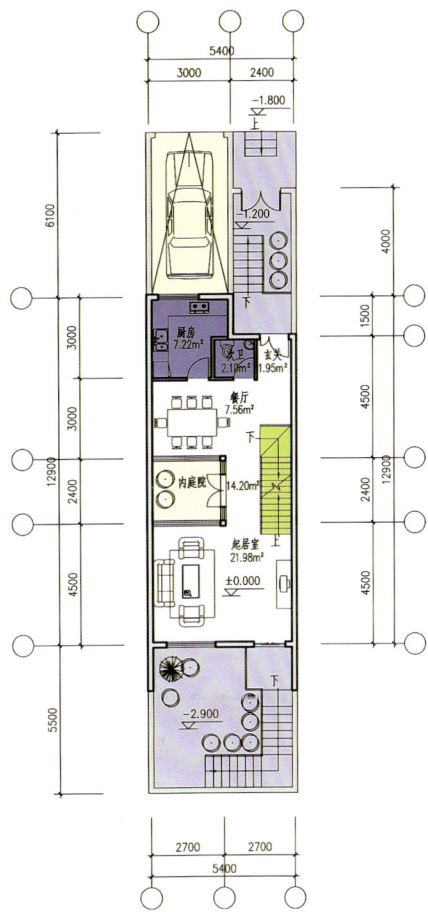

厨房
7.22m²

洗卫
2.11m²

金美
1.95m²

餐厅
7.56m²

内庭院
14.20m²

起居室
21.98m²
±0.000

−2.900

−1.800

−1.200

5400
3000 2400

6100

3000

3000

4500

5500

4000

1500

4500

2400

4500

12900

12900

2700 2700
5400

8

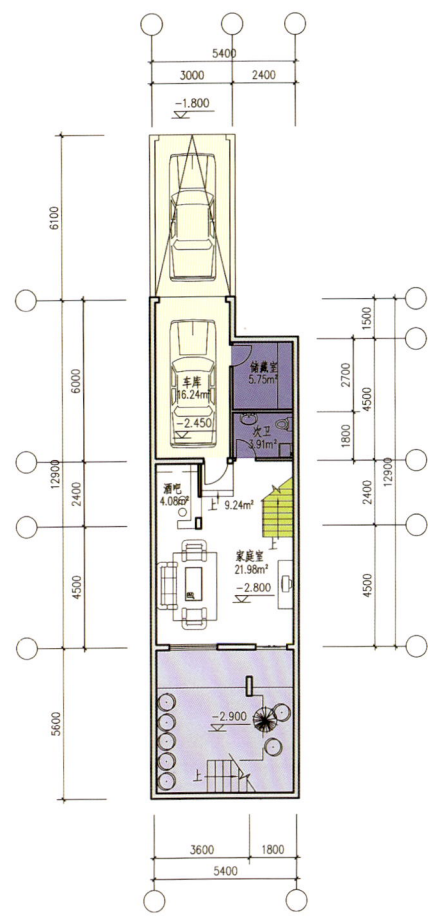

−1.800

车库
16.24m²
−2.450

储藏室
5.75m²

灯卫
1.91m²

酒吧
4.08m²

9.24m²

家庭室
21.98m²
−2.800

−2.900

5400
3000 2400

6100

6000

2400

4500

5600

1500

2700

1800

2400

4500

12900

12900

3600 1800
5400

10

9

11

12

13

14

15

16

12. 双拼户型立面示意图
13. 双拼户型平面图图
14/15/16. 建筑立面与局部

亿城天筑

开 发 商：北京亿城房地产开发有限公司
项 目 地 址：北京西南四环花乡桥南
建筑规划设计：北京中联环建文建筑设计有限公司
建 筑 面 积：83 690 m²
容 积 率：1.05
总 住 户 数：518套

1. 别墅立面
2. 园林景观
3. 总平面图

亿城天筑项目位于北京西南四环花乡桥南500米处，总建筑面积8万多平方米，住宅部分容积率1.05。项目在花园洋房舒适居住的基础上创新，依照别墅的尺度和品质，在地中海式风情建筑体内，创造出诸多符合中国传统生活习性的空间布局，主推两种洋房格局：4-5层退台花园洋房和9层观景电梯洋房，518套人性化自然居所。

亿城天筑推出新品花园soho，在极具个性的LOFT空间，肆意挥霍想象，住户可以在这里布置像家一样温馨随意的办公室，可以设计自己的商务会客厅，或者可以一层办公，搭建隔层用于栖居……楼层挑高5米，缔造楼中楼的mini house，打破传统模式，解构创造力至上的新经济时代，精致办公，物尽其用。通过夹层巧妙实现动静分离，缔造双层双厅的复式空间，一层空间，两层内涵，错跃式结构自然分出不同生活。选择工作，还是栖居？或者将二者融于一体，在这里实现如此简单。高空间，充裕采光，真实自然，自由畅想和感性刺激不用创造，让创意在不同高度唯美绽放。

The area of construction: 83 690 m²
The rate of volume: 1.05
The number of dwellings: 518sets

Tianzhu Yeland is located in 500 meters south of North-west Sihuan Hua Village Bridge of which the total construction area is over 80,000 square meters and dwelling volume rate is 1.05. The innovation of this project is based on the comfortable living condition of foreign-style buildings. In accordance with the standards and quality of foreign-style villa and the Mediterranean-style architecture, it arranges the space that accord with Chinese traditional habit of living. The recommendation of two patterns of foreign-style houses:4th -5th floors are western style houses with back yard and 9th floors are the house with viewing lift. There are 518 sets of human natural houses.

Tianzhu Yeland launches new residents with Soho style which aims to squander imagination in LOFT space of personality, so that residents can decorate their houses as warm and casual office, besides, they can design their own business reception room or one floor for business and the next floor for living. The height of the storey is five meters which creates extra mini house in the dwelling. This design breaks through the traditional style to seek for the innovation of new economic era, and make full use of the material and space. It ingeniously separates dynamic and static by sandwiching to create mixed and double-storey space which can be divided into different lives. For working or living, or blend them together? Here it seems easy to make it. Enough space, plenty of light, reality, nature, freedom and emotional stimulation have made the innovation flow.

3

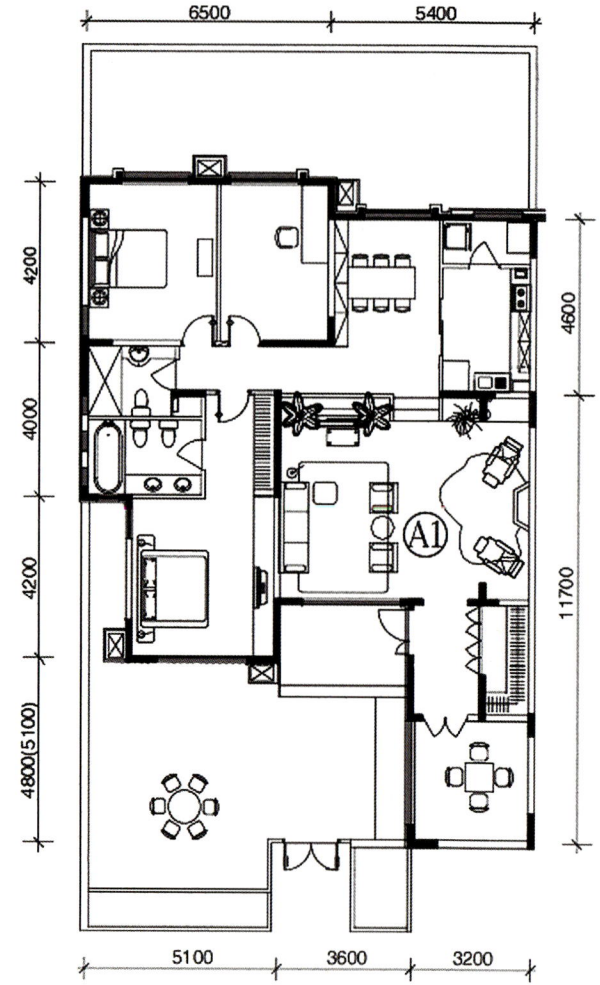

5

4

6

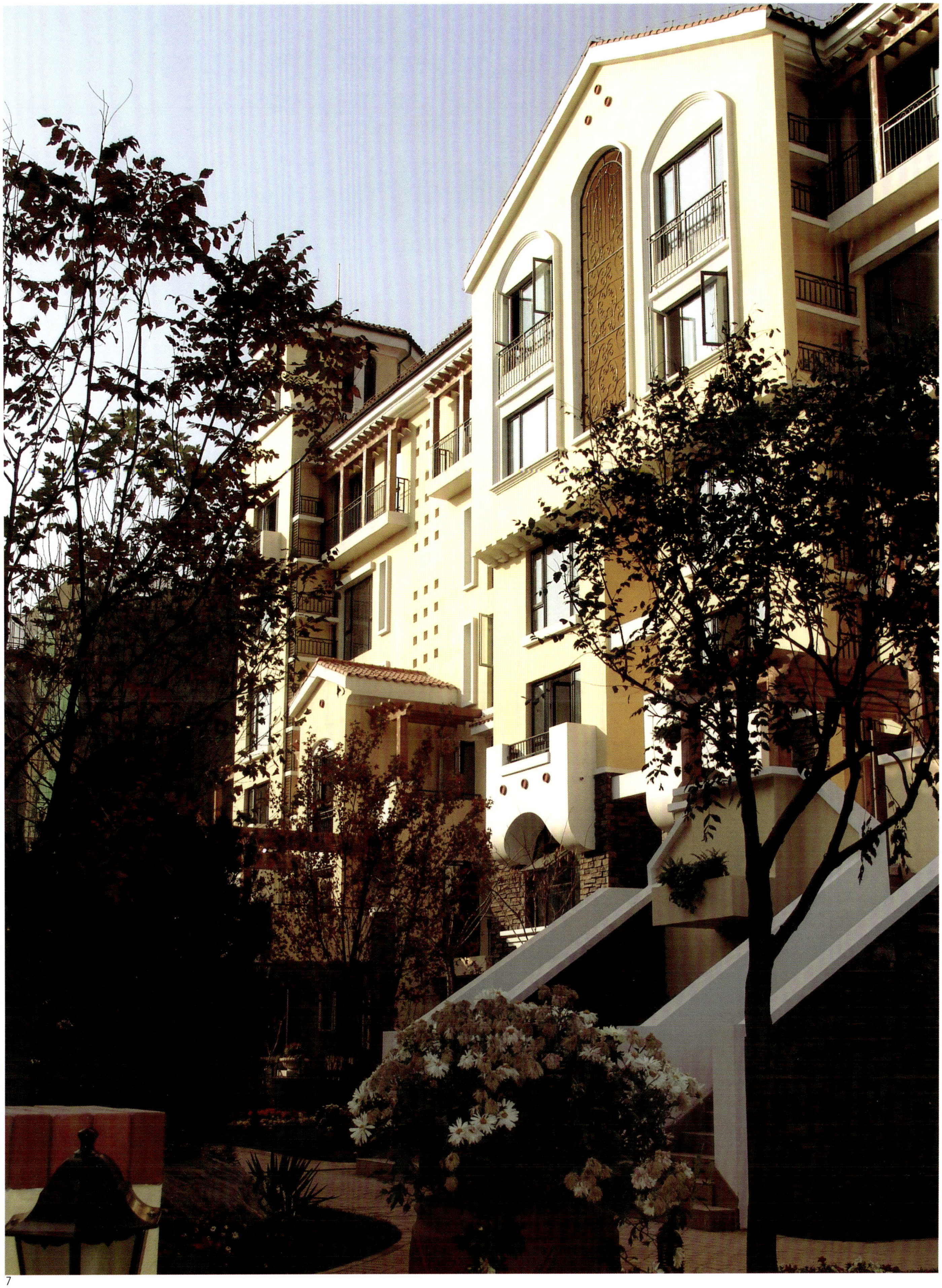

4/6/7. 别墅实景图

5. 户型平面图

7

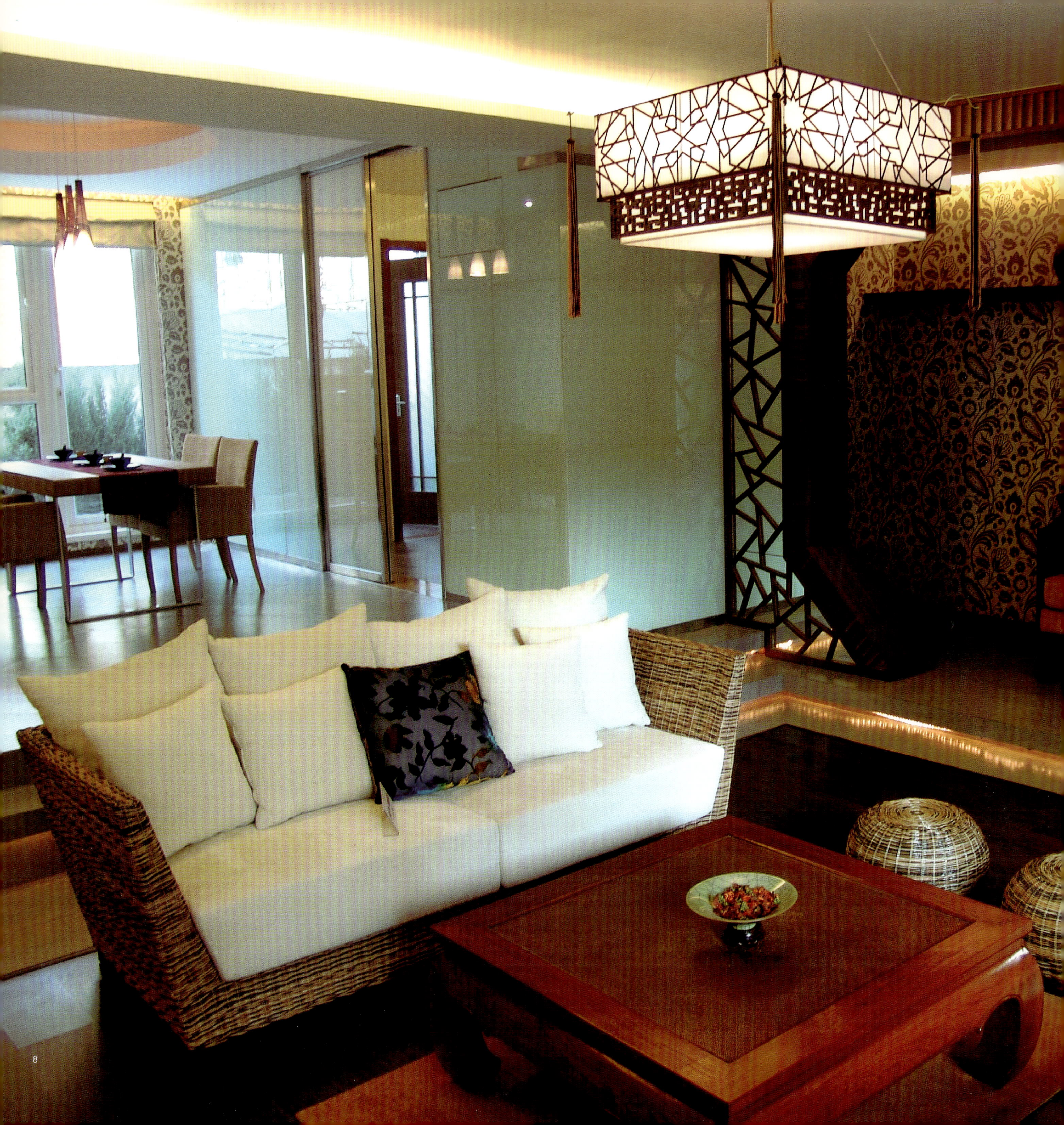

9

10

8/9. 客厅实景图
10. 卧室实景图

1

香草天空(康城二期)湖景叠拼别墅

开　发　商：北京银信兴业房地产开发有限公司
建 筑 设 计：澳大利亚五合国际·澳洲易道
占 地 面 积：103 000 m²
总建筑面积：110 000 m²
绿 化 率：36%
容 积 率：1.19
总 户 数：384套

2

香草天空(格调别墅)位于"大康城"成熟一期西侧,总占地10.3公顷,其产品形式为低层低密度叠拼别墅,上叠有艺术阁楼和风情露台、下叠有私家庭院及情趣地下空间,小区内建有豪华泳池会所。小区南侧紧邻27000平方米代征城市湖景绿地、7000余平方米人工湖面、景观大道环绕其中、郊野公园及周边无际绿林带构筑纯粹格调景观意象。

香草天空整体规划以南部代征湖景绿地及50米宽自然林带为景观依托,流畅、柔和的纹理充分表达了自由、浪漫的设计思想。为营造更多自然的、原生的格调空间,小区内布以大面积步行栈道。小区绿化环境以带状绿地为主,曲径通幽的景观打破传统兵营式排列,与区内自然步行系统构成业主的主要室外活动空间;朝向人性化、邻里空间尺度佳、前庭后院绿茵阵阵。

香草天空的建筑呈现出一派新地域主义风格,与传统建筑相比,除吸收其精华的工艺设计元素与功能构件外,重点创新表现在对屋顶及阳台等部位的大胆突破——俊朗开阔的屋顶造型与内敛深灰使建筑整体大器中透着完美与精致;外墙现代解构主义的进退错落风格,令建筑体态生动丰富,层次感渐近,格调生活的不同场景可得到充分演绎;内敛深灰与赭石面砖的对比色运用,在视觉上形成强烈印象,但又不失与屋顶色彩的协调,建筑整体表现极具尊贵与文化感,充分彰显格调建筑翌翌光辉。香草天空在外立面的设计上着力突出居住建筑的空间尺度,居住的自然与舒适。

香草天空独揽大自然圣境,人工湖水面如一天然巨玉镶嵌于香草绿地之心,作为大面积风景板块的视觉中心,湖岸有山石、草坡入水,并配以叠水景致;园区植被依地势而建,或为苍翠、或为葱郁的绿植跌宕出纯粹而精致的自然空间。郁郁葱葱的毛白杨林带、曲线优美的林中香径、朴野自然的山石跌水……和谐的风景元素构成了湖景绿地舒缓有度的空间节奏。香草天空完整一体化的景观大空间设计表现为现代而简约的构成风格;整体景观由一条主轴绿谷与湖景绿地串联而成,自然形成入口门区景观区、绿轴景观区、庭院景观区等相对独立而又紧密相联的景观区域;景观体系层次丰富、功能合理,构成了主题明确、过渡自然、内容丰盈的景观意象。塑造了人与人,人与自然共生的和谐篇章。

香草天空户型细分为18种,建筑面积为181-315平方米。户型设计充分彰显格调人士另类个性,在空间的设计上极具人性化——情调空间与氛围的营造是其最显著的特点(品位人士可将此空间设计为私家香薰室);整体空间布置完全按照人性化、舒适化、科学化的原则设计,同时室内与室外空间的呼应与融合亦得以充分表现:室内屋顶露天阳台、首层庭院空间的自然过渡使格调人士与外部自然更亲密接触。

松

双

经

沥

东

路

三元乳品厂东路

市双桥乳品厂

施工

双 纬 南 27.36 路

D 17#楼 6   D 18#楼 6   19#楼 D 6

A 12#楼 4   13#楼 A 4   B 16#楼 4

A 10#楼 4   11#楼 A 4   B 15#楼 4

A 8#楼 4   9#楼 A 46.63   B 14#楼 4

C 5#楼 4   C 6#楼 4   7#楼 C 4

C 3#楼 4   4#楼 C 4

A 1#楼   2#楼 A

22#楼   20#楼   21#楼

会所

140

4

Developer: Beijing Yinxinn Xingye Real Estate Development Co., Ltd.

Floor area: 103 000 m^2

Gross floor area: 110 000 m^2

Greening rate: 36%

Volume rate: 1.19

Total units: 384

Vanilla Sky (villa of fascination) is located in the west of the completed 1st phase of Cannes, covering an area of 10.3 hectare. Its product includes low-rise and low-density cityhouses, superposed decks of which are artistic attics and scenic terraces, and inferior one are private courtyards as well as basements. Besides, there are luxurious swimming pool clubs. In the south the residential quater close next to the city lake view and green area of 27,000 square meters, man-made lake of more than 7,000 square meters. Pure fascinating landscape is created by Feature Avenue surrounding it, the country park as well as boundless greening belt nearby.

Based on the lake view and greening area in the south and the

natural tree's line of 50 meters' wide, such smooth and tender grain fully express our design of freedom and romance. To create more natural and original style, large areas of the residential quarter are spread with walking trestles, breaking through traditional regimented array, forming main outdoor space for owners. Also, you can enjoy personalized direction, proper scale and the greenness in the front and back yards.

Construction of Vanilla Sky turns on a style of new regionalism. Spacious roofs and dark grey tone is the most important innovation of the constructions apart from the essential crafts design elements and function components that have been taken into them. Houses in picturesque disorder of deconstructionism have been adopted and enriched, polishing different circumstances of elegant life. And contrasting colors of dark grey and burnt sienna brick can impress you generally to show culture of Vanilla Sky. In terms of the outside elevation, space scale is emphasized to offer natural and comfortable living conditions for you.

As well as natural view, Vanilla Sky also possesses a man-made

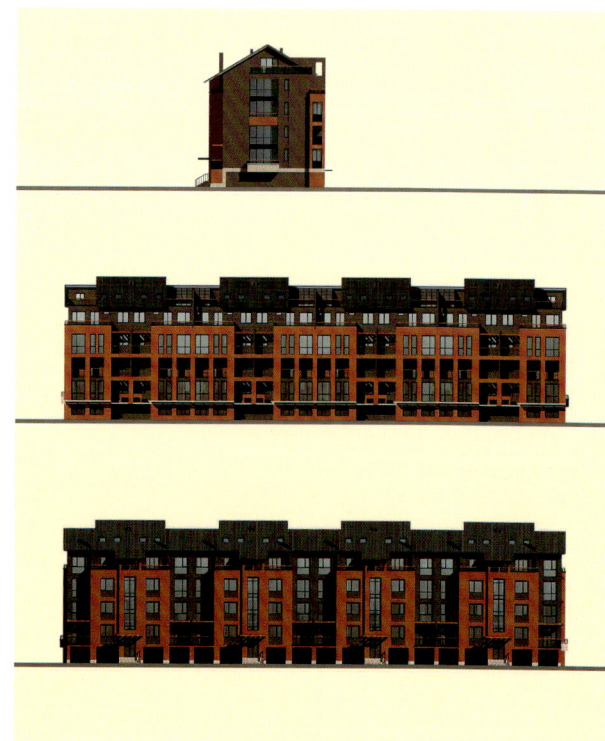

5

141

lake which is inserted in the center of the greening area like a mirror. Harmonious scenic element such as hill stones on the bank, water along the grass slope, green plants show you pure and finished natural space. General feature space under the integration of Vanilla Sky is full of modern and brief style: a middle-axis green valley connecting the lake view naturally makes the entry feature, the green axis landscape and courtyard feature which are relatively independent. Rich layers, reasonable functions demonstrate a clear feature image, harmony between people and nature.

There are 18 housing types with areas of 18 square meters to 315 square meters. And the design obviously expresses designers' special personalities. Then, the distinctive feature of the personalized space is the elegant space (also can be designed as private Aromatic room). Arrangement of space is set comfortably and reasonably which mergence of indoor and outdoor space is fully embodied, such as the indoor ceiling, patios, initiate courtyard allows people to get along with nature more closely.

4. 联排别墅效果
5. 联排立面图
6. 会所平面图
7. 别墅人视效果

6

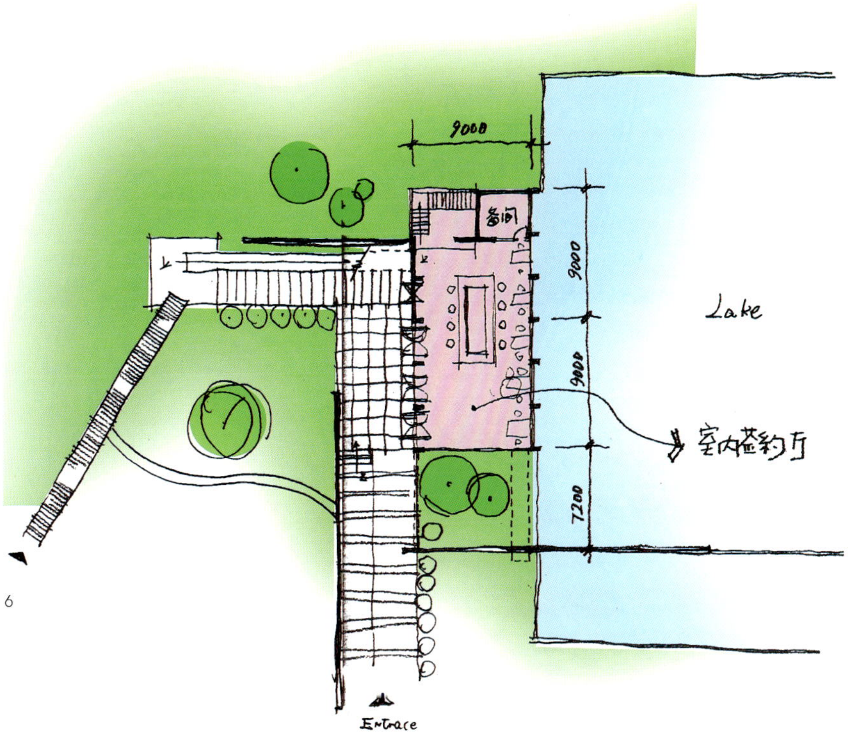

7

8

8. 湖景别墅效果
9. 组团景观广场

10

11

12

13

14

10/11/12/13/14/15/16. 别墅实景

Developer: Chengdu Vanke Real Estate Co.,Ltd.

Project address: 6,East Pihe Road, Xindu District,Chengdu City

Total area: 425 000 m^2

Total floor area: about 260 000 m^2

Volume rate: 1.01

Area interval: Western houses, 90 to 200 m^2

terrace houses, about220-260 m^2

1. 水边小景
2. 宅间绿化
3. 小区景观

万科·双水岸

开 发 商：成都万科房地产有限公司

项 目 地 址：成都市新都区毗河东路6号

总占地面积：425 000 m^2

总建筑面积：约260 000 m^2

容 积 率：1.01

面 积 区 间： 洋房：约90-200 m^2　　联排：约220-260 m^2

　　万科·双水岸坐落于成都北部新城区域，周围一片风景聚集地，那里有成都市七大绿色走廊之一——毗河风景带、50多万平方米的北湖公园、十里飘香的桂湖公园、占地近50万平方米的成都市植物园、即将扩建到3300多亩的熊猫基地和18万平米的泥巴沱风景区、还有规划中"熊猫小镇"。被这1河2湖3景区层层围绕的低密度亲水大社区万科·双水岸，堪称绿色氧吧。

　　万科·双水岸以不超过1.6的低容积率，独享1300米毗河原生态景观，原汁原味的西班牙风情建筑，温暖舒适的社区环境……

这幸福，你可以轻松拥有。

　　从未来居住的舒适度出发，双水岸坚持把大面积洋房、连排建筑在社区最核心的地块，享受优越风光。留出大部分为生态景观用地。

　　对社区水系的精心打造，是万科·双水岸的一大特点。小区水系全部建成后，未来将考虑将毗河水引进小区内，通过人工湿地处理净化后，由内河主河道输送至人工湖，再经人工湿地汇入

河中，形成社区活水循环。社区沿毗河的1300米河岸，万科精心打造了13万平米毗河景观绿化带，尽享精彩生活。

　　万科·双水岸4期"水上的院子"位于社区中心地带，以院落为社区的基本单元，分布于水域周边，形成一个个独立的空间。大开放小围合的建筑规划，开阔的空间带来居住的舒适感和主动的交流；合理的尺度带来心理的归属感和领域感。连排、洋房、电梯等多元建筑形态与主题园林相呼应，丰富的户型选择与阳光露台巧妙组合，令生活丰盛，心灵纯净。

Vanke•Shuangshui Bank is located in the new town area of West Chengdu, surrounded by a lot of landscape areas, such as one of the seven greening corridors in Chengdu-Pihe Landscape Area, 800-acres-large Beihu Park, Guihu Park full of flowery flavor, Chengdu Botanical Garden, the Panda Foundation which is to be expanded to more than 3300 acres and the 180,000-square-meters-large Nibatuo Landscape Area, and also the Town of Panda being planned. Large community Vanke•Shuangshui Bank, with low density, closed to the waters, and surrounded layer by layer by Pihe River, the 2 lakes and the 3 landscape areas, can almost be regarded as a green oxygenic bar.

Vanke presents you the Pihe River landscape greening belt of 13 square meters along the 1,300-meter-long bank of the river, providing you with wonderful life here. You can easily enjoy the happiness of living in this warm and comfortable community with such originally ecological river feature, low volume rate within 1.6 as well as the original Spanish style construction.

And for your comfortable residence in the future, Shuangshui Bank insists on large areas of western-style houses and townhouses being built in the core of the community so that you can enjoy the beautiful scenery.

One great feature of Vanke•Shuangshui Bank is to mould the water system of the community meticulously. The Pihe River is considered to be introduced into the block after the water system is completed. After processed, the river water will be transported through the major inner river course to the man-made lake, then be converged back into the river through the man-made wetlands, finally forming a Flowing water cycle in the community.

The 4th phase of Vanke•Shuangshui Bank is in the central area of the community, with courtyards as its basic unit spreading around the waters, forming each independent space. Broad space due to openess on the whole and enclosure in detail in the construction plan makes you comfortable when living in and communicate with others positively, meanwhile, the reasonable scale brings you sense of belonging. Multiple construction forms such as town houses, western-style houses, elevators, etc. echo with the theme-garden, and abundant choices of housing types are ingenuously combined with the sunshine balconies to enrich your life and clear your mind.

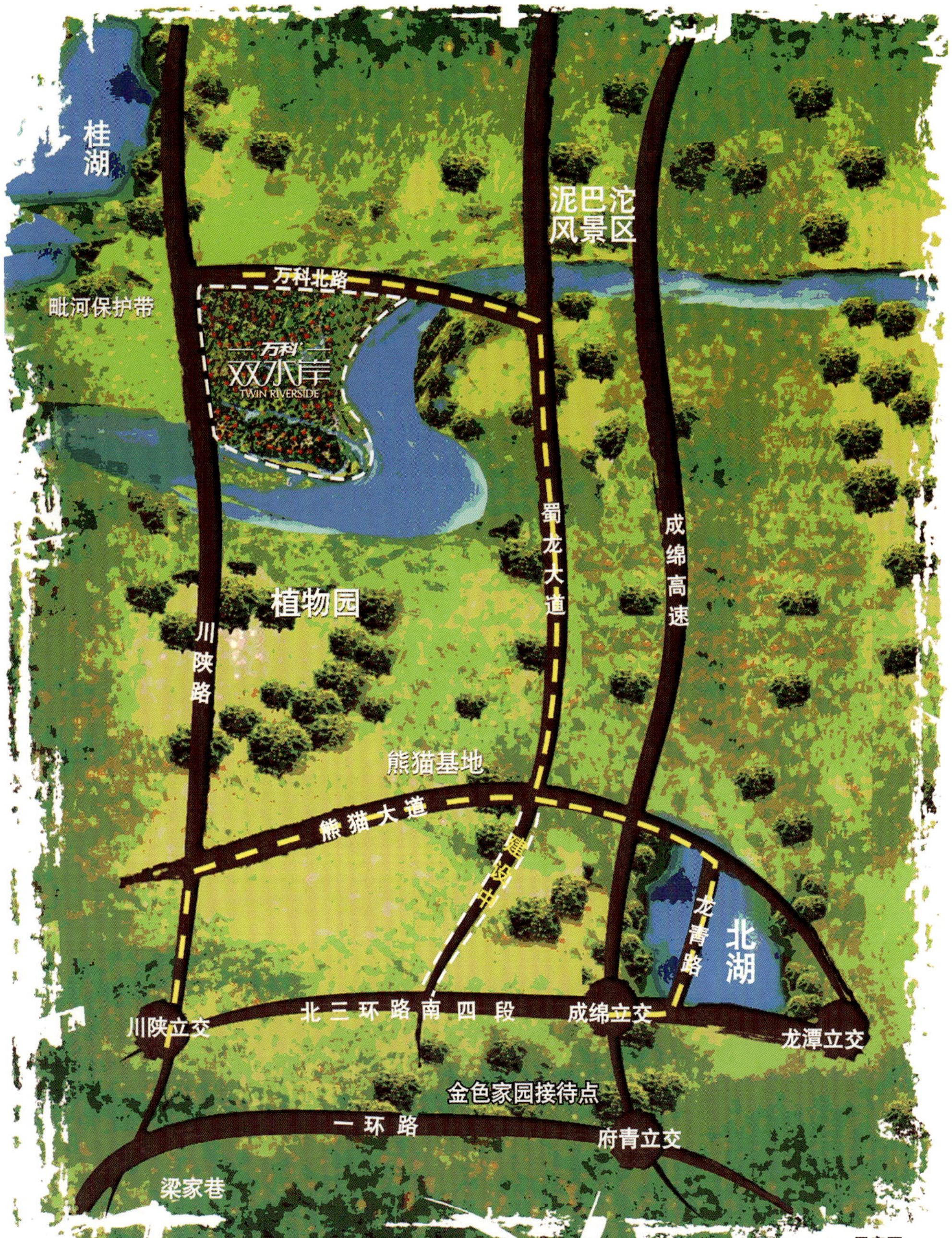

桂湖

泥巴沱风景区

毗河保护带

万科北路

万科
双水岸
TWIN RIVERSIDE

植物园

川陕路

蜀龙大道

成绵高速

熊猫基地

熊猫大道

星级井

龙青路

北湖

川陕立交

北三环路南四段

成绵立交

龙潭立交

金色家园接待点

一环路

府青立交

梁家巷

示意图

4

5

6

1

2

孔雀城三期

开　发　商：京御房地产开发有限公司
项 目 地 址：河北永定河与106国道交点西南
景 观 设 计：香港泛亚国际
规 划 设 计：美国RTKL
　　　　　　翰时国际
总占地面积：271 500 m²
总建筑面积：361 563 m²
绿 化 率：40%
容 积 率：0.49

　　孔雀城位于京开高速与永定河交汇处西南，总占地27.15公顷。项目地处未来城市试验区核心区域，是以独栋、独栋合院、联排别墅、园墅和花墅为主力产品的田园小镇；休闲运动、田园风情、别墅生活、小镇氛围，让孔雀城创造自然、和谐、美丽的生活方式，为都市精英打造真正的居住梦想。

　　项目采用加州线型规划，充分结合项目的原生地貌、河岸特色等，南加州风格中又融于清新的田园生活气息，原生的杨树林大道中间布有水景与整个社区完美的结合在一起，让家人能有更舒适的休闲空间。项目目前一期已经顺利入住，社区超市、医疗中心、健身房等配套已经开放；二期已经售罄，三期已盛装上市，联排、园墅、花墅临风待赏，让业主提前十年享受别墅生活。

　　该项目规划设计定义为魅力小镇，整体的建筑风格延续了一、二期的南加州风格，是一、二期田园别墅与城市的过渡，立面设计清爽、明快，有层次的叠落露台，提供观景平台，使建筑内外的景观相互融合，为业主营造一种自由、休闲的小镇生活。规划整体布局借鉴欧洲小镇住宅的型式，围合的组团空间成为小镇空间的基本单元，庭院理念设计营造既保持传统文化又独具西方舒适性的生活氛围，并通过建筑体量围合出相对私密的尺度及宜人的组团院落。

3

4

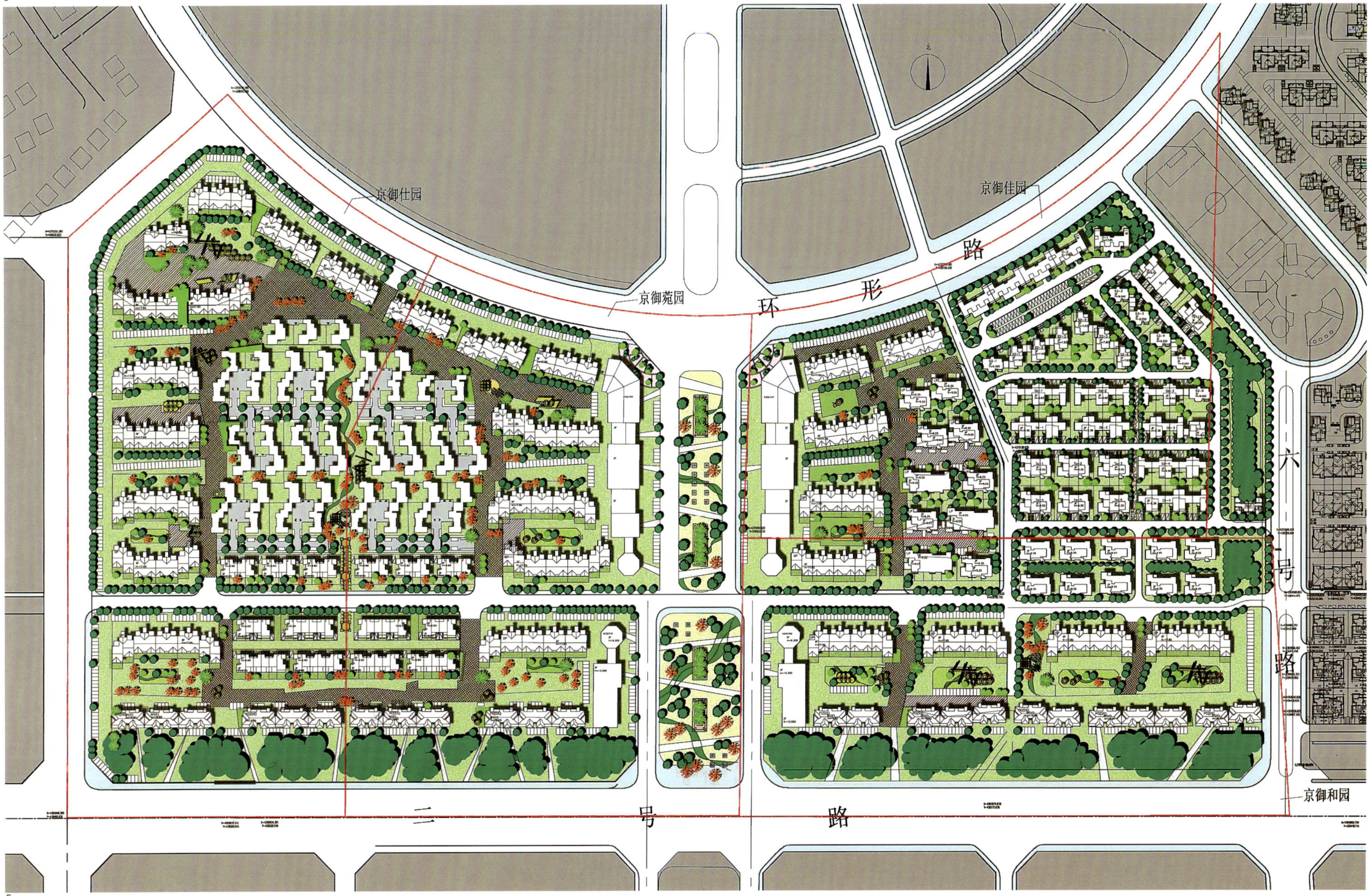

京御仕园

京御佳园

京御苑园

环 形 路

京御和园

六 号 路

二 号 路

5

6

7

Developer: Jingyu Real Estate Co., Ltd

Project Address: southwest of the intersection between the Yongding River and 106 National Highway

Landscape Design: Hong Kong Pan-Asia International

Planning Design: RTKL, A&S International Design

Area: 271 500 m²

Area of structure: 361,563 m²

Greening rate: 40%

Volume rate: 0.49

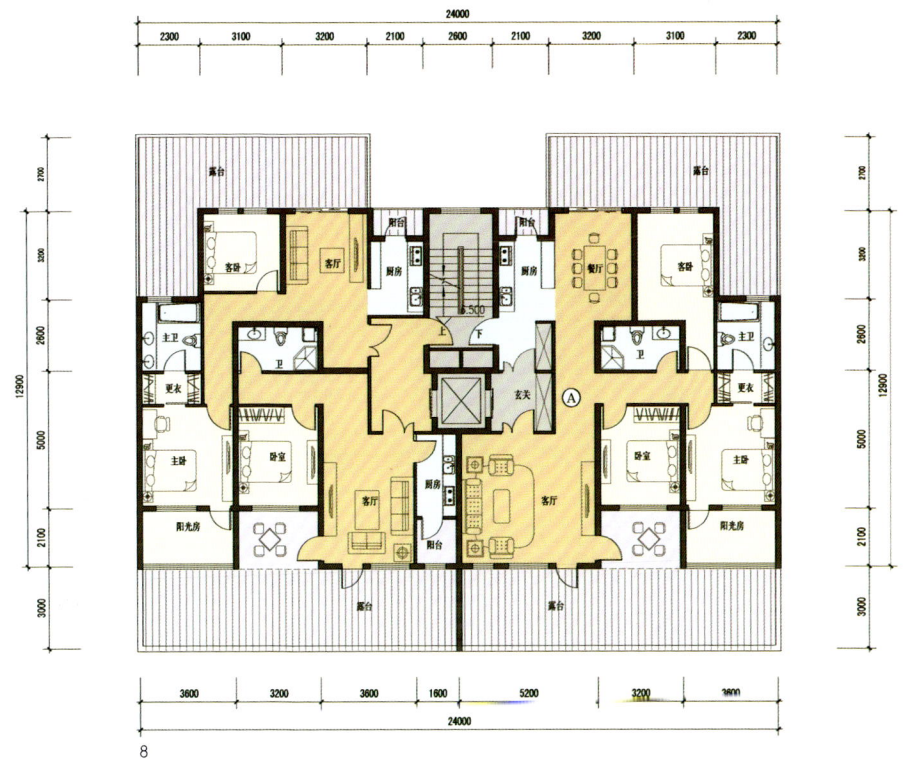

8

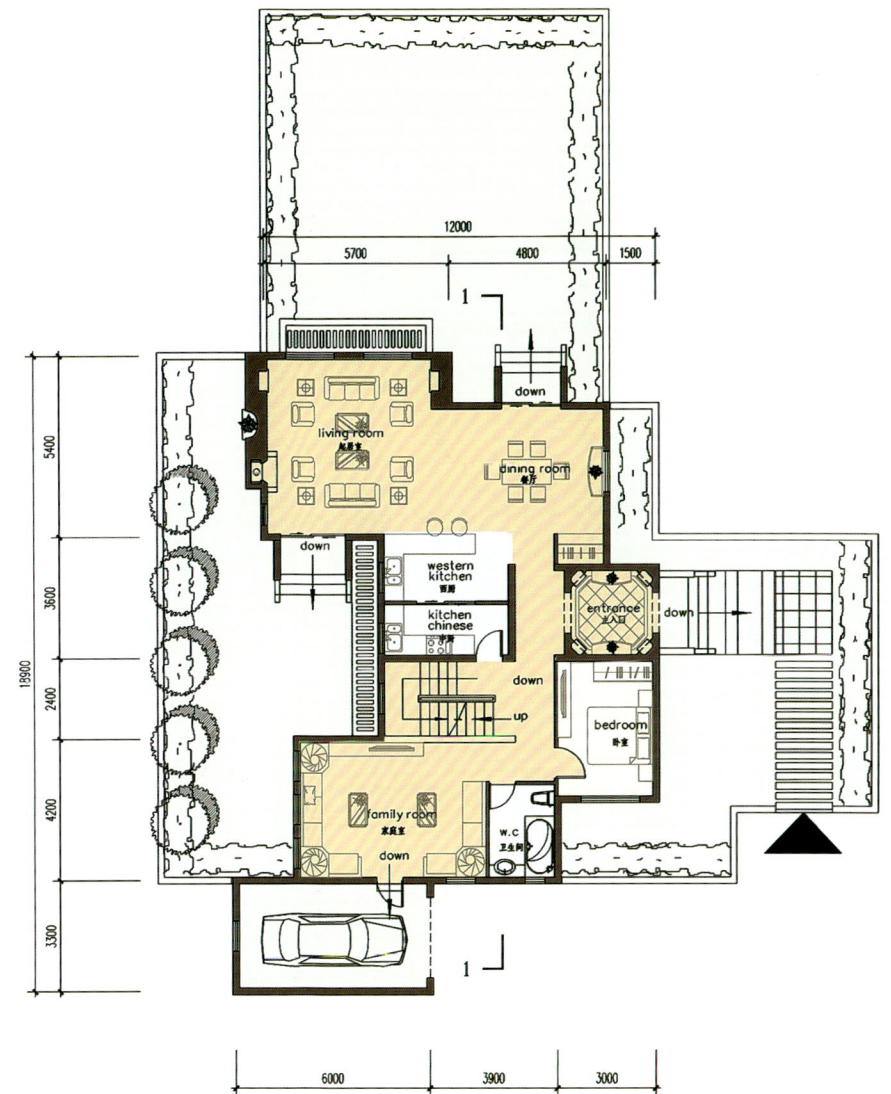

6. 入口实景图

7. 总体鸟瞰图

8. 叠拼平面图

9. 院墅平面图

9

The development is located southwest of the intersection between the Yongding River and 106 National Highway with total land coverage of 27.15ha. The development is situated in the center area of future city experiment block. It is a rural town which is based on single-family, single-family courtyard, the scheduled villas, villas and villas with garden flowers. Recreational sports, pastoral life, villa life and small town atmosphere make Peacock City create the natural, harmonious and beautiful life style. All this will let city elites read their dream.

The development adopts the California line-type and fully integrated the original landform, riparian characteristics and so on. Southern Californian style merges in fresh pastoral life atmosphere. There has waterscape in the middle of Yang's native forest road which is perfectly combined with the whole community. And this will give a more comfortable recreation space for the families. At present, development(phase Ⅰ) has been a successful move and facilities like community stores, medical centers and gymnasiums have already been opened to the dwellers. Development (phase Ⅱ) have been sold out and phase Ⅲ have been grandly listed. We are waiting for your visit to the scheduled villas, villas and villas with garden flowers. You can surely feel the villa life 10 years in advance.

The programming of the development was defined as a charming town. The whole architectural style continued the Southern Californian style of development phase Ⅰ and phase Ⅱ. Also, it continued the transition between villa in rurality style and cities. The facade design of the development is fresh and lively. Its balconies which have several levels provide you a viewing platform. The landscape within and outside the building combines together and give you a leisure town life. The arrangement of the programming draws on the European pattern of residential and its basic units are composed of roped space group. The design not only set out the traditional culture but also the comfortable western life style. Moreover, the building group will give you a individual space and a pleasant courtyard.

10. 建筑实景
11. 立面构成
12. 阳台一角
13. 室内

10

155

11

12

13

Developer: Dongguan Shanshui Town Development Co., Ltd.
Project address: Chang'an Town, Dongguan City
Conctruction designing company: Shenzhen European Construction Planning Co., Ltd.
Floor area: 73 800 m²
Volume rate: 1.09
Greening rate: 50%

1. 西班牙海岸山城风貌
2/3. 丰富的建筑立面

中惠山水名城

开 发 商：东莞市中惠山水名城开发有限公司
项 目 地 址：东莞市长安镇
建筑设计公司：深圳市欧普建筑设计有限公司
占 地 面 积：73 800 m²
容 积 率：1.09
绿 化 率：50%

中惠山水名城二期位于东莞长安市中心区一块东南向的陡峭坡地上。是一处地势北高南低的原生坡地，建筑依山形地势而规划，北靠莲花山，东临莲花水库与广东省最大的鹭鸟保护区。周边有莲花别墅、莲花山庄、长安乡村高尔夫球会、台商会馆等高档配套，自然环境极其优越，是一处罕有的养生山水宝地。

山水名城小区由Townhouse 和高层建筑组成，小区结合北高南低的自然地势，运用"南北向、大间距、低密度"的规划布局，全力营造一种只属于少数名门望族典藏的极致尊崇生活。

整个二期工程只有59套豪华别墅，并由三种户型组成：12米面宽、15米进深的双拼别墅，13米宽、12米进深的"硅谷别墅"；还有9米和7.5米面宽、14米进深的联排别墅。全部别墅皆为四层建筑，面积由413平方米起，至598平方米。所有别墅主要以3幢为一组，也有部分是2幢为一组，4幢为一组则占极少数。所以，59幢别墅中，竟有百分之七十五是三面环花园的单边端户。这是同类楼盘中罕见的，也是此项目的特色之一。

项目的建筑风格取材自西班牙南部海岸山城。利用天然山坡，为每一幢别墅营造富于山地特色的景观及园地条件，加上"原汁原味"的建筑细部，令Ronda小镇的风貌呈现在旺中带静的长安市中心一角。

二期地块内能见度最高的地方放置了一条景观轴线，令区内居民可以在会所及"楼王楼后"西北角的交通节点（也是整个一、二期别墅区地块的中心点）观赏有著名山城三藩市 Lombard Street 韵味的盘山景观小径，或利用其登上山顶公园。整个项目尽量扩大私家花园的空间。又为了便于管理及保持各家花园的私密性，各户花园之间均不设公共空间，而是以挡土墙直接分隔花园。

为表达山地别墅的韵味，三类户型均统一采用较轻松的西班牙住宅建筑风格，含适量古典元素以体现豪华及工艺素质，满足当地高端物业市场对别墅的要求。

Second phase of Zhonghui Shanshui Town is located on a steep slope land of the central area of Chang'an Town, Dongguan City. It is an original slope that high in the north side and low in the south side. The construction is built against the Lianhua Mountain in the north, and faces the Lianhua reservior and the largest egret sanctuary of Guangdong Province in the east. Around it are the Lotusvillas, Chang'an Country Golf Club, Taiwanese clan associations, and the excellent natural environment makes it a advantegeous residential area for living.

The residential quater consits of townhouses and high-rise buildings. The natural terrian of the residential quater is planned to mould a luxurious life for the minority noble groups with most efforts.

There are only 59 sets of luxurious villas included in the whole second phase'project, which consists of 3 housing types: Semi-detached house of 12 metrs'wide and 15 meters'deep, Silicon Valley villas of 13 meters'wide and 12 meters'deep, and, townhouses of 9 and 7.5 meters'wide and 14 meters deep. All are four-storey buildings with areas ranging from 413 m^2 to 598 m^2, and with every three as one group, partly evert two as one, minimumly every four as one. Therefore, 75% out of the 59 villas are unilateral houses facing gardens in three directions, which is a unique feature of the project.

In the construction style of the project, taking example from South Spanish coastal town, we make good use of the natural slope to create special hill landscapes and gardens. Then with the archetypal details, the Ronda Town turns on a quiet corner in the prosperous city ceter.

There's a feature axis set in the most visible area in the section of the second phase which allows residents to watch the famous hill town, Lambard Street of Shanfan City, the scenic path or climb up to the garden on top along it. Space of private gardens of the whole project is expanded while no public space is set between each garden. Instead, each is seperated by bin—wall, which is also easy to manage.

Light Spanish residential construction style is adopted in all of the three housing types to express hill villas, and proper classic element is included to embody its luxury and craft quality, meeting the demand of local high—end property market for the villas.

4. 总平面图
5/6. 别墅实景图

4

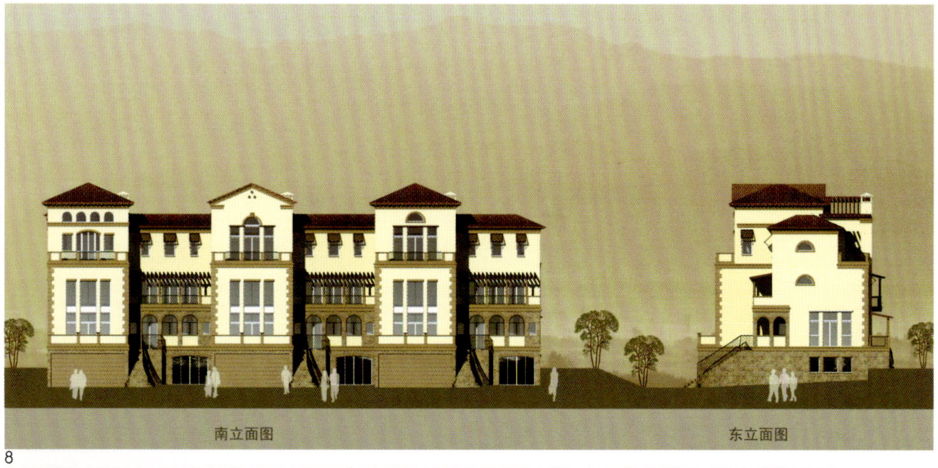

南立面图　　　　　　　　　　　　　　　　　　　东立面图

8

8. C户型南立面及东立面图

9/12. 别墅实景图

10. D户型北立面及西立面图

11. 南入户南立面及东立面图

9

北立面图　　　　　　　西立面图

10

南立面图　　　　　　　东立面图

11

12

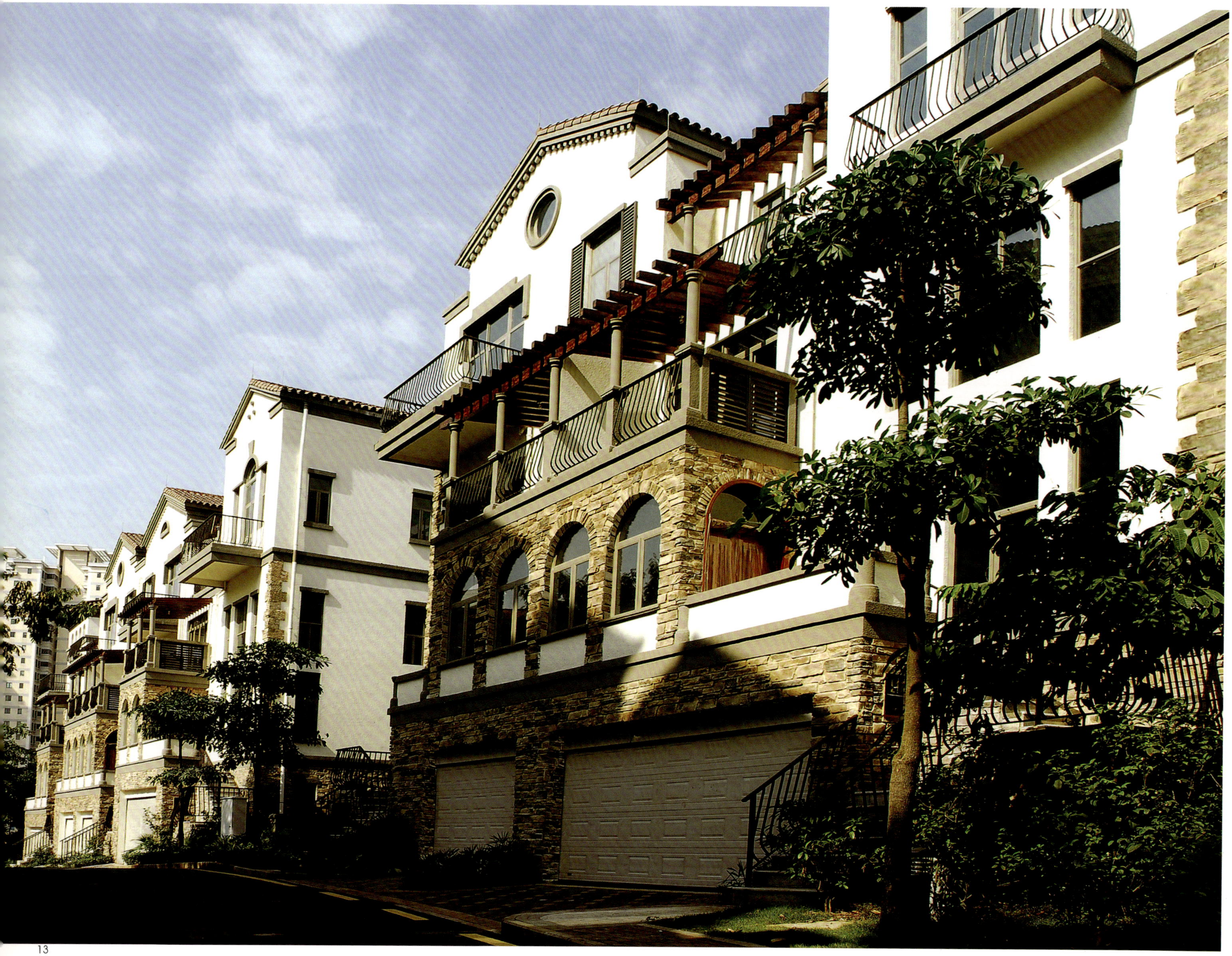
13

14

15

13/14/15. 别墅实景图

総用地面積：323 400 m²
総建筑面積：412 700 m²
住宅建筑面積：400 000 m²
公共建筑面積：2 800 m²
総居住户数：2871户
総人口数：9188人
建筑平均层数：5.2层
人口毛密度：284人/公顷
绿化率：45.3%
容积率：1.28
停车位：978辆
户均停车位：0.34

1/2. 效果图
3. 总平面图

桂林市梧桐·墅

开 发 商：桂林袭汇房地产开发有限责任公司
　　　　　桂林翔鹏房地产开发有限公司
项目地址：桂林市叠彩区灵川县龙头岭经济开发区
规划设计：深圳市国际印象建筑设计有限公司

桂林市梧桐·墅项目规划方案设计项目规划地块邻近桂林市灵川县城龙头岭，距离桂林市中心约15公里，距离桂黄公路主干道约300米，用地紧邻灵川县镇政府。场地西临湘贵铁路，东接银渠路，北接灵青公路，南接朝阳路，东西长约450米，南北长约1100米，形状方正，地形较为平整。建设场地地处桂林百里生态长廊，地理环境优越，植被丰厚，视野开阔，空气清新；同时，场地临近灵川县镇政府，市政配套较齐全，建设基础良好。建设场地周边已规划和已建设的城市道路网功能明确，较为均匀地接入到建设场地的各个部位，为住户出行提供便利快捷的交通。

规划原则

1.基于对生态旅游型城市住宅的深入思考，将对居住空间的组织溶入到城市和自然景观中去，脱离传统住宅区单体排列的巢臼，把城市空间和环境景观作为设计的出发点，围绕生态、和谐、便捷、舒适的生活主题，以东西向的步行景观轴线和南北向的观赏景观水系纵横交叉为中心，一动一静，一玩一赏，来调动全区的互动与平衡。

2.强调小区景观的均好性，通过以组团为基础单位的划分及管理，合理的营造集多样化、趣味性、私密性为一体的组团空间管理模式：各个组团依托统一的运做方式，生活资料，却又相互间彼此独立，自行管理，完全拉开私密空间与公共空间的距离，真正做到统一运做，区域管理，公私分级，内外有别的立体生活模式。

3.在景观最优化的前提下，注重提高土地利用率，以及结构体系的合理性，获得最大程度的经济性。

4.规划设计还遵循以下几个原则：

①生态——尽量多地保留原有生态环境，尊重自然景观，以承袭地域文脉，寻求场地精神的根源性；
②环保——绿色、生态、人性化设计；
③休闲——结合自然环境条件，提供多样化的休闲娱乐方式，以达到良好的居家生活效果；
④情趣——休闲舒服的氛围，精致宜人的景观，体贴人性的管理；
⑤发展——良好的社会效应、环境效应和经济效应。

总体布局

　　总体布局根据地块现有的城市道路系统和原有地块地貌特征，采用依山顺水、因地制宜、合理开发、时序开发的原则，将地块分为别墅区、洋房区、多层区和保留山体公园等四大部分。别墅区利用设计地块北部的天然丘陵为基础，围绕原有的山体水域顺应等高线排部，层次丰富，布局灵活。

　　洋房区以保留的山体公园为核心成带状布局，景观优良，单独成区，同时把别墅区低密度空间向多层区的相对高密度空间顺应过度，打造由南向北逐步放低的小区天际线，为小区整体空间形态的丰富性提供条件。

　　多层区主要布置在设计场地上较为平整的地段上，采用了BBS模式的院落式布局，按不同的方式形成对外半封闭、对景观面全开放的组团空间，每组团200—300户规模。组团内利用建筑间的错位组合，形成相互错落又相互联成一体的景观内院和游休憩绿地，大部分住宅均能享受三个面的组团绿地。

　　规划地块西南角和西北角分别设计有18班小学一处和9班幼儿园两处，不但解决本地块相关服务，同时为后续开发创造良好的配套条件。

4

5

　　本设计强调景观环境朝向以及交通安全的"均好性"。一方面考虑从中心景观带到组团、宅间绿地等各景观节点尽可能多地覆盖和照顾大多数住户，另一方面单体分布的位置亦和其拥有的临窗景观密切相关。设计中将各景观元素精心组织，按照生态原则层层递进、渗透，并形成节点间的有机沟通，各住户亦在此形成丰富变化的环境空间中真正得益。半围合式的组团更加增添了安全感和归属感。

总体布局舒展开阔，区域明显、主次有序，其原则分以下几点：

①保护和优化现有生态景观形态。
②结合地形，布置建筑组团，强调建筑与环境的融合。
③核心设计中央绿化景观以及组团绿地以提供休闲娱乐的最佳环境。
④交通组织合理，出入口位置恰当，各级别道路清楚连续。
⑤满足功能需求，符合各项技术规范。
⑥分期开发设计合理，时序体系脉络清晰。

交通系统

通过总体规划分析，将总体交通系统作统一调整，将道路功能结构进行明确划分、等级制定，并保留有开发的余地。

主要入口道路设计为长约100米的景观林荫大道，为集中的步行商业街，是城市环境向小区环境过渡的载体；组团间道路设计为6米的林荫双车道，可做临时停车之用；组团内车行环线宽4米，兼做消防环线，每个独立组团都是一进一出的两个口，管理高效便捷；景观步行道路系统贯穿于整个小区。

绿化系统

根据生态学原理，组织了系统的、稳定的绿化系统。同时，依据公共绿地的使用对象、服务范围、赏玩特性等不同目的，来划分绿地等级，作到目的明确，层次清晰。

一方面大面积的保留原有林地，并对其进行维护修葺，以突显本土地域特色；一方面精心打造贯穿规划地块东西的步行景观大道和串联南北的景观水系，做好小区的形象工程。各独立组团独特设计属于自己的组团绿地，配套以老人儿童活动场地，休憩草地，山水凉亭等景观小品。

户型设计

根据地势、朝向、分区、密集程度等因素，规划中设计了适用于各种家庭结构的户型，形式丰富，产品多样，以适应不同人群的使用需求。

别墅户型分双拼、小独栋、豪华别墅等形式，每户均配有单独车库，私家花园，面积从250平方米到480平方米不等。

花园洋房为一梯四户的上下叠拼形式，一二层为一户，配备独立入户花园，享受私家园林情趣；三四层为一户，则设计有大露台，屋顶花园，享受于阳光美景之中。

6

7

多层住宅分经济适用型、标准型、宽景豪华型等多种样式，底层半架空为车库，南向大房间下设计为汽车停车，北向小房间下设计为自行车停车，各单元刷卡入户，实行封闭式管理，以提升小区整体素质。

建筑风格

本项目建筑风格应业主要求，以地中海风情为主题，打造广西的样板生活情景住区，并代表桂林地区新兴的城市形象。

第一期开发，以法国波尔多地区的浪漫情调为主线情节展开，充分运用活泼多元的建筑元素，错落有致的屋面变化以及典雅、温馨的建筑色彩，再配合丰富动感的园林景观设计，共同营造出整个住区高雅、舒缓、情趣的居住氛围和生活理念。

4. 商业街入口效果图
5/6. 别墅效果图
7. 洋房效果图
8. 双拼效果图

8

Developer: Guilin Xihui Real Estate Development Co., Ltd
Guilin Xiangpeng Real Estate Development Co., Ltd

Project address: Longtouling Economic Development Zone, Lingchuan County, Diecai District, Guilin City

Conceptual design: Shenzhen Int-impress Architectural Design Co., Ltd

Total plot area: 323,400 square meters

Gross floor area: 412,700 square meters

Residential area: 400,000 square meters

Public area: 12,800 square meters

Total units: 2,871 square meters

Total population: 9,188

Average floors: 5.2

Residential density: 284 per hectare

Greening rate: 45.3%

Volume rate: 1.28

Parking space: 978

Average parking space: 0.34

Project of Guilin Parasol Villa is close to Longtou Hill, Lingchuan County, Guilin City, next to Lingchuan Government 15 kilometers away from downtown area of the city and 300 meters away from Guihuang Main Road. In its west is Xianggui Railroad, Yinqu Road in the east, Lingqing Road in the north and Zhaoyang Road in the south. It is about 450 meters in the west-east direction, about 1,100 meters in the south-north direction, quite a square and even land. The area is right in the Guilin 100-meter-long ecological corridors with abundant vegetation, broad views and fresh air. It's being next to the Lingchuan Government equips it with complete utilities and good construction foundation. Planned surrounding field and net of city road are clearly and evenly connected to each part of the construction to offer residents convenient traffic conditions to go out..

Principle of the plan

1. The organization of residential space, based on the consideration on city residence of ecological and travelling type, is merged into the city as well as the natural feature to get rid of traditional residential unit array. Then out of the starting point of the design, the space and features are filled with theme of ecology, harmony, convenience and comfort. Finally, the cross of the dynamic west-east walking feature axis and the static south-north water feature system, one for fun and one for watching, have balanced the whole community and its interaction.

2. Uniformity of the features is emphasized by the division and management of the groups, reasonably creating a multiple, funny and private group space managing pattern. United operating way and living data but respectively independent from each other to self-manage can totally lengthen distance between private and public space, which makes a stereoscopic life style.

3. Upon the condition of the optimization of the feature, we pay

attention to improve the use of the land and the reasonability of the structure to gain profit on the largest scale.

4. Principles of the design are as follow:

①Ecology-----Ecological environment remained as much t as possible in respect for the nature, succeeding the regional culture and the root of the field.;

②Environment protection-----Green, ecological and humanized design;

③Leisure----Multiple entertainment combined with natural condition to seek best residential life;

④Style----Leisure atmosphere, nice landscapes and considerate management;

⑤Development----favorable social, environmental and economic effect.

General layout

According to the present city road system and original terrain features, and in the principle of taking advantage of the mountain and the water to develop the layout reasonably in right timing, we divide the section into four parts as villa area, western houses area, multi-layer areas, and mountain park remained.

The villa area is designed based on the natural hill in the north, surrounding the remaining hills and waters and subjecting to the array of the contour line, rich in layers and flexible in arrangement.

Western house area takes the mountain park as its core; such ribbon arrangement possesses excellent features. Being independent, it makes a transition from the low density space of the villa area to the comparatively high density space and draws the skyline lowing from the south to the north gradually, finally provides for the general space of the residential quarter with abundance.

Multi-layer is mainly arranged in an even sector of the field, adopting a courtyard layout of BBS mode, and forming semi-closed from outside and open feature group space in different ways. Each group includes 200 to 300. Inside the group, residents can enjoy staggered inner feature courtyards, swimming pools and resting green land.

There are a primary school with 18 classes and kindergartens with 9 classes in the south-west and north-west corners of the planning sectors, not only solving relevant local land sector service, but also providing good condition for later development.

This design emphasizes on the features' environmental orientation and homogeneous traffic safety. On the one hand we offer more residents more access to central feature belt, groups, residential green land. On the other hand individual location is closely related to the window features. Each element is elaborately organized with the ecological principle to progress and filter layer by layer, forming communicative chances between the plots so that every resident can benefit from such variable environment space. Semi-enclosure groups also add more to sense of safety and belonging.

Relaxing and broad layout, with distinctive region and order, and principle as follow:

①To protect and optimized present ecological feature shape

②To combine the terrain, arrange architectural groups and to make emphasize on the integration of construction and environment

③Core design, the central greening feature as well as the green land offer best environment for entertainment

④Reasonable organization, proper exit and entrance location, clear and continuous road of every grade

⑤To meet the demand for function and to obey every technological standards

⑥To develop in different phases and to design reasonably with time order system in clear nexus.

Transport system

According to the general analysis, we adjust the overall transport system, precisely divide the road function structure, set grades and preserve all the alternatives for development.

11/12/14. 钟楼实景图
13. 入口实景图

12

13

14

Main entry road is designed as a feature boulevard about 100 meters' long, it's also a commercial waking street which is a conveyor of the transition from the city surrounding to the residential one. Between each group there are 6-meter-long two lane streets which are also for temporary parking. And inside the groups are 4-meter-long ring roads for vehicles which are also for fire-fighting. Finally, each independent group gets one exit and one entry, convenient and effective for management while feature walking road system spread through out the whole residential quarter.

Greening system

On the basis of ecological principles, systematic and stable greening system has been set up. Meanwhile, for different purposes such as who use the public greening land, he service range, and entertaining, etc. we divide the greening land into several ranks to make our goal clear.

Large areas of the original forest are remained and well repaired to

15. 会所景观
16/17. 文化墙实景

15

feature local region which the residential quarter image project being improved by elaborately molding the feature walking avenue which is planned to divide the wet and the east and the water feature system connecting the north and the south. Each group owns its independent green land with activity fields for the elderly and the children , the resting grass and summer shelter.

Housing design

Taking factors like terrain, orientation, division and density scale into consideration, we have designed housing types of various kind of structure, rich in their forms, various in the production, to meet demand of different groups.

There are housing types like semi-detached houses, detached houses, and luxurious houses, etc., each of which are equipped with individual garage, private garden, with area of anywhere between 250 to 480 square meters.

Garden western houses are in the form of four units as one. First and second floor are as one residence with individual indoor garden, allowing you to enjoy fun for gardens. Third and fourth are as one with patios and roof gardens where you can be bathed in the sunshine and the beautiful landscape.

High-rise residential buildings are divided in to economic, standard, and wide views types, which are set to be garages in the bottom layers. Large south-oriented rooms are designed as car parking space, small north-oriented rooms as bicycle space. Closure management of entering the units by punching the cards can improve the general quality of the residential quarter.

Architectural style

On the request of the owners, Mediterranean style has been adopted as the architectural style to mould sample living residential quarter in Guangxi and the project is representing newly risen Guilin.

In the first phase's development we unfold with the main clue of romance tone from France Bordeaux area, fully using active and multiple architectural elements to give more changes and warm colors to the house along with dynamic garden feature design. Successfully, we have created an elegant, comfortable and tonal atmosphere in the residential quarter as well as introducing such life concept to you.

16

17

174

1. 泳池景观
2. 会所局部
3. 围合景观

1

深圳东海·万豪广场

开 发 商：深圳东海集团有限公司
项 目 地 址：深圳市南山区填海五区
建 筑 设 计：深圳市水木清建筑设计事务所
总用地面积：10 875 m²
总建筑面积：29 164 m²
容 积 率：1.914
总 户 数：58

本项目设计获深圳市第十二届优秀工程勘察设计住宅三等奖。

优点：

1. 园林式居住模式，平台花园空间设计细腻，从设计上提高了居住品质。

2. 户型方正实用，南北通透。

3. 住户与商业人流分布合理，实用方便。造型新颖，虽有古典元素，却极具现代风味。

4. 商业空间活泼，极富特色。入口广场别具匠心，提升了区域环境质量。

5. 造型设计细腻深入，整体形象佳。

东海万豪广场位于华侨城南成熟地段不可复制的红树湾片区，坐拥沙河及名商两大高尔夫球场，深圳大学、南山外国语小学、南山高新中学等文教配套。屋苑前是广阔的深圳湾和延绵15公里的滨海生态公园。

万豪广场汇豪门气派与高尚品味于一身，单元面积从80至185平方米（复式），一楼复式花园形成了窗窗有景、移步换景的动人景致，缤纷植物飘然而至；花园交相呼应，相映生辉，形成一副立体的旖旎画卷，让家中的每个角落都成为赏心悦目的优裕空间。屋苑内水景园林、瀑布泳池等精彩纷呈，商业裙楼荟萃完善配套设施，提供充裕停车位，发展商提供24小时全天候酒店式优质物业管理，地铁科技园东站步行数分钟即达。

Project location: Tianhai Zone Five, Nanshan District, Shenzhen

Developer: Shenzhen East Pacific Group Ltd.

Architectural design: Shenzhen Thinker Design & Consult Pty Ltd.

Total plot area: 10 875 m²

Gross building area: 29 164 m²

Plot ratio: 1.914

Total units: 58

The design of this project has won a third prize of 12th Shenzhen excellent prospective and designing residence.

Merits:

1. Garden living pattern, terrace garden space with delicate design which improves living quality.

2. Square and practical house types which is spacious in both south and north direction.

3. Reasonable distributed residents and business crowds. Originally designed architectures with both classical elements and modern style.

4. Active and special commercial space. The unique entry plaza promotes the environmental quality of the zone.

5. Intensive modeling design and good general image.

East Pacific Marriott Plaza is located in mangrove creek area which is unique in the mature section of south Huaqiao Town, possessing the two large golf courts, Shahe and Mingshang, and educational institute like Shenzhen University, Nanshan Foreign Language School and Nanshan Gaoxin Middle School, etc. In front of it is the broad Shenzhen Bay and seafront Ecological Park that extends for 15 kilometers.

4

5

Marriott Plaza takes luxurious style and noble taste into it. Areas of the units various from 80 to 185 square meters (duplex), duplex gardens on the first floor present a view of landscape can be seen in every widow and landscape changes as people walk. There are abundant kinds of plants and gardens echoes to each other, forming a spatial and romantic picture scroll which filled the whole house with enjoyable atmosphere. Inside the house there are water feature garden, waterfalls and swimming pools and so on. Business podium buildings are equipped with necessary facilities and plenty of parking spaces are provided. The developer provides 24-hour good quality hotel-form property management. Finally, you can get to here within minutes' walk from Metro Station East Science Park.

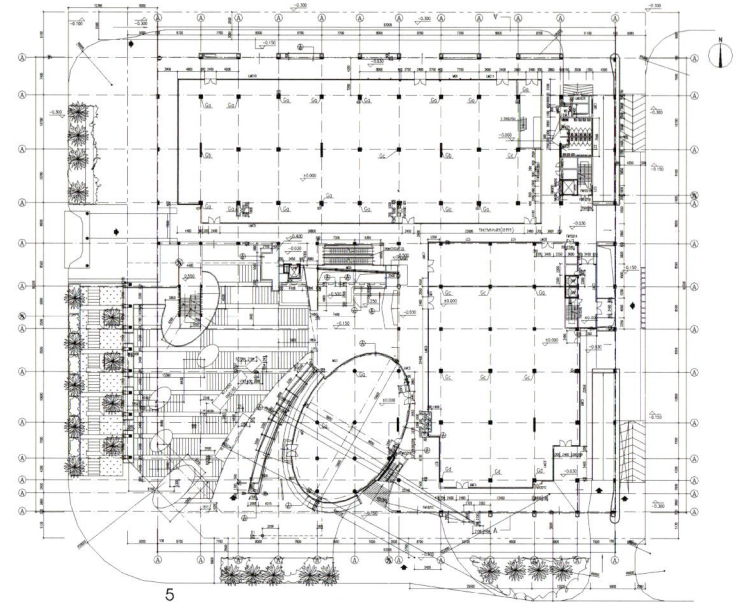

6

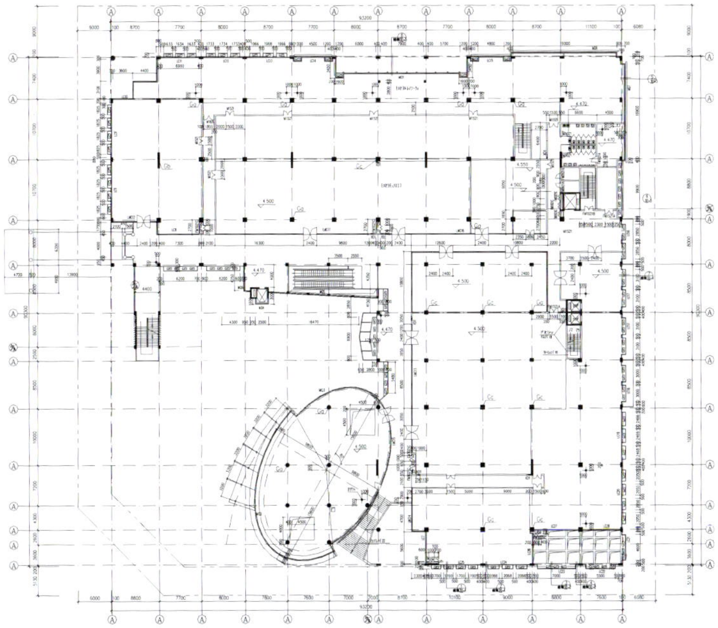

7

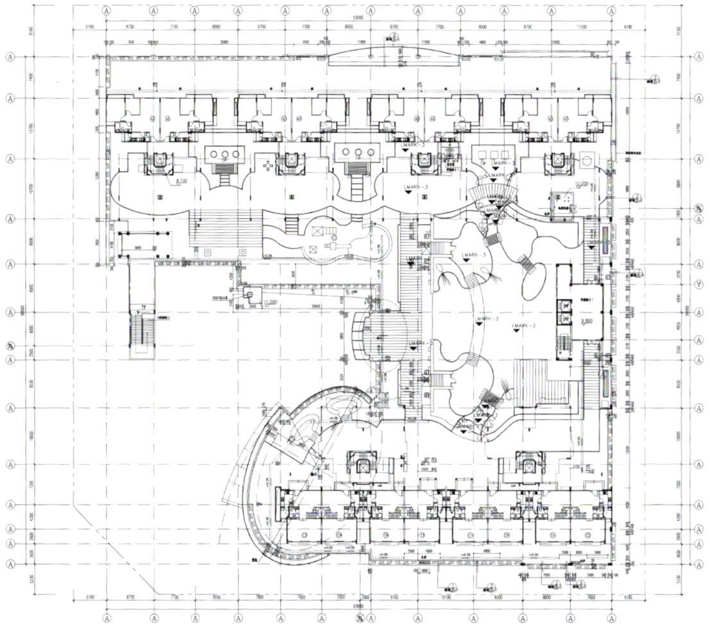

8

4. 总平面图
5. 一层平面图
6. 中庭景观
7. 二层平面图
8. 三层平面图
9. 会所

9

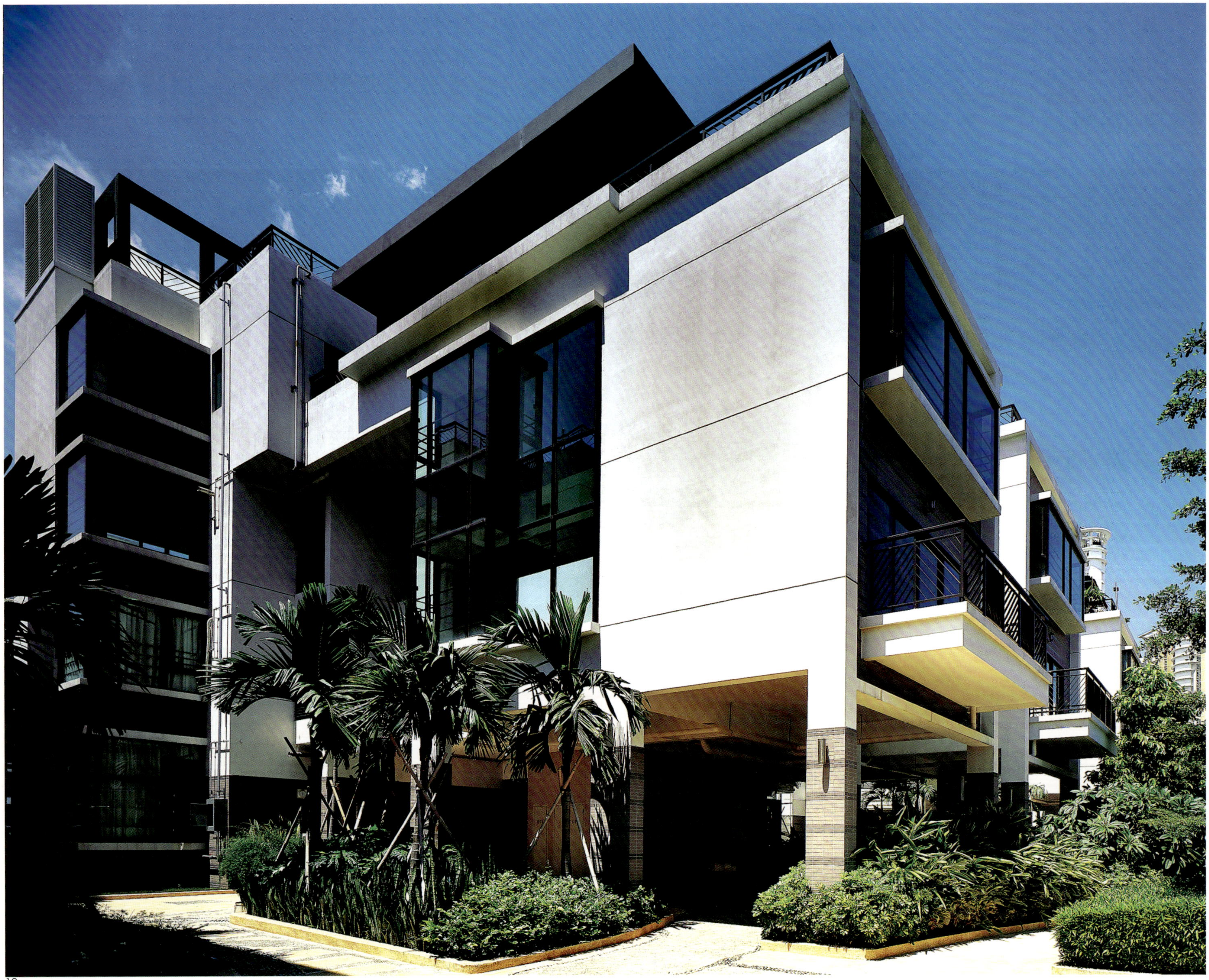

10

12

11

10. 建筑立面
11. 建筑透视
12. 架空层

小高层住宅

Small high-rise
Residence

小高层住宅

1

占 地 面 积：47 882 m²
总建筑面积：81 506 m²
总 户 数：689
停 车 位：400
容 积 率：1.5
绿 化 率：65%

1. 庭院效果图
2/3. 建筑局部
4. 实景图
5. 鸟瞰图

中海日辉台

开 发 商：深圳中海地产有限公司
项 目 地 址：深圳龙岗坂田
景 观 设 计：泛亚易道景观设计有限公司
建 筑 设 计：深圳华森建筑与工程设计顾问有限公司
方 案 设 计：张晖 严庆平
结 构 形 成：异形柱框架

2

3

4

5

6. 庭院透视图
7. 入口
8. 总平面图
9/10. 效果图

Project Address: Bantian Road Longgang District Shenzhen

Developer: Shenzhen Zhonghai Real Estate Development Co., Ltd

Design of Landscape: Fanyayidao Landscape design Co., Ltd

Design of Building: Shenzhen Huasen Aichitectural and Engineering Designing Consultant Ltd

Scheme design: Zhanghui, Yan Tianping

Formation of the structure: Special-Shaped Column Frame Structure

Area Covered: 47 882 m²

Total Building Coverage: 81 506 m²

Total units: 689

Parking spaces: 400

Afforesting rate: 65%

Plot ratio: 1.5

8

The project Zhonghai Rihuitai is located on the joint point of Banxuegang Avenue and Beier Road in Shenzhen and the north of it is Beier Road, the west of it is Banxuegang Avenue. It is a Hawaii style building group with the area of land to be used of 47881.78 m^2 and a floor area ratio of 1.5.The type of the buildings combines the multi-storey apartment with the high-rise apartment and the main type of flat based on small three-bedroom. The main target is for domestic sales. The primary is the first-time home buyers and the secondary is second home buyers .It aims at the white-collars of IT sector who is young and with a high level of knowledge. The project Rihuitai is the middle quality excellent residential district developed by the Shenzhen Zhonghai Real Estate Development Co. Ltd. It is consist of 7 multi-storey and high-rise buildings, a kindergarten of 9 classes and several shops along the street on the ground floor.

The shape of the land is Irregular polygon and relatively lack of the condition of surrounding landscape.

The buildings are arranged along the boundary of the land for used. That creates a large inner yard enclosed by the buildings. Great effort was made to create an attractive living environment for the dwellers with a more beautiful landscape of gardening .The main flat types of flat are the two-room apartment and the three-room apartment of which the building coverage ranges between 80 m^2 to 110 m^2.The design is so elaborate that it creates the reasonable functions under the limit of land ,and better ventilation and illumination as well as a good orientation of landscape. The exterior wall of the building is

9

covered with tiles in warm colors matched with the dark colors and the light ones. Stone finish was designed on the bottom part of the buildings and some key facade. That conjoining with design of cornice and the frame makes the buildings elegant and unique. With the landscape elements of tropical style such as palm trees, wooden bridge and pavilion made of grass,, the project look unique and become the model of delicate small apartments.

10

1/2/4. 一期建筑立面局部
3/5. 一期建筑立面中景

1

武汉万科西半岛

开　发　商：武汉市万科房地产有限公司

项 目 地 址：武汉东西湖区东西湖区环湖路以东、金银湖西岸

建筑设计公司：加拿大AEL建筑景观设计有限公司

总 用 地 面 积：201 798 m²

总 建 筑 面 积：207 197 m²

容　积　率：1.3

绿　化　率：39.8%

总　户　数：840

停　车　位：437

2

　　万科西半岛项目位于武汉市区西北方向，属汉口东西湖区金银湖生态旅游区内，距汉口西北广场车行距离15公里(20分钟)，距汉口火车站12公里，距万科在东西湖区已开发完成的另一大型人居项目万科四季花城5公里。

　　西半岛有着典型的江南水体地貌，优越的生态环境，多样的自然物种，稀缺的双半岛地形，西半岛的用地约36.3公倾，其中约20.2公倾地建房子，另外的2/5是代征地，大部分用作绿化、留给身体、眼睛、休息；建筑在大面积绿地、湖面的环绕中，在这里，建筑是一种点缀，自然才是主角。西半岛建筑以地中海附近国家建筑为蓝本，广泛运用了地中海的元素，立面丰富、精

致，产品天然、质朴、手工，这里的建筑非工业化社会中的流水线产品可比，它的味道不像速溶咖啡，简单而雷同。"手工"感与"不可复制"性是它的最大特点，使每间房子都透出思考的痕迹。以接近定制的手法，营造了一种自然的、健康的、悠闲的、富有艺术感的岛居生活。

　　规划形成过程：加拿大AEL建筑景观设计有限公司邀请了国内外的知名设计师对西半岛思考研究西半岛地块的规划，西半岛的规划定稿是芬兰佩加萨尔米宁设计师，中南建筑设计院ATK的作品，芬兰有千湖之国的称号，对这样水域丰富的地块，他们驾轻就熟。

　　金银湖主体水域延伸到地块内部，将地块天然分隔成2个半岛，双半岛地形，设计分别把它称作左岛、右岛。西半岛地块被湖水、绿地环绕，在这里，建筑是一种点缀，优越的自然环境才是主角。2005年开始开发左岛一区、左岛二区13栋多层、销售大厅规划用地面积61238.4平方米，建筑面积62220平方米，560户左右。左岛由岛居退台美墅、岛居情景洋房、小高层、别墅四种产品类型组成；退台多层，面积在80—140平方米之间，有少量复式产品，面积在160平方米左右；小高层面积在125—150平方米左右，别墅面积在200平方米左右。

亲水公园

　　小区在改造中保留了大部分原始地貌，标高和植物，保护原生态；并于湖堤处设计如树池、木栈道、亲水平台、雕塑等景观，让您可以参与其中；

运动公园

　　从环湖路过来，映入眼帘的首先是大面积的绿地，万科拿出了红线内的地和代征地做大面积绿化，通过自然的推坡、葱郁的林木、天然材料的装饰，形成一道天然的环境屏障，将社区自然的围合。与四季花城的假日广场不同，西半岛的运动公园更强调功能上的作用，充分考虑了各个年龄层的需求。建设了符合国家标准的2个篮球场、1个网球场、儿童游乐园、慢跑径、散步道等。它其实也是一个运动会所，处在自然之中的一个充满了负离子的会所。锈蚀钢板墙，雕塑元素，随着时间的流逝变化着，与西半岛一起生长；晨曦飞鸟雕塑是社区里会说话的风景，传承西半岛追求自然、生态的精神。在满眼的绿色中，在湖风吹拂下，大人在湖边打球，小孩在坡地上玩耍，老人在曲曲折折的步道上闲庭信步，这种景象不是谁都可以享受到。您看到的也将不是单纯的建筑，而是生动的生活场景，构造了丰富的城市界面。

6

7

Project location: West Bank of Golden & Silver Lake, East of Huanhu Road, Dongxihu District, Wuhan

Developer: Wuhan Vanke Real Estate Co., Ltd.

Architectural design: Canada AEL Architectural Landscape Design Co., Ltd.

Gross plot area: 201 798 m²

Gross building area: 207 197 m²

Plot area: 1.3

Afforesting: 39.8%

Total units: 840

Parking spaces: 437

Project Vanke•The Peninsula, is located in the northwest of downtown area of Wuhan. It belongs to the Golden & Silver Lake Ecological Tourism Zone, Dongxihu District, Hankou, 15 kilometers (for 20 minutes by bicycle) away from Hankou's Xibeihu Plaza, 12 kilometers away from Hankou Railway Station and 5 kilometers away from another Vanke's residential project Wonderland, which has been completed in the same district.

The peninsula gets a classic terrain of Jiangnan water system, excellent ecological environment, various natural species and rare terrain of double peninsulas. The west peninsula covers an area of 36.6 hectares, 20.2 of which are used to build houses, and another 2/5 is eminent domain, mostly used for greening to people's health. The construction is surrounded by large areas of green land and lakes. Here, construction is some kind of decoration while natural plays the main role. The peninsula is constructed taking the ones in the nearby countries of Mediterranean Sea as blue prints, so large number of Mediterranean elements is used. Compared to the assembly-line product in non-industrial society, it is abundant and delicate in elevation, natural and man-made built. It's simple but not alike others and it's not like instant coffee. Manual sense and being unduplicated are its best features, allowing every house demonstrates the tide of thinking. It is nearly customized to create a island life that is natural, healthy, leisure and full of sense of art.

Process of the formation of the plan: Canada AEL Architectural Landscape Design Company Limited has invited famous designers at home and abroad to study the plan for the terrain of the section of the Peninsula, which is finalized by Finland designer Peja Salminen. Finland is honored as a country with thousands of lakes, so they can handle well such a land with plenty of waters. It's also designed by Central-South Architectural Design Institute ATK Studio.

Main waters of the Golden & Silver Lake extend into inside the section, naturally dividing it into 2 peninsulas which is respectively referred to as the left peninsula and right peninsula. Block A and more than 13 high-rise dwelling buildings of the left peninsula as well as the sales hall are planed with an area of 61,238.4 square meters and the building area is 62,220 square meters containing about 560 residents. The left peninsula is made up of four types, island-living setback villa, and western houses with sceneries, small high-rise dwelling buildings and villas. Areas of setback multistoried buildings are between 80 and 140 square meters, small high-rise dwelling buildings are between 125 and 150 square meters, villas around 200 square meters and few duplex product are around 160 square meters.

6/7. 二期效果图
8/9. 二期建成实景

8

9

10

11

12

13

14

15

16

Waterfront Park:

Most part of the original terrain, true height and plants are preserved in the remold of the residential quarter for the purpose of protecting the original ecology. Besides, landscape like there are trees and ponds, trestles, waterfront platforms, sculptures are set alongside the lake so that you can enjoy them.

Spots Park:

What you can see first coming from Huanhu Road are large areas of green lands. Land within the red line and the eminent domain is used to be large areas green lands, and by the decoration of the natural slope, verdant woods and the natural materials, a natural environment barricade is formed, enclosing the community. Different from the Holiday Plaza of Wonderland, the sports park of the Peninsula is emphasized more on function, taking full consideration of demand of people at all age levels. There are 2 basketball courts, 1 tennis courts, a children amusement park, jogging tracks and paseos are constructed according to national standards. In fact, it is also a club for sports which is naturally full of anion. Corroded Steel panel walls, sculpture elements have been accompanying the Peninsula to grow as time goes by. Birds in the morning and the sculptures are lively scenery of the community, succeeding the pursuit of nature and ecology. With green full of your eyes, wind from the lakes breezing you gently, you can play balls on the bank of the lake, children can play on the slopes and old people can walk on the winding paths. Such precious scene can be shared here, and what you can see is no longer pure architects, but lively life which make an ample city interface.

17

18

19

14. 三期建筑全景效果

15/19. 三期公寓效果

16/17. 三期联排别墅效果

18. 三期建设背景中的会所实景

1

总规划用地面积：127 000 m²
总建筑面积：240 000 m²
容　积　率：2
建筑密度：16%
绿化率：35%
总　户　数：约2400

1. 南向效果图
2. 临街单栋效果图
3. 一期鸟瞰图
4. 总平面图
5. 花园看18层建筑效果
6. 花园看11层建筑效果

中海金沙馨园

开　发　商：广州中海地产有限公司

项目地址：广州市白云区金沙洲居住新城环洲三路南段

规划建筑设计：广州瀚华建筑设计有限公司

景观设计：广州市晴川环境艺术有限公司

2

二带：沿江绿化带，充分利用江岸线打造南方滨水区亲水性强的公共绿地供居民休闲娱乐，构筑完善的生态绿地系统网。

三片六区：利用北环高速公路、广佛公路及金沙洲大桥等有利条件，结合自然地形差异和道路骨架形态，划分六大居住区间。

经典元素　舒适家园

虽是限价项目，中海金沙馨园沿袭中海地产多年经验累积而成的经典建筑元素，发扬均好性强的产品特点，以实用的户型设计、良好的对流通风规划、通透明亮的设计，用心打造一个和谐、美丽、宜居的精品住区，实现真正的"限价房的价格，商品房的享受"。

实用户型

中海地产致力于开发中高档住宅，拥有丰富的经验，积累了一大批经过市场检验、成熟合理的户型并应用到中海金沙馨园的设计中。"大面宽、短进深"、南北通透对流、方正间隔、超高实用率、双阳台、L型阳台、双卫生间、动静分区等合理布局，体贴细节成就舒适家居。

对流通风

中海金沙馨园总体规划充分考虑营造小区"风环境"。由南至北两进院落的布局有利于引导顺畅的南北对流通风，结合岭南建筑的特点，设置风雨走廊与架空层，不但加大了园林绿化的空间，还在低空位置形成均匀流畅的通风。其次，前后交错相连的设计，使小区内形成良好的"穿堂风"，保证小区的各个空间均有流畅均匀的通风条件，从而营造了良好的小区"风环境"。

通透明亮

良好的采光使中海金沙馨园的建筑，户内和户外都给人通透、明亮的感觉。电梯厅与走廊通透设计，节约能源的同时，引导光线和自然风穿行在电梯筒和厅房各窗户间，形成极佳采光和对流效果，使室外室内均有良好的"风环境"和"光环境"；顶部局部开敞的阳光半地下车库，将绿化引入地库，营造地库良好的自然通风采光条件。

超宽楼距

城市土地寸土寸金，70米超宽楼距带来市区罕有的视觉盛宴，开阔天地纷呈生活乐趣。一杯咖啡的时间，从都市繁华回归生态低密度舒适家园，高尚社区享受，尽可怡然自得。

超大园林

除了金沙洲规划的城市级生态公园、城市体育公园及沿江滨水休闲区，中海金沙馨园一期还规划了逾2万平方米的大型生态社区园林让家在绿波中靠岸。超高绿化覆盖率、依地势均匀分布绿化带，处处萦绕清新绿意。

3

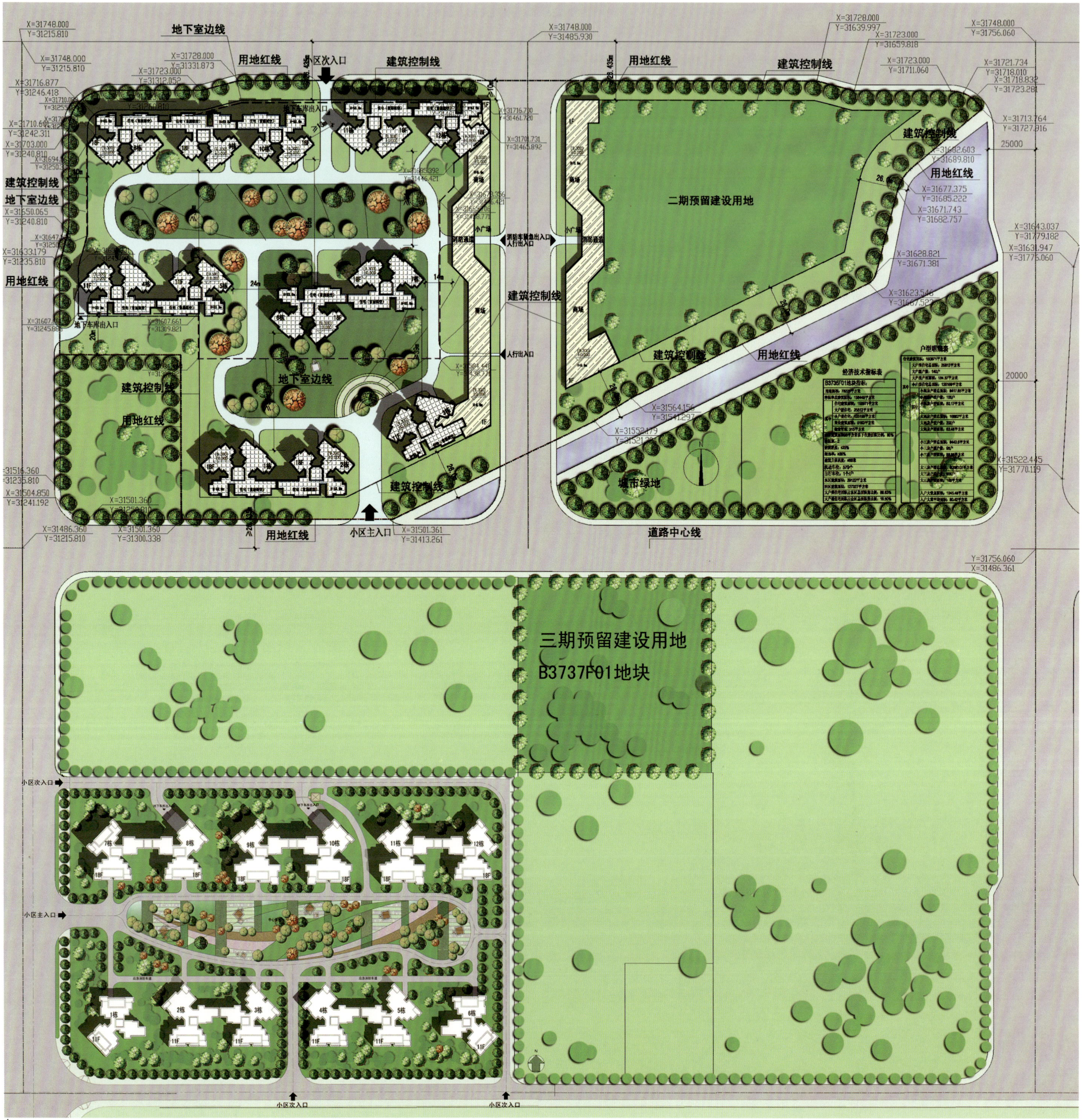
地下室边线
用地红线
小区次入口
建筑控制线
地下车库出入口
消防车道
小广场
消防车辆出入口
人行出入口
建筑控制线
地下室边线
小区主入口
二期预留建设用地
用地红线
建筑控制线
经济技术指标表
户型面积表
城市绿地
道路中心线
25000
20000
X=31748.000
Y=31215.810
X=31748.000
Y=31485.930
X=31728.000
Y=31639.997
X=31748.000
Y=31756.060
X=31486.360
Y=31215.810
X=31501.360
Y=31300.338
X=31501.361
Y=31413.261
Y=31756.060
X=31486.361
三期预留建设用地
B3737F01地块
小区次入口
小区主入口
小区次入口
小区次入口

4

Project address: north Huanzhou Third Road, Jinshazhou New Town, in Baiyun District Guangzhou

Developer: Guangzhou Zhonghai real estate

Planning constructive design: Guangzhou Haohua architect design Co,td

Landscape design: Guangzhou Qingchuan environment & artist Co.td

Overall planning covering area: 127 000 m^2

Overall constructive area: 240 000 m^2

Floor area ratio: 2

Building depotition: 16%

Greening rate: 35%

Overall households amounts: about 2400

Zhonghai jinsha xin residential square is located in F District Jinshazhou, to the west and east side of jinshazhou, huanzhou second road and huanzhou third road, which the main line linked north and south ,on the north side of shafeng road. The project can be divided into three ,the first and the second are the low density ecological living community, the third stage of the project is the only one land use with commercial equipment in F District in Jinshazhou. The covering of the project is nearly 130 thousand square metres, and its floor area ratio is 2.0.Surrudings plan to have large number of city planning public green land. It belong to In addition to two nursery schools in planning ,two primary schools, one middle school, it will build into a the rare excellent well-equipped low-

5

6

7

8

9

10

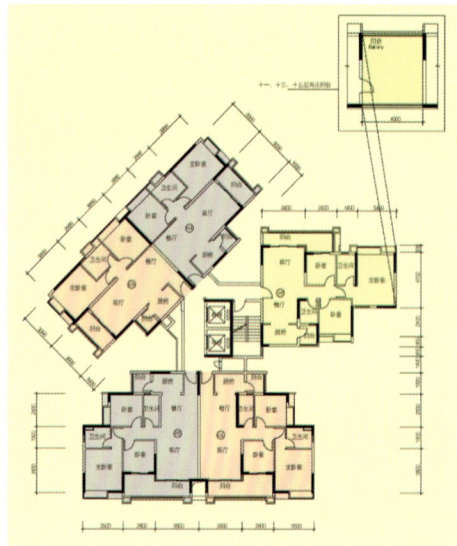

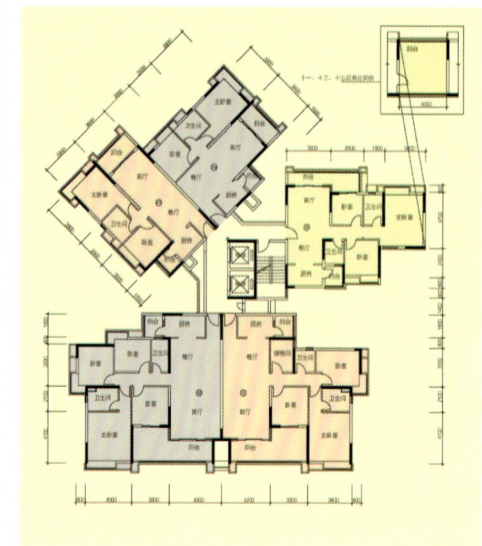

greening zone along east river mainstream plants stretched into the human living.

Inside the community, it place six living parks make the

City ecological garden and natural lake cater to the living habbits "loving mountain and water" of people in guangzhou , forming "the city is surrounded water, and people are surrouded by green". What's more , Jinshazhou was located in the center of Guangfu Metropolitan Area . There is no doubt that it will have a brilliant future.

"two–three films as six zones" scientific regional planning

The greening systerm of Jinshazhou systerm are made up of the riverside landscape greening zone, highway proofing greening zone ,city ecological park and corridors greening zones, which place in term of natural.

11

density ecological living community in Guangzhou. Now the first stage of the project, the last theatrical excellent units is putting forward.

Jinshazhou is located in the center of Guangfu metropolitan area, in the northwestward of Guangzhou, to the east side of baisha river, on the north, west, south side of Nanhai District, Foshan. It own a ideal natural ecological environment and advanced transportation condition In the newest planning , Jinshazhou fix its position as the living new town in Guangfu metropolitan area. it will be built into a living center that is most modern and most excellent management with largest scale and natural character in west Guangzhou in the future.

"The city is surrounded water, and people are surrounded by green" Ecological living new town

Landscapes in the jinshazhou bring fresh air. the government spent huge sum of money planning the riverside park, riverside sports park forming the greening landscape spool thread along river. In arm of lake and wold was planed to built a city-level ecological park in the west, which work in concert with the

12

13

14

One center: center of modern trade and finance, entertainments, located in jinshazhou bridge; integrate shopping, entertainment, leisure , financial service.

Two belt: along green belt, make full use of waterfront along rivers, create a water-loving public green for leisure and entertainment, build full-fledged

Two-three films as six zones: Make use of North Central Expressway, Guangfu Road , Jinshazhou Bridge, combine natural topography difference and road framework dividing into six living spaces.

classical elements confortable home

Although it is a fixed price project , it fellow the classical architect elements that accumulated by several years promoting its even and good character With its practical design and good convection and draught, it try to build a harmony and beauty and lodgeable residential square "Cost the price of a fix price house, but live the commercial residential building." has came true.

practical style

It devoted to develop medium-to-high grade apartments. It has a rich experience which accumulated a large number of rational and well-developed types of flat that go through market test , which are adopt into design. The considerate details make the living homelike: rooms with short depth and large breadth, penetrating from north to south, upright and foursquare partition, super utility ratio, double balcony, L style balcony, double toilet, zones divided into activity and tranquility.

Convection and draught

The overall planning of Zhonghai jinsha xin town consider fully creating "wind environment" for community. Layout that enter courtyard twice from north to south is good for leading draught and convection from north to south , combined the constructive character of south of the

16

Five Ridges, story built on stilts and hallway that go through wind and rain ,which not only enlarge the greening area, but also form the fluent and even draught in low altitude. What's more, design that have front-and-back shift and link forming "draught", which guarantee fluency and even aeration in every space of the (residential)quarters creating "wind environment" in the (residential) quarters.

Penetrating and well-illuminated

Good lighting make people feel penetrating and well-illuminated outdoors and indoors. Design about elevator hall and corridors, while energy-saving ,can lead lights and natural winds go through every windows in halls and rooms forming optimum lighting and convection effects. Good "wind environment" and "light environment" come into being outdoors and indoors. The half-underground garage that have open sunshine in coping part invite greening into basement creating

natural lighting and aeration condition.

super wide building distance

An inch of land, an inch of gold.70 metres wide building distance bring a visual feast to city presenting living pleasure. Just the time spend on a coffee can return from the prosperity city to the low density comfortable homeland. Just enjoy yourself in the high-taste community.

super gardening

in addition to the planning ecological park in city level, city sports park and rest area along the river, the first stage of the project also has planed a large scale ecological community that has over 20 thousand square metres along the bank. Superelevate greening covering rate, the even greening covering zone adapted the terrain, You may enjoy a fresh and green surrunings.

1

占 地 面 积：1 261 920 m²
总建筑面积：1 300 000 m²
容 积 率：2.5
绿 化 率：28%
总 户 数：1200
停 车 位：地下98

1. 庭院透视图
2. 景观小景
3. 中庭鸟瞰实景图

南京东郊小镇

开 发 商：百胜麒麟（南京）建设发展有限公司
项 目 地 址：南京市江宁区江宁区麒麟镇东郊小镇第1、2、3、4街区
景观设计单位：老圃造园工程股份有限公司
建筑设计单位：澳大利亚道克设计（深圳）咨询有限公司

2

　　东郊小镇坐落于南京市区东郊的紫金山东麓，隶属于江宁区麒麟镇。地块北接沪宁高速公路，西临锦绣花园已建楼盘，南至宁杭公路，东至南京二环高速公路，整个项目与建设中的仙西新市区相连，项目距麒麟镇不足3公里，距中山门不到12公里，至新街口约15公里。项目用地紧邻以沪宁高速及二环构成的南京东郊的绿色生态网架，项目内为丘陵地形，起伏变化较大，现状多为农田，尚有部分村落及农户集居地。整个项目红线内宗地总占地为120多万平方米，规划总建筑面积为130万平方米，计划分5-6年开发，年开发量在20万平方米左右。项目一期第一街区南至宁杭路，东至晨光村小学，北至用地红线，西至区内道路。用地面积为75222.6平方米，南北长304-307米，东西宽约220米。一期项目绿化率为43.5%。容积率为1.302。第一街区规划总建筑面积

为10万平方米左右，其中住宅约为96000平方米，商业用房约为1400平方米，一期共计36幢房屋，除了19号楼为7层挑高高层住宅以外，其余的均为6层普通多层住宅（21号楼为5层样板房），住宅合计为963户。项目现已开发至第四期。

　　东郊小镇位于南京三大风景区环抱之中，西临钟山风景区，东傍汤山风景区，北接佛教圣地栖霞山，景色怡人，空气清新，为未来的业主创造美丽而又悠闲的生活环境．

　　东郊小镇南临宁杭公路，北接沪宁高速公路，与建设中的仙西新市区相连，快速交通网络完善；小镇周围有55、309、314、游5、太作、南汤等公交线路，第一街区交付初期会开通上下班时间的社

区巴士，新公交线路的延伸也在洽谈中；南京地铁2号线计划2009年开通至马群，白水桥立交已建成通车，正常情况下开车到新街口也就30到35分钟。

　　东郊小镇位于南京东郊麒麟镇，麒麟镇被定位为南京大都市近郊以居住区开发为特色，制造业占有一定地位的卫星小城镇．小镇临近仙林大学城，充分享受大学城的规划利好．项目占地120多万平方米，总建筑面积超过100万平方米，初步规划总人口3.5万。按照新市政的规模和功能要求，将分期建设幼儿园、小学、公交车站和社区大型生活、商业、文化、体育等服务设施。

Project location: eastern suburb town block 1,2,3,4 Kirin town jiangning district Nanjing city

Developer: Baisheng Kirin (Nanjing) Construction Development Co., Ltd.

Landscape design: Old Farmer Landscape Architecture Co., Ltd.

The area of land: 1 261 920 m^2

The total area of construction: 1 300 000 m^2

Volume rate: 2.5

Greening rate: 28%

The number of dwellings: 1 200

Parking: underground 98

Eastern suburb town is located in Nanjing western Kirin town which is orientated to be outskirts of Nanjing metropolis that make a feature of residential area development and manufacture. This town is close to Xianlin University City and takes full advantage of the good plan. The area of the project is 1916 acre and the total area of the construction is more than 1,000,000m^2, and the population of preliminary scheme is 35,000. According to the new municipal scale and functional requirements, the construction will be phased to kindergartens, primary school, bus station, large-scale community life, business, culture and sports service facilities.

3

4. 中庭休闲景观
5. 楼宇间水景

Eastern suburb town lies in east of Purple Mountain which belongs to Kirin town Jiangning district. It lies north of Nanjing-Hangzhou Highway and to the west of Fairview Park that has been completed, to the south of Nanjing-Hangzhou Highway and Nanjing second ring highway is in the east. The whole project is connected to the new Xianxi city center which is under construction in the west, within 3 kilometers away from Qilin Town and 12 kilometers away from Zhonshan Gate, about 15 kilometers from Xinjiekou. The project is also closely next to the ecological net consisting of the Hu-Ning Motorway and the second ring road. There are a lot of hills inside the project with various changes in terrain, most of which are farmland and parts of villages. The total project covers an area of 1,912 acreages with the red line, and its total floor area is planned to be, 130,000 square meters, first block of which is planned to be 10,000 square meters including residential block of 96,000 square meters and commercial houses of 1,400 meters. There are 36 houses with 963 resident units in total in the first phase, most of them are all 6-floor common resident buildings (No.21 is a 5-floor sample) except that No.19 is a 7-floor high-rise resident building. So far it has been developed to the fourth phase.

Eastern suburb Town is cradled in the arms of Nanjing's three scenic spots, Clock Town Scenic Spot in the west, Tandshan Mountain Scenic Spot in the east, and Buddhist Holy Land Xiashan Mountain. Such beautiful scenery and fresh air allow prospect owners to create a beautiful and leisure living environment.

Quick and perfect transportation web is here with Ning-Hang Road in the south of the town, Hu-Ning Motorway in the north and the new Xianxi City being under construction. Around the town there are bus routes No.55, No.309, No.314, No.5, Taizuo, Nangtang, etc. And in the beginning period of the delivery of the first block there are community buses during the time of on and off work while the extend part of the new bus route under negotiation; then the Nanjing Line 2 is to be operated to Maqun in 2009; finaaly, the Baishui Bridge has been completed which allows trip by car to Xinkoujie with 30 to 35 minutes under general circumstances.

5

6/7/8/9. 建筑立面

7

8

9

10

11

10. 阳台景观

11. 入口外望

12. 样板房客厅实景

13/14. 样板房餐厅实景

12

13

14

1

2

武汉·大华南湖公园世家二期

开 发 商：大华集团
项 目 地 点：武汉市洪山区南湖风景区
建筑规划设计：澳洲高臣建筑事务所
基 地 面 积：175 000 m²
总建筑面积：326 845 m²
容 积 率：1.67
绿 化 率：32.0 %
总 户 数：2318 户

"大华南湖公园世家"项目位于武汉市洪山区南湖风景区内，具有得天独厚的地理位置优势。本项目分三期开发，二期地块位于整个用地的中部，用地面积175000平方米，东面隔出版城路与一期工程相邻，北面为武梁路，西面为石牌岭路，南面为南湖新城路，拟建70栋建筑，其中有9栋商住楼、59栋住宅楼、2栋社区服务楼，1所小学，9处地下室，总计建筑面积326845.58平方米。设计立足于高起点、高品位，注重居住文化、居住素质的提升，营造一种全新理念的面向未来的居住方式及生活理念。

用地概况：用地较为周正，东西向长度由358—450米，南北向纵深约430米，地势北高南低，坡度平缓，地面高程约为

21.0—22.4米。根据周边及本用地的条件分析，本地段适于开发建设中高档的住宅小区，并可利用临路、临河的优势布置商业用房，提供社区服务并可以充分挖掘其经济效益。

总体布局：总体采用纵向行列式布局方式，建筑物均按南北向布置，保证住宅每户均有良好的通风采光和朝向景观；强调绿化主轴和景观视觉通道为主线景观轴，并形成空间轴线，强调景观的共享性及空间的序列性。依照主线景观轴划分组团，形成一个完整且具有层次的空间序列，并通过建筑的排列、景观小品、入口广场等空间及景观上的设计，体现组团的特点和可识别性。商业部分尽量利用商业价值高的沿街面。结合城市规划要点要

求，适当让出空间做为小区的入口广场，形成小区空间序列的起始；将内部景观空间向外延伸形成景观点，并结合入口广场的景观环境设计，形成居住小区的一个入口景观，成为整个建筑群在沿街面上的一个景观亮点。

绿化系统：绿化系统是本工程的主要特色，以生命绿带和景观视觉走廊为核心贯穿整个用地区域。本期设计在此主轴规划下，着重组团入口景观、广场空间、内部绿化空间，各组团之间形成内部绿化空间节点，设置老年人和儿童的活动场地，满足功能要求架设的木质铺地强调景观空间的节点，从而活泼整个小区的环境景观设计。

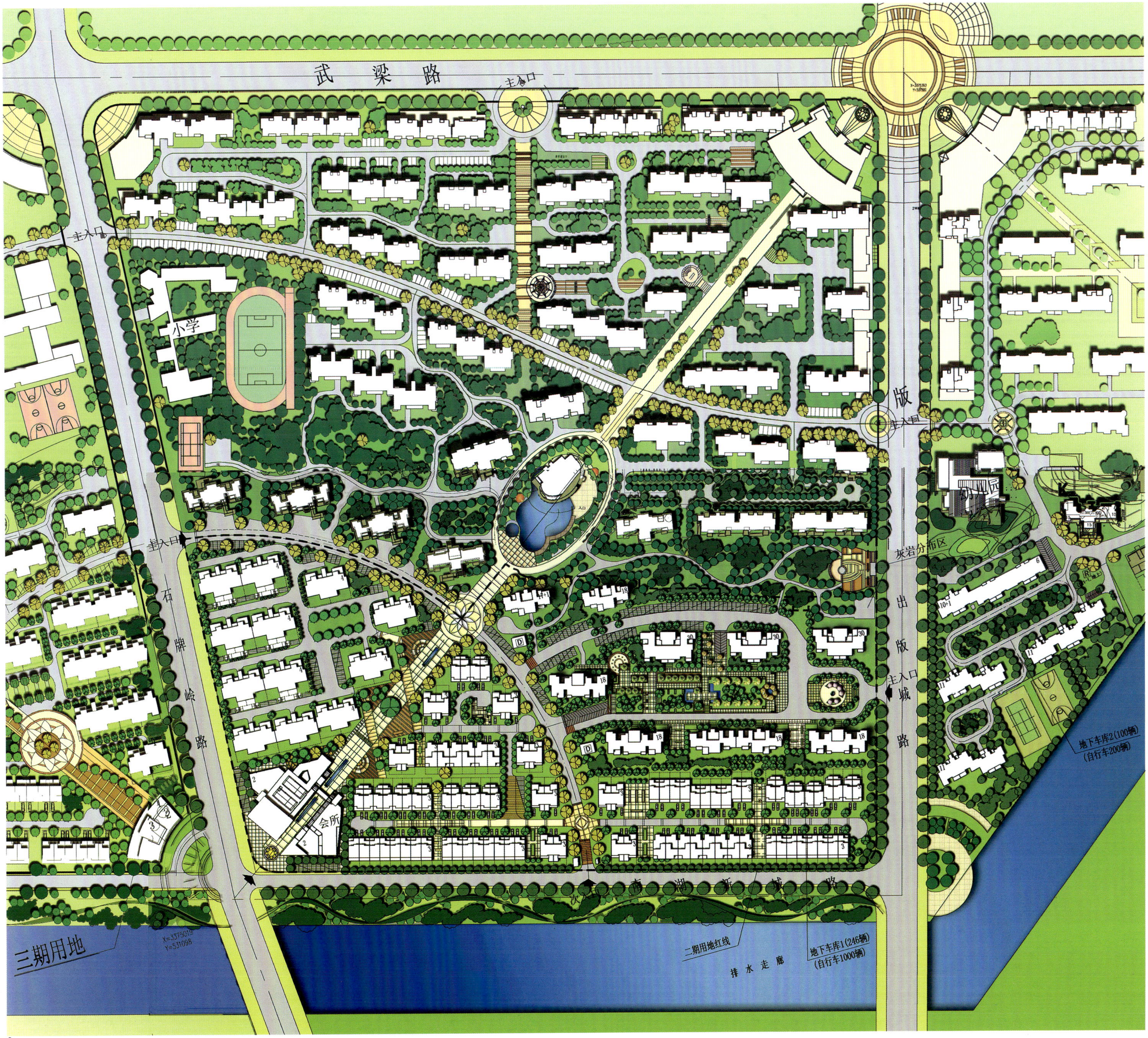
武梁路
主入口
小学
版
出版城路
石牌岭路
会所
幼儿园
灰岩分布区
湖新城路
三期用地
二期用地红线
排水走廊
地下车库1(246辆)
(自行车1000辆)
地下车库2(100辆)
(自行车200辆)

3

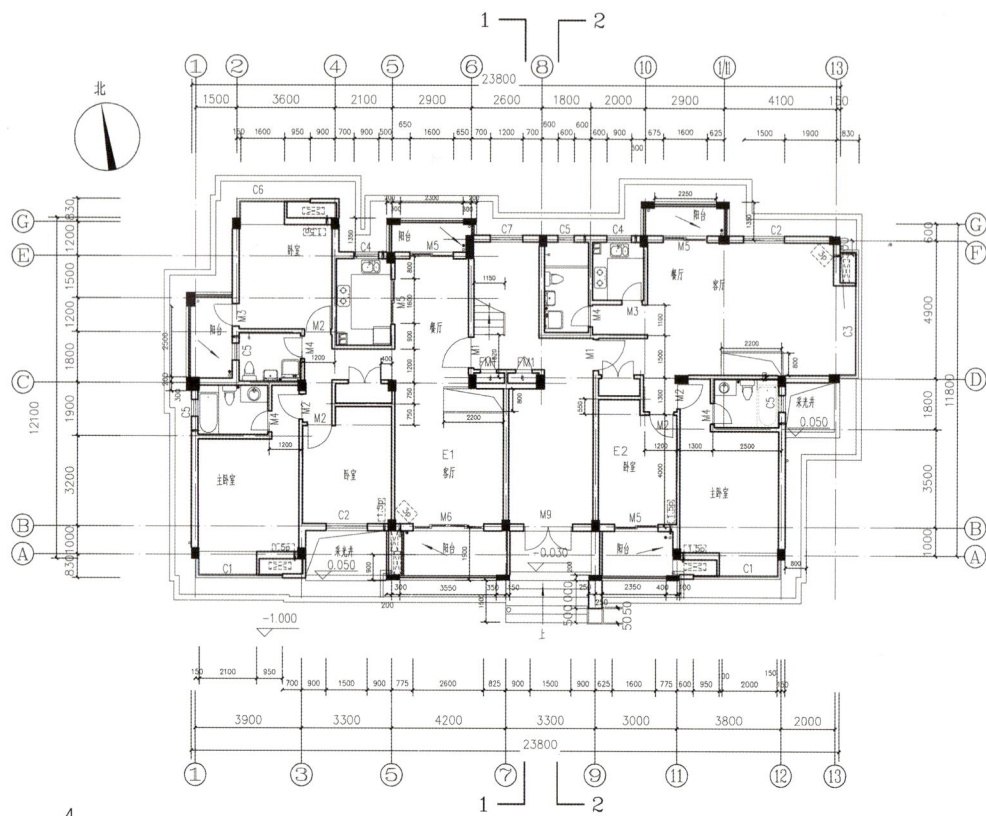

4

平面设计：住宅单元户型设计平面方正实用，采光通风良好，交通厅均可直接采光。户型有一房、二房、三房、四房和叠拼多种形式，建筑面积由50.90—226.76 m²，以80—96 m²二房和119—129 m²三房为主导房型；户型平面设计上，引入了"+1"的设计概念，即在二房和三房的基本户型中增加一间小面积多用途房，提升了居住档次；为客户提供多样选择而促进销售。

建筑造型：强调稳重、细腻，以深沉的色调和细部装饰、挑檐、线脚的精心设计，利用色彩、材质的变化，以及坡屋顶的跌落、变化，达到清新、自然的效果，使小区个性鲜明，独具特色，具有独特的 "海派" 情调。在商业的细部上精致而富有内涵，以端庄又不失华丽的新古典式建筑风格演绎异域风情。

4. 多层标准层平面图
5. 多层实景1
6. 多层实景2
7. 多层大堂入口

5

6

Floor area: 175 000 m^2
Building area: 326 845 m^2
Greening rate: 32.0%
Volume rate: 1.67
The number of dwellings: 2318

Project of "Dahua Lapark" is in Nanhu Scenic Area, a unique and beneficial geographic location in Hongshan District, Wuhan City. It has been devied into 3 phases to develop, second sector of them is in the middle of the whole area with a floor area of 175,000m^2 and in its North, Shipailing Road in its West and Nanhu New Town Road is in its South. It is to build 70 blocks, 9 of which are business living blocks, 59 are residential blocks, 2 community service buildings, a primary school and 9 basements, total floor area is 326,845m^2. A whole bright new living rationale is created based on the attention to the improvement of living culture and quality.

Shape of the land: its length of Westeast is 358m to 450m and 430m of Northsouth, and it is higher in the North and lower in the South with a gentle slope measured 21.0 to 22.4 meters high. Analyse from its surroundins and the use of local land, this section can be developed as the high-level residential blocks under construction now. The advantage of being next to roads and river also can be beneficial to arrange commercial housings, serve the community and fully bring up its economic efficiency.

General layout:longitudinal procession layout is adopted and each construction is set to be North and South towards to promise that all residents live in cool and bright houses with nice views; greening main axis and passages with feature views is empasized, forming a space axis commonly shared. A whole neat space order comes from the groups that devided through the main feature axis. And group feature and how it is recognized is embodied through the order design, the features, entry plaza, all things in space and features.

7

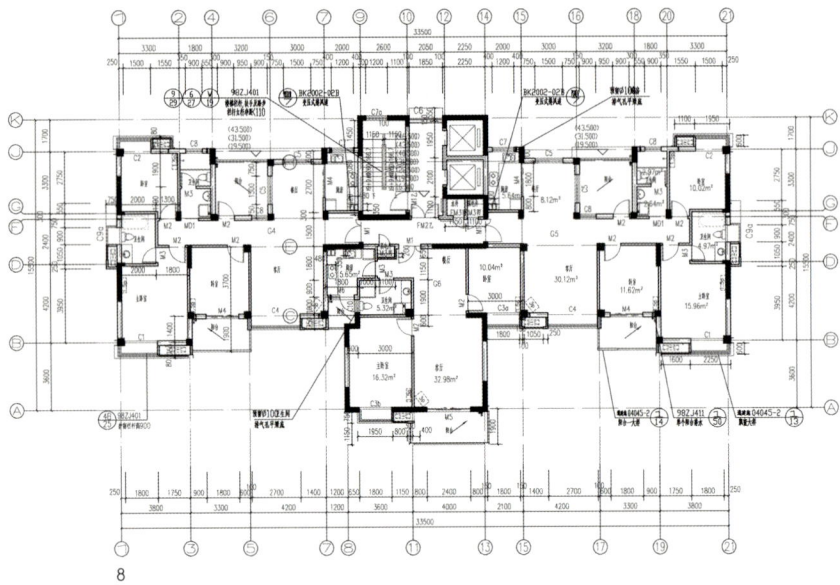

8

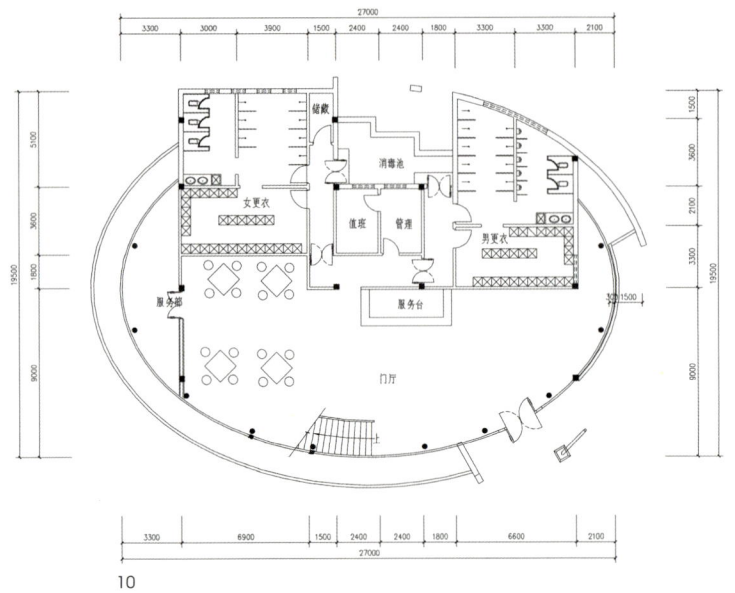

10

9

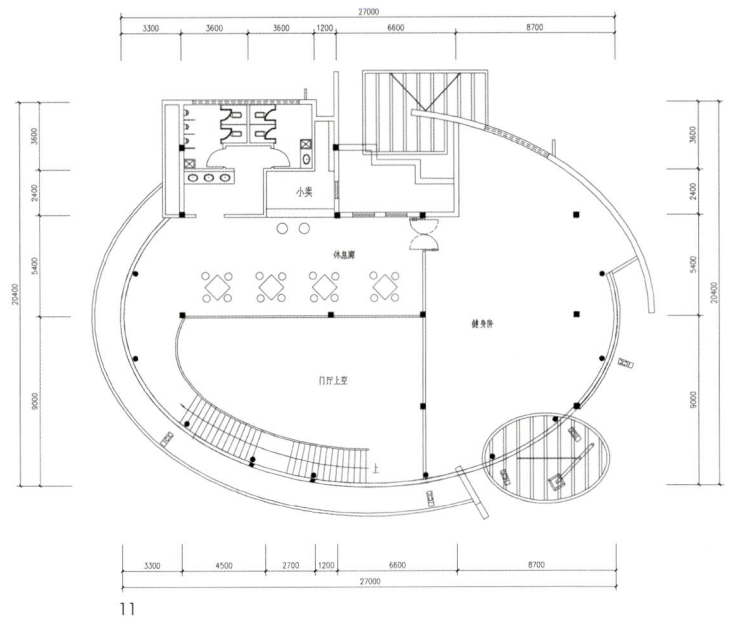

11

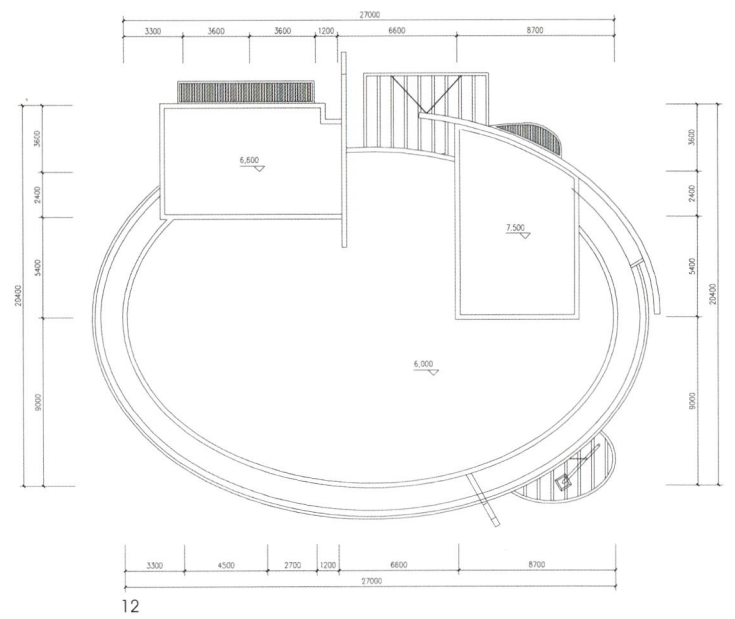

12

8. 高层标准层平面图
9. 高层实景
10. 会所三层平面图
11. 会所二层平面图
12. 会所一层平面图
13/14/15. 会所景观

13

14

15

According to the requirement of city plan, commercial parts make best of the high-valued street side to make room for entry plaza of the block, making it a start of the space. The entry feature becomes the brigtht spots of the whole construction groups due to its extending to outside, combining the entry plaza feature and the inner feature space. The greening system: main characteristic of the project with living greening belt and visual feature corrdors as its core penetrating through the whole area. To make the environment of the whole block alive, great efforts have been made, such as the greening sapce among entry feature, the plaza space and the inner greening space, the setting of eldly and children activity field, wooden floor meeting the function demand to emphaziing to knot of space.

Graphic design: practical, bright and cool residential units, all their traffic halls can get enough sunlight directly. There are housing types with from one room, two rooms, three rooms, and four romm to cityhouse, floor areas from 50.92m^2 to 226.76m^2, 80 to 96 m^2 and ones with 3 rooms of 119 to 129 m^2 is leading the housing types. In terms of graphic designs, ratiaonale of "+1" is introduced, that is, a small multi-purpose romm is added to the basic type of the ones with two rooms and three rooms. And finally, promoting sales to provide customers with various choices.

Construction modelling: steadiness, equsitivenss. Dark color tones, detailed decoration and delecate design of overhanging eave and moldings, the use of variation in colors and material, and declining, changes of the slope roof together form a fresh, natural and special block.

16

17

18

19

20

21

22

23

24

25

26

27

28

1

A派公寓

开 发 商：北京浩隆房地产开发有限责任公司
园林设计：易道(北京)
建筑设计：翰时国际
占地面积：18 182 m²
建筑面积：64 600 m²
绿 化 率：31%
容 积 率：2.7

2

　　A派公寓由4栋住宅楼及两栋配套楼组成，其中1栋16层的观景公寓，其余三栋是CBD内罕见的9层板式公寓，这四栋楼构成围合式社区，社区中间是4000多平米的共享绿化空间。

　　建筑外立面塑造的美式现代风格，外立面材质选用砖红色搭配灰色高级面砖及局部的白色涂料，将建筑元素如凸窗、景观阳台通过现代建筑手法组织起来，营造出居住兼具时尚气息的建筑形象。品质感非常强，色彩温暖又不失稳重。

　　社区园林是本项目的一大亮点，中心集中绿地达4000多平米，由国际一流的园林设计公司易道（北京）担纲，营造公园般优美、静谧的社区园林环境。园林中着重利用草坡、林地、水系及植物群落，合理组配成有机生态循环系统，保证社区园林的生

均由开发商精心购置，这些植物的科学搭配尤其是水生植物的配置使社区园林成为一个非常难得的生态组合。

　　在园林材料的选择方面，也同样注重自然、生态。以大面积的林地草坡为主。景点用材基本为天然石料，木料……以树、草、木、石，这几种最淳朴的设计语言描绘城市人对自然真实的向往与殷切的归属感。

　　社区园林主体设计元素为两条线状元素及多处点状元素相互交融、有机组合而成。两条线状元素由弯曲的水系及迂回的小路组成：叠水、石溪、喷水池塘组成丰富多变、具有灵性的水体生态系统；而自然曲折的青石板小路又环绕林间，顺着这条充满自然情趣的小路，可以体验畅游林地的愉悦。这些点状元素似乐章

顶木亭、旱喷广场、草地剧场形成自然、充满生机活力却又不失多功能用途的活动空间。入口池塘处为大块的山石形成的案堤、木台，在此展眺园林。童趣广场是专门为孩子们提供的游乐空间，设置安全、参与性强的游乐设施。被一片相对区域较大的水面包围的是临水茶座，与山顶木亭遥相呼应。劳累一天后，或者在夏季的傍晚，这里伴着活泼的喷泉、听着潺潺的水声，恬静地品一杯红茶，实在是很惬意。山顶的木亭是社区园林的一个制高点，树影掩映中潺潺叠水顺石而下，可以远眺斜阳，而近处又身处树林一般，享受鸟语花香。旱喷广场及草地剧场是园区中两个不同尺度的集聚空间，在这儿与邻居、朋友轻松的聊天，是社区园林中交流的好场地。整体来看，社区的景观设计使得线状与点状元素合理连贯地被安排成一个科学有机的整体，在保证功能性的同时使人享受在公园般的闲情雅趣。

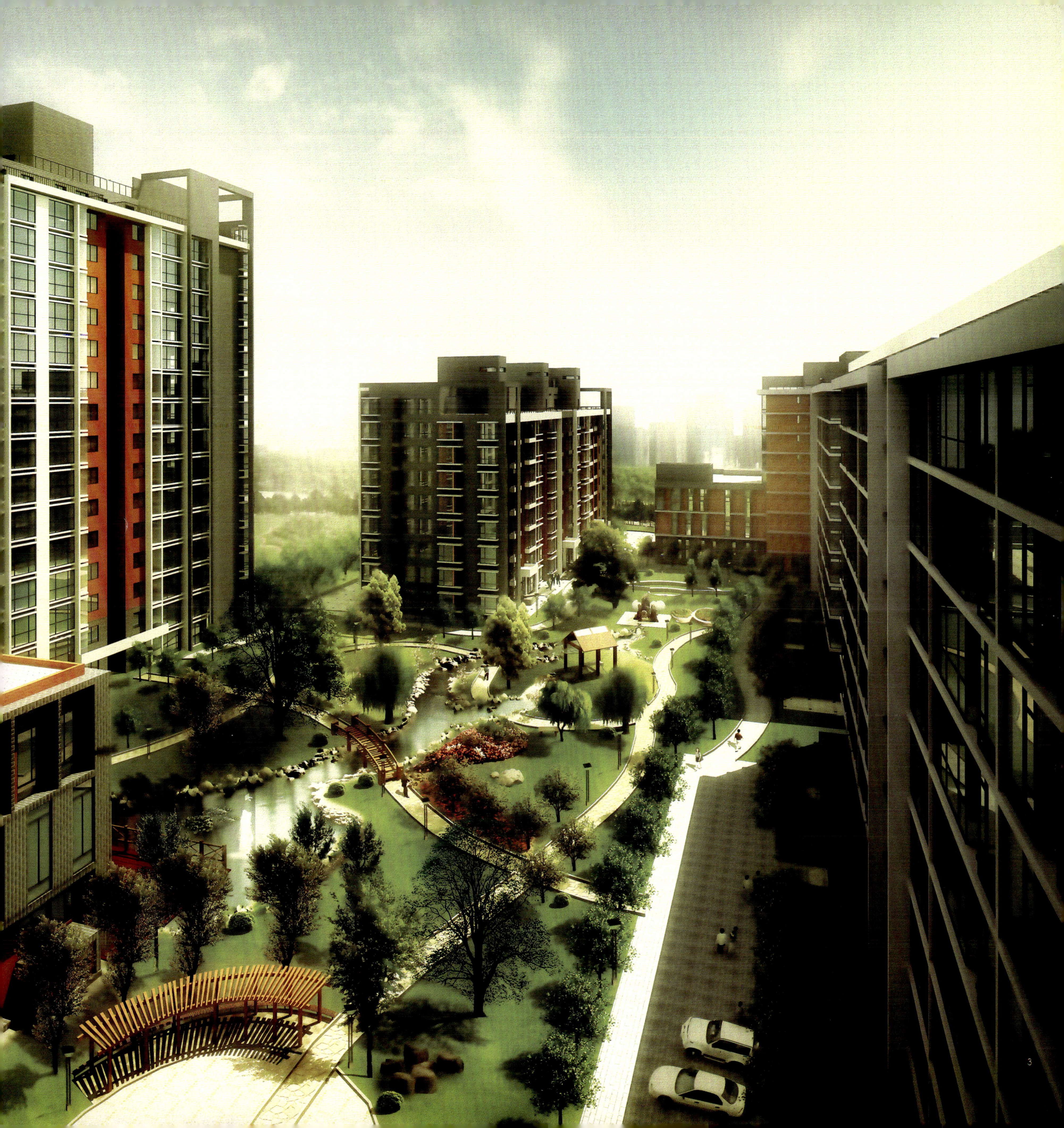

Developer: Beijing Haolong Real Estate Development Co.,Ltd.

Landscape Design: Yidao (Beijing)

Architectural Design: A&S International Design

Area: 18,182 square meters

Building Area: 64,600 square meters

Green Rate: 30%

Volume: 2.7

A PIE is composed of four dwelling buildings and two matching buildings. One is a sixteen-viewing block of flats, and other three apartmenrs are the rare nine-floors plate type of apartment in CBD. The floor buildings constitute a surrounded community. In the middle of the community, there is a public green space covering about 4,000 aquare meters.

The building facade shape the modern American style. the materials of facade is made up of the super bricks with brick and grey, and the part of white paint. All of these make the architactural elements such as comvex window and balcony organized by the landscape of modern construction, creating a constructive image with living and fashion. It has a high sense of quality, and the tones is just OK. The community garden is a major bright spot in this project. There are more than 4,000 square meters of green space in the center Yidao (Beijing), for creating a beautiful and tranquil environment of the community garden as a park.. It is placed stress on the use of grassland, woodland, water and plant community, in order to build a circulatory system with organic ecological group reasonably. The system make the ecology of the community garden and the sustainable development possiable. As you see, there are several magnificent trees in the garden which are purchased carefully by

4. 鸟瞰图
5. 总平面图

5

the developers. The scientific collocation of these plants, particulary the collocation of the aquatic plants, make the community garden become a very rare combination of ecology. On the choice of the material of community, nature and ecology is also paid high attention to. Relying mainly on a large of woodland and grassland, the basic materials of the scenery are natural stone, wood… Using trees, grass, wood, stone—the most simple language to depict the townfolks' real yearning for nature and the high sense of belonging.

The main elements of the community garden design are made up of two linear elements and multiple points elements. Two linear elements consist of crooked hydrographic net and winding path. Overlapping water, stone river and spray pond make the water ecological system more changeable and attractive. Along the crooked stale path which is full of flavor and circle the woodland, you can enjoy the pleasure of visiting the woodland. The multiple points spread like a movement from east to west—from the pond of entry, the teahouse clubs of abutting on the river, the amusement park, the peak wooden booth, the dry fountain square to the theatre of grassland, forming a moving space with nature vitality, and more functional use too. There are embankments and tables which are made of the big rocks in the pond of entry. Here overlook the garden . The amusement park is a space providing pleasure specially for children. And the playing equipments in it are safe and with a high participation. The one that is surrounded by a large area of water ralatively is the teahouse clubs of abutting on the river. The teahouse and the peak wooden booth coordinate with each other from afar After working all day long, or evening in the summer, it is very satisfactory for you to sit tranquilly for a cup of tea, seeing the lively fountains and listening to the sound of running water. The peak wooden booth is a commanding point of community garden. In the shade of the trees, the overlapped water is gone gurgling and babbling along the stone. You can overlook the setting sun, and just like dwelling in the forest to enjoy the cheep and the flowery fragrance. The dry square and the theatre of grassland are the two different scales of assembling space in the community. You can sit here and have a relaxing chat with neighbors or friends . It is a good place to communicate in the community garden. Overall, the landscape design of the community make the linear and the point elements formed a scientific and organic whole reasonably and coherently. With guaranteeing the functionality, it make sure that people can enjoy the leisurely and comfortably just like in a park at the same time.

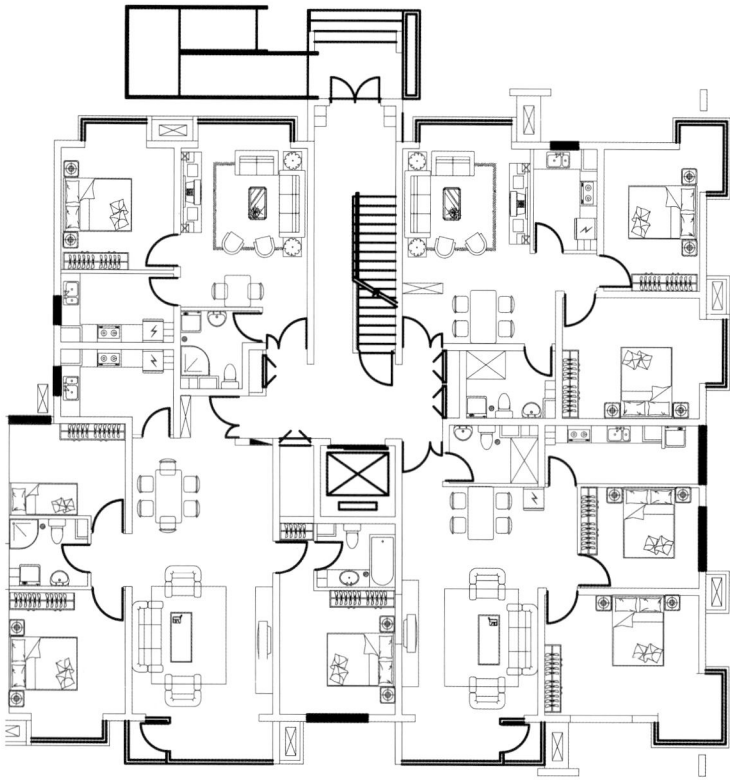

6

7

8

9

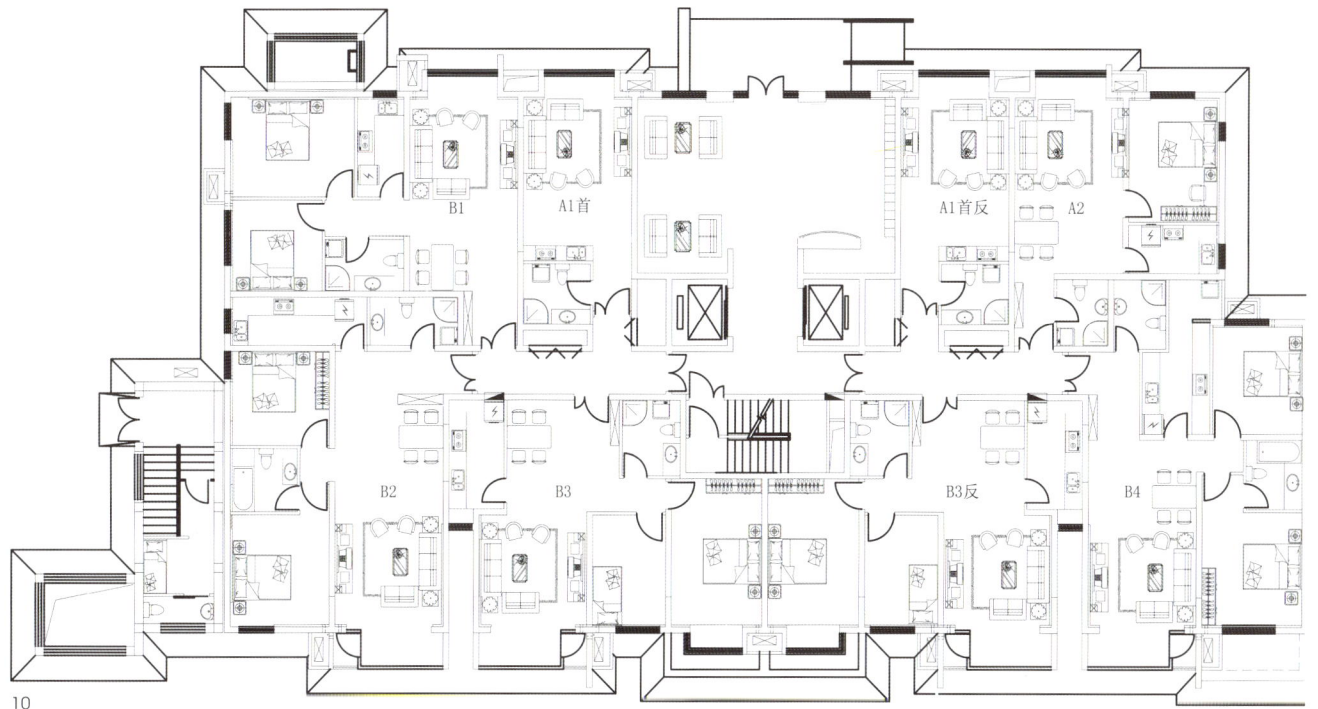

6/7. 2、3栋户型平面图

8/9. 2、3栋户型平面图

10. 1栋户型大样

11. 公共空间效果

B1　A1首　　A1首反　A2

B2　　B3　　　　B3反　B4

10

11

12

13

12/13. 4栋户型平面图
14/15/16. 建筑效果图

14

15

16

中海·南湖一号

开　发　商：长春中海地产有限公司
项　目　地　址：长春朝阳区南湖大路与工农大路交汇处
景　观　设　计：泛亚国际
规划建筑设计：北京构易建筑设计有限公司
占　地　面　积：65 000 m²
总建筑面积：86 000 m²
容　积　率：1.32
绿　化　率：35%
住　户　数：291套
地　下　车　位：498个

中海·南湖一号项目位于吉林省长春市，地块呈三角状，东北临工农大路，西北临城市最大园林公园南湖，南临南湖大路，周边建筑形态多为著名高校和重点中学，本项目所在区域为长春城市形象最为突出的地区，地理位置很优越，历史的印痕和天然形成的自然景观也最集中，是很理想的居住场所。在建筑风格特点上，适合发展传统的建筑形态和文化气息比较浓郁的地域建筑形式。整个规划人车分流，车辆不进入小区。建筑风格属于英式维多利亚风格，融合了安女王风格和哥特式风格的一些元素特点和符号，表现为凸窗、角楼、尖卷、陡峭的坡屋顶等建筑符号。

景观设计强调和建筑设计风格的一致性，注重品质的塑造，同时也强调园内景和园外景的互融与共生。小区绿化讲究"点、线、面"相结合的方式。利用借景、对景、收放空间变化手段来实现景观的多层次性、有趣性、均好性、可识别性。以创造生态型居住环境为规划目标，贯彻"尊重自然"与"可持续发展"思想，在保持原有的地形地貌的前提下，贯彻生态原则、文化原则与效益原则，力求塑造一个具有环境优雅文化内涵丰富、经济效益显著和个性鲜明的花园式经典居住空间。居住区的景观设计不是单纯意义上的绿化设计，它包括绿化、铺装、标志系统、景观照明、景观水景、景观小品六大方面。而这六大方面最后应达到整个景观系统的生态、功能和动观的效应。

设计以满足人们对居住环境所要求的舒适性、健康性、生态性、安全性和经济性为出发点，创造出一个布局合理、功能齐备、交通便捷、环境优美的现代住宅区。讲究人与环境的融合，建筑与整体规划布局的融合，居住行为与休闲、娱乐的融合，建筑与绿化的融合，并充分考虑城市社区的标志性建筑物，使居住者有强烈的归属感和自豪感。

院落作为社区活动与景观的中心可识别性的利用，在院落的边缘围合着房屋，房屋和院落保持着距离，院落之间的路径编织成网络，路径使得院落的空间和表层可以被感知和经历，在院落里是绿色的空间，也是周围小环境的调节中枢，并且通过不同形式的广场、草地、小径、花架、喷泉、座椅、植物、儿童游乐场等景观要素形成可识别的空间场。

　　结合现有场地的特点，环境设计从基本的景观要素入手，将小型的活动场地如水景、常绿树树池、草坪、花架、花房等加以不同形式的排列组合，用简单理性的手法解决了"带状绿地"在平面构图中的困难性，在排列组合的过程中，注重了环境与建筑的对位关系，环境设计中最大限度地减少了硬质材料的品种数量，最大限度地增加了常绿植物的种植量，使环境色彩朴素沉静，以保持色彩浓重的建筑物的完整性。

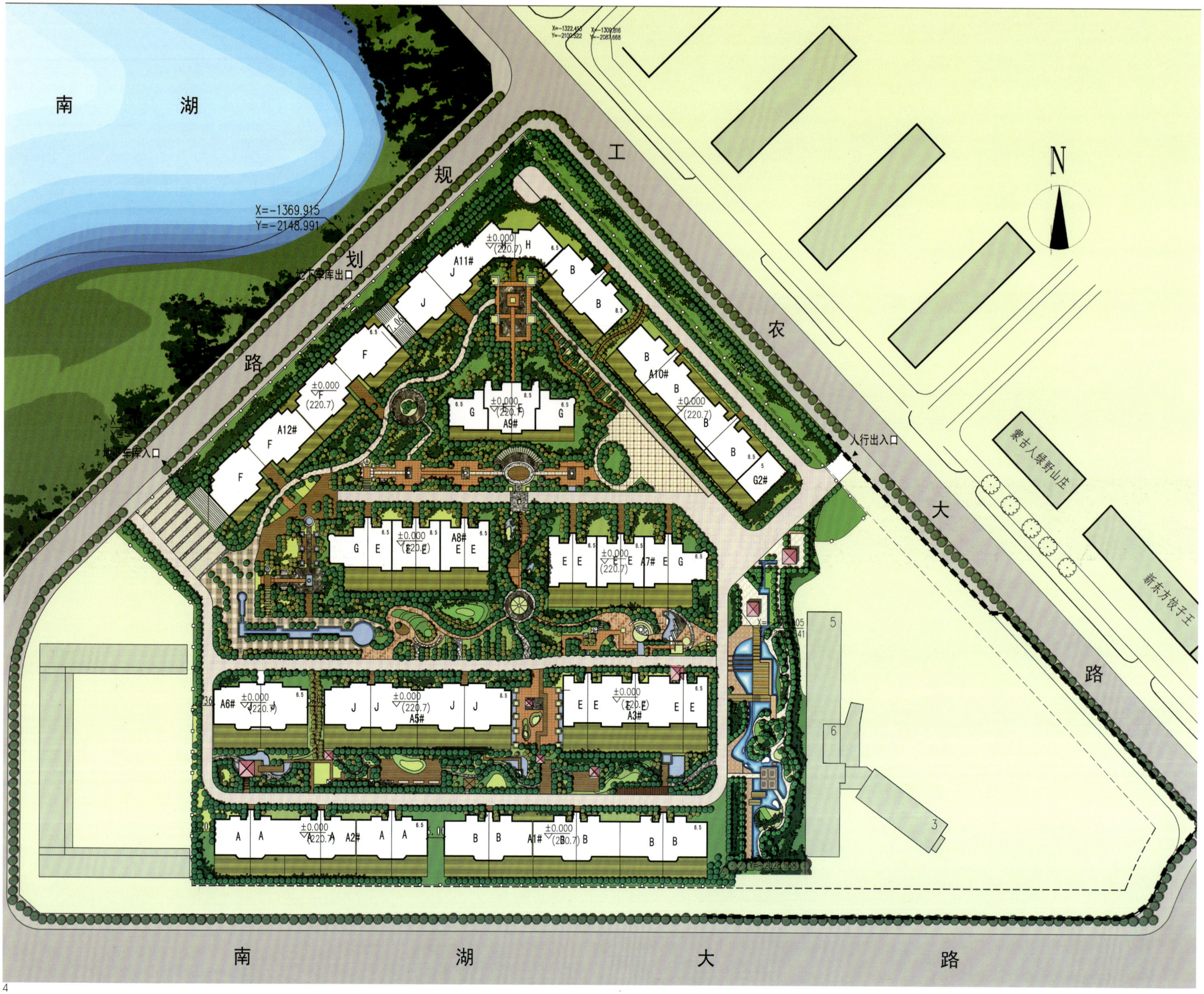

南 湖

规 划 路

工 农 路

大 路

N

X=-1369.915
Y=-2148.991

南 湖 大 路

4

Floor area: 65 000 m²

Building area: 86 000 m²

Greening rate: 35%

Volume rate: 1.32

The number of dwellings: 291

Zhonghai·South Lake No.1 item is located in Changchun City, Jilin Province, block show a triangle shape, northeast approaches the Worker-peasant Road, northwest approaches the largest garden prak South Lake of the city, south approaches South Lake Road, most architectural form neatly are famous colleges and Key Middle School, these areas from this project are most prominent for the image of Changchun City, which location is very superior, focusing on historical and natural-printsof natural landscape, is an ideal living place. In the features of

architecture style, suitable for such a architectural form with traditional form and relatively strong, cultural atmosphere. The whole planning devides human and vehicle, vehicles are banned to the district, the style belongs to British Victorian Style, blending in several elements and symbols between the Quene's Style and Gothic style , performing as complex windows, conner towe, steep sloping roof such symbols.

What landscape design and architectural design stressed is the consistency with the style, focusing on improving quality, meanwhile emphasized that blend between inner scene and outer scene. The District Greening adopts a new way combined with "point, line, face". To make full use of comparing to borrow scenery, opposite scenery and such changeable method, which to realize multilevel, interesting of scenery. In order to aim the planning on creating ecological living environment, carrying out the concept of respect "nature" and "sustainable development" on the basis of keeping original terrain, carrying out ecological principles, cultural principles and effective principles, make it classical living space which has elegant environment, significant economic effect and clear garden. Landscape of residential areas is not just so-called greening design, it includes greening pavement, the symbol of system, landscape lighting, waterscape, scenery pieces 6 puints. From the six points it reaches to the dynamic concept about ecology function and development of the whole system.

In order to satisfy people's comfortable life, ecological life, security and economic , design creates a modern department which has reasonable layout. various functions, convenient transport and beautiful environment. Focusing on combination of human and environment, the combination of architecture and the whole layout, the combination living behavior and entertainment, the combination of architecture greening, what's more, considering the symbol architecture in the city, giving the residents strong scene of belonging and sense of proud.

5

6

4. 总平面图
5. 入口效果图
6. 入口手绘图
7. 入口平面图

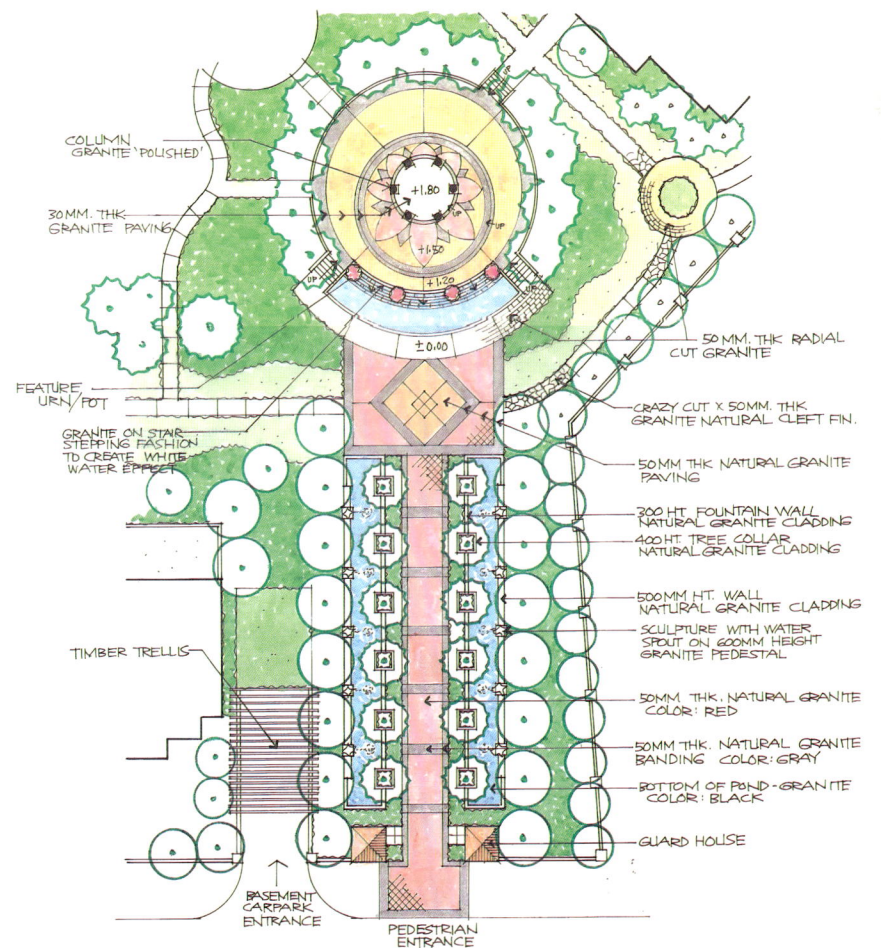

COLUMN GRANITE 'POLISHED'

30 MM. THK. GRANITE PAVING

FEATURE URN/POT

GRANITE ON STAIR STEPPING FASHION TO CREATE WHITE WATER EFFECT

TIMBER TRELLIS

50 MM. THK. RADIAL CUT GRANITE

CRAZY CUT X 50MM. THK. GRANITE NATURAL CLEFT FIN.

50 MM THK NATURAL GRANITE PAVING

300 HT FOUNTAIN WALL NATURAL GRANITE CLADDING

400 HT TREE COLLAR NATURAL GRANITE CLADDING

500 MM HT. WALL NATURAL GRANITE CLADDING

SCULPTURE WITH WATER SPOUT ON 600MM HEIGHT GRANITE PEDESTAL

50MM. THK. NATURAL GRANITE COLOR: RED

50 MM. THK. NATURAL GRANITE BANDING COLOR: GRAY

BOTTOM OF POND - GRANITE COLOR: BLACK

GUARD HOUSE

BASEMENT CARPARK ENTRANCE

PEDESTRIAN ENTRANCE

7

8

9

8/9/12. 效果图
10. 户型A平面图
11. 户型B平面图

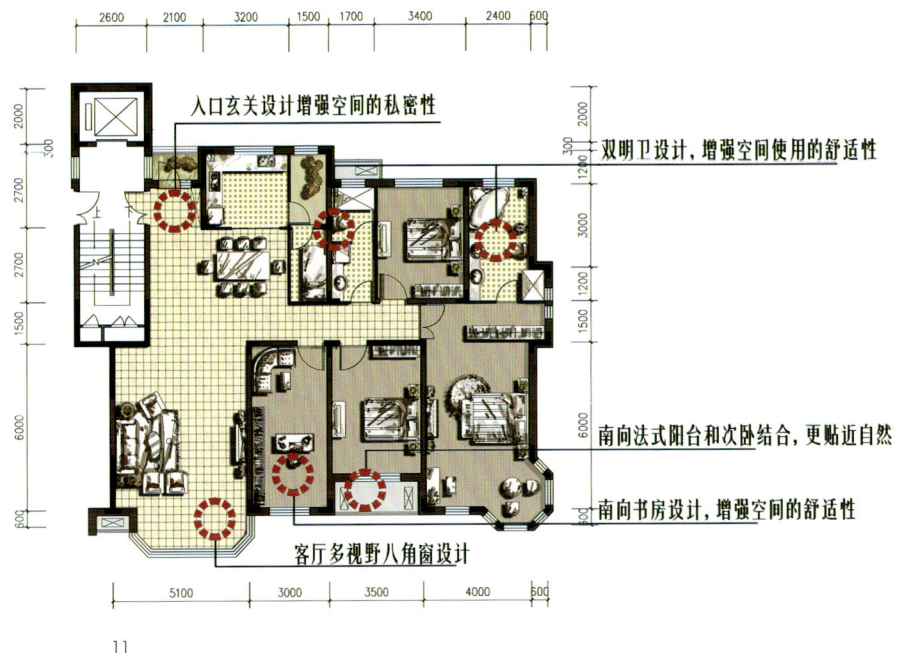

左图 (10):

2600　2100　3300　3500　2400　600

2700　3200　3000　1500　5700　1500

入口玄关设计增强空间的私密性

五件套明卫空间,增加主卧豪华感

主卧空间和书房空间可拆合

5.1米舒适大客厅,方显豪华感

多角度 观景窗设计

5100　3500　4000　600

10

右图 (11):

2600　2100　3200　1500　1700　3400　2400　600

2000　300　2700　2700　1500　6000　600

1200　3000　1200　1500　6000

入口玄关设计增强空间的私密性

双明卫设计,增强空间使用的舒适性

南向法式阳台和次卧结合,更贴近自然

南向书房设计,增强空间的舒适性

客厅多视野八角窗设计

5100　3000　3500　4000　600

11

12

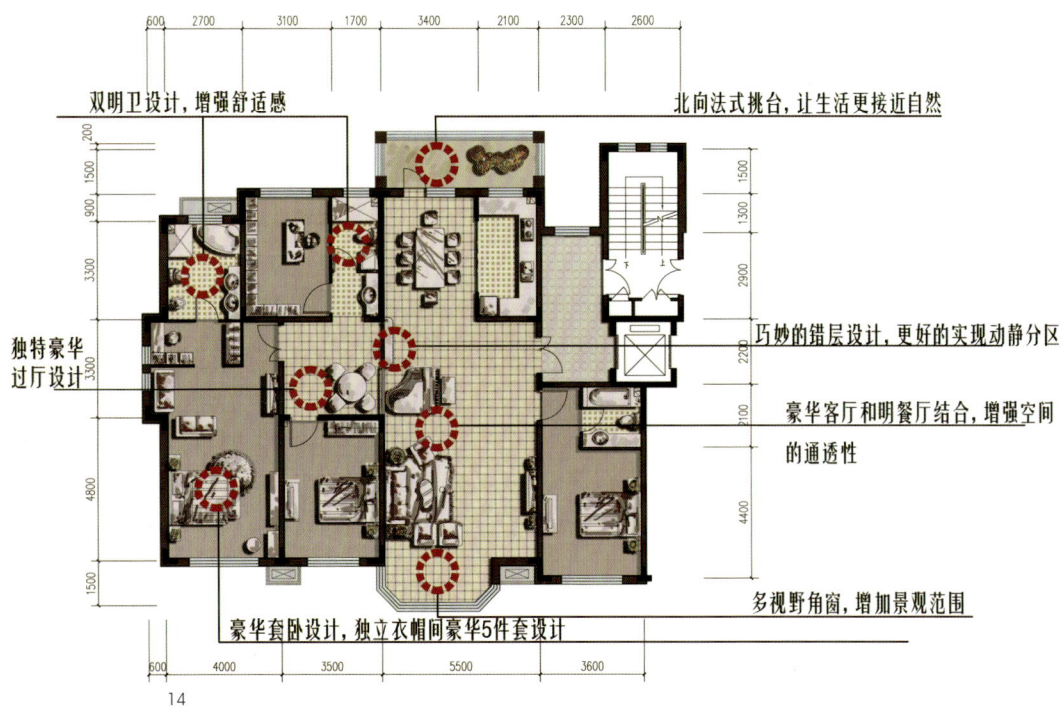

Courtyard as the realized using of community activities and middle landscape at the edge of yard round with houses, the houses and yard keep with distance the routh between the yard is waved to a net, the route shows the yard space and surface can be sensed and experienced. The green space in the yard ,also as surrounding environment, and important point by all diferent forms square, ground, path, mere form ,fountain, chair plants, amusement park,etc. Such element of landscape informed to a realized ground space.

Combined with the features of present ground, environment design put a basis of foundamental elements of landscape, making small scale activities ground to blend differen forms of phenomenon such as waterscape, the green pod, lawn ,mereform ,flower house and so on, by simple way overcoming the difficulties about the face of banding Green land. From the process of rank and combine, focusing on the relation on environment and architecture. During the design , to be a large extent, decreasing the quangtities of stiff materials, increasing plant quantities of green plants, showing the color of environment simple and calm in order to keep a whole feeling on colorful architecture.

双明卫设计,增强舒适感

北向法式挑台,让生活更接近自然

独特豪华
过厅设计

巧妙的错层设计,更好的实现动静分区

豪华客厅和明餐厅结合,增强空间
的通透性

多视野角窗,增加景观范围

豪华套卧设计,独立衣帽间豪华5件套设计

14

八角窗设计，扩大书房的视野感受

使用空间的全明设计

双明卫设计，增强使用的舒适性

15

16

17

1

山东德州高地世纪城

开 发 商：德州高地置业有限公司

用地面积：218 294 m²

总建筑面积：278 448 m²

容 积 率：1.28

绿 化 率：23.7%

1. 时代建筑造型
2. 区位分析图
3. 总平面图

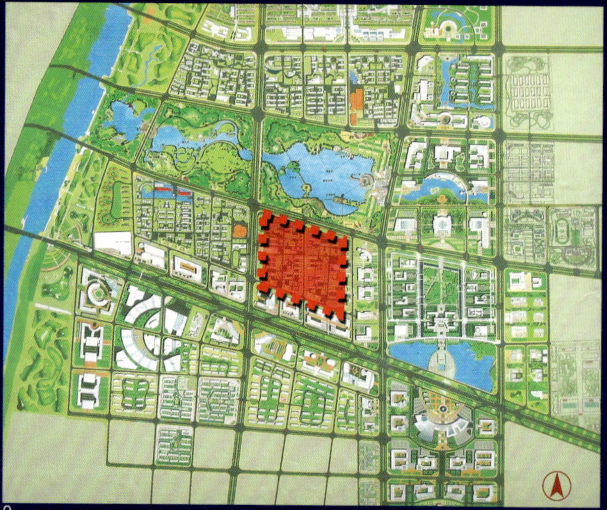

2

山东德州高地世纪城位于山东德州市河东新城，北临东方红路，西接杏园路，南至东风北街，在新城的规划中位置非常重要。设计强调中央景观轴线的同时将5层的叠加式复式、5层的阳光花房、6层住宅及11层的小高层住宅相互组合，形成错落有致的组团空间，在保证全部住宅拥具备朝向的均好性的同时保证了绝大部分住宅的景观均好性。并且利用建筑单体间的不同围合方式，形成疏密有致的大小庭院，使每个组团中的单体能拥有一个公共的宅前大花园。同时，设计中特别设置了若干大型的公共集中绿地，其中可配置缓步健身道、活动广场等健康体育设施，为小区用户提供了广阔的外部活动场所，对用户间形成相对稳定的院落归属感和熟络的社交关系提供良好的外部环境，从更深层次满足人们相互了解交往的心理需求，提供一个更具人性魅力的居住空间。

小区的西部组团包括了商业综合楼、小学、和两房、三房、阳光花房等几种住宅产品。在组团级的集中绿地附近充分利用了中央的景观，又丰富了组团内部的空间及立面效果；小区的东北部组团包含了超市及以三房和叠加式复式为主的住宅产品。几个部分通过组团的集中绿地来分开，保证了各自区域功能及使用的完整；小区的东南组团包含农贸市场，三房的多层住宅为主的产品。组团内部的住宅除拥有良好的朝向之外，更享有超宽的楼间距，使得组团级的绿地空间能与用户楼前的绿化融为一体，充分体现了"以人为本"的设计理念。

立面选型：多层住宅的立面设计上运用了一些新古典主义的手法，颜色以相对素雅为主，结合砖、石材、涂料的不同表现力，创造出既简约又不失稳重，既年轻而又不轻浮的建筑风格。商业的立面设计则采用比较地域风情化的手法，吸取了沿海的建筑风格，将塔楼、柱廊、天窗等元素融合运用。

绿化系统设计：根据地势基本平坦的特点，对地块进行局部改造，采用绿化小径，绿化堆山，绿色停车坪等手法，使小区内的绿化景观高低起伏、错落有致，形成一系列舒适宜人的绿化休闲空间。小区西、北部各有一条25m和30m宽的绿化带，形成小区的绿化屏障，将外部的噪声污染隔离，提升了小区的居住品质。小区的绿化系统均有两个以上集中公共绿地，东南区更有一个非常开阔的集中绿地，建筑与园林完美地结合。整个小区的绿化可配置各种运动、休闲场所，在丰富住户休闲生活的同时，也丰富了绿化景观。西区的小学结合运动场布置了各种绿化带，与西边25m宽的市政绿化，一起营造出一个花繁叶茂的现代园林式小学。整个小区的绿化，有主有次、有静有动，层次分明，并加以文化教育的内容，使人们在休闲娱乐之余还可以得到精神素质上的进一步提升。

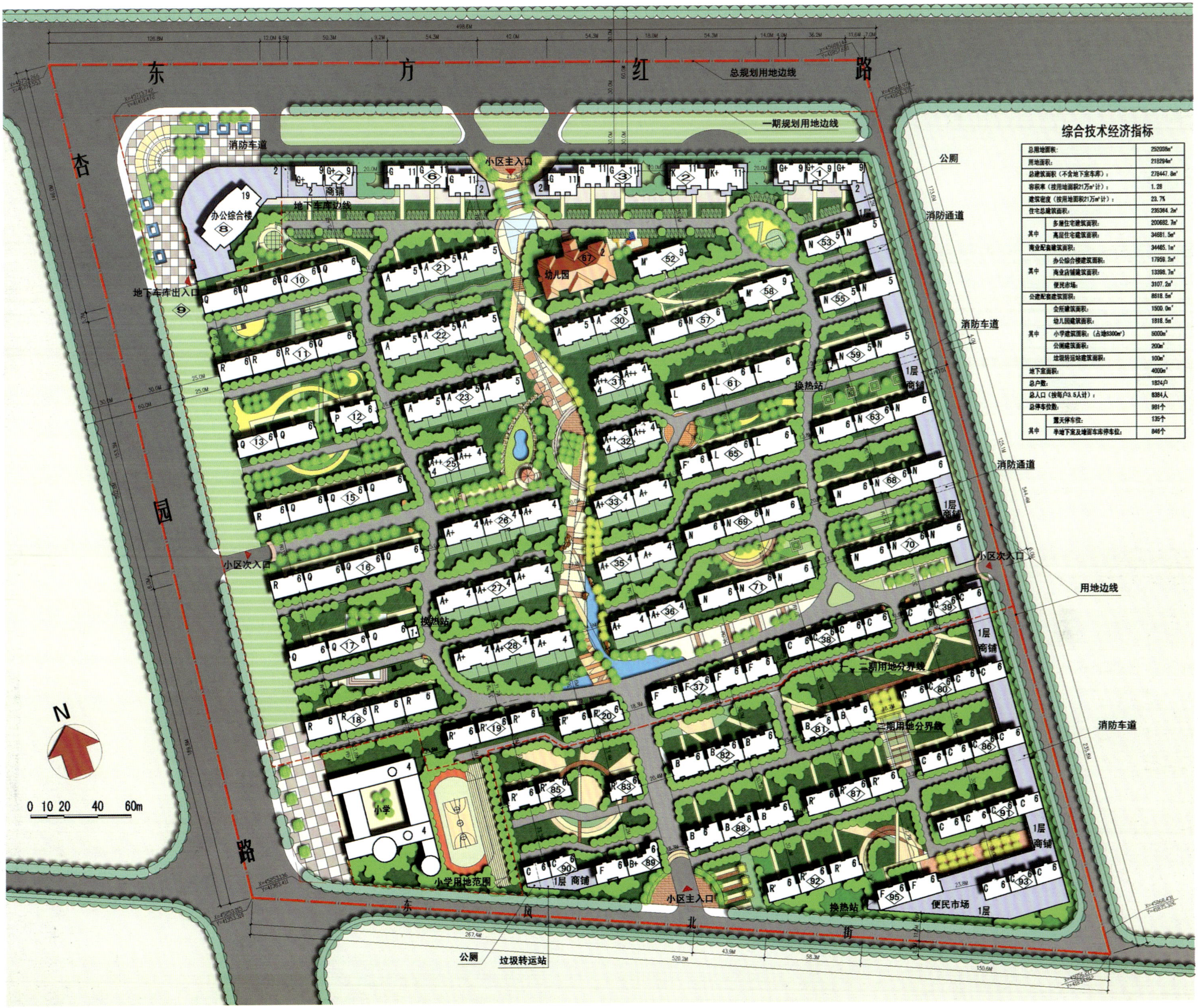

东 方 红 路

杏 园 路

东 风 北 街

N

0 10 20 40 60m

消防车道
办公综合楼
地下车库边线
地下车库出入口
小区主入口
商铺
一期规划用地边线
总规划用地边线
公厕
消防通道
消防车道
换热站
1层商铺
小区次入口
用地边线
二期用地分界线
1层商铺
消防车道
二期用地分界线
消防通道
幼儿园
小学
小学用地范围
小区主入口
换热站
便民市场 1层
1层商铺
公厕 垃圾转运站

236

3

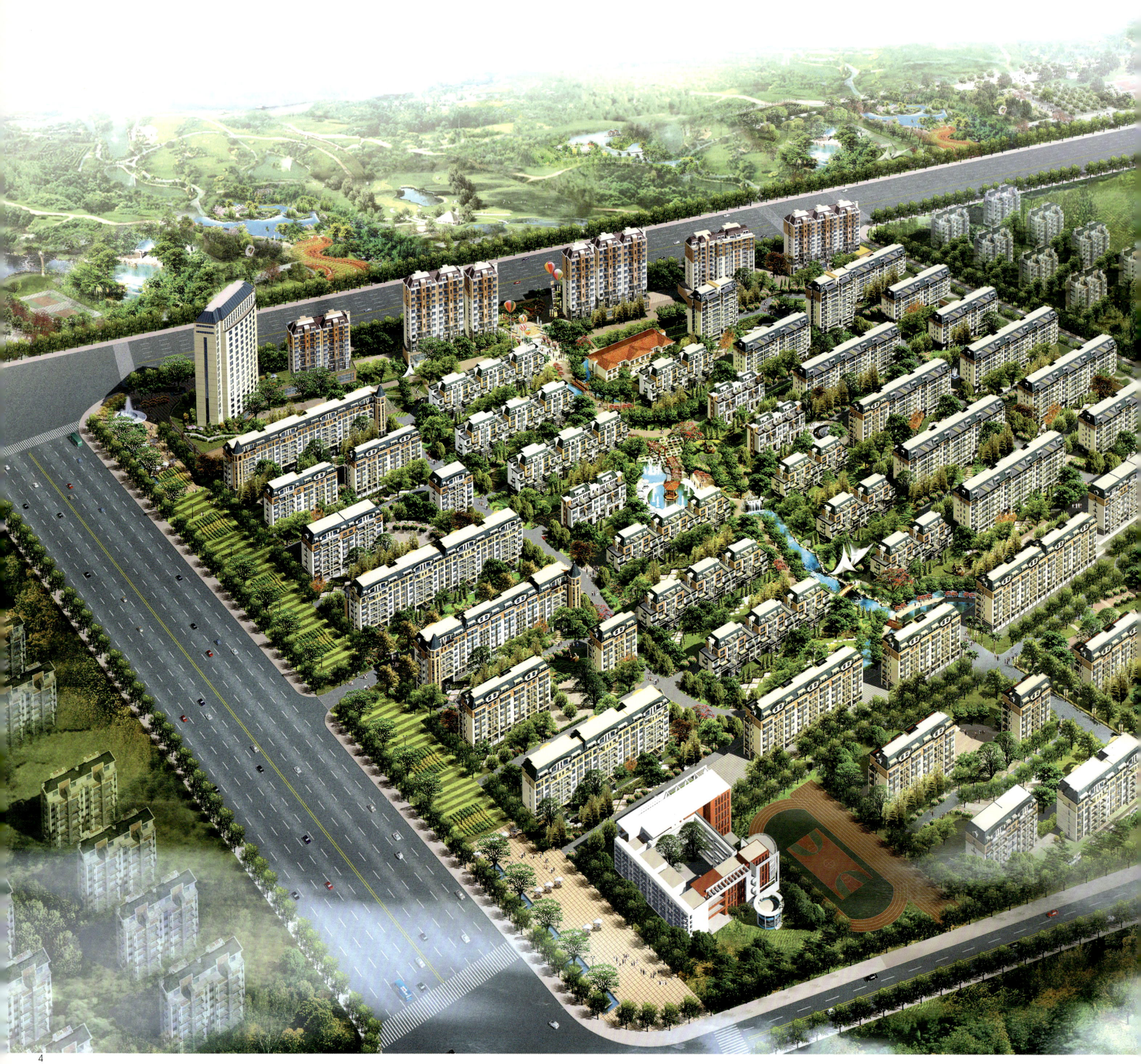

4

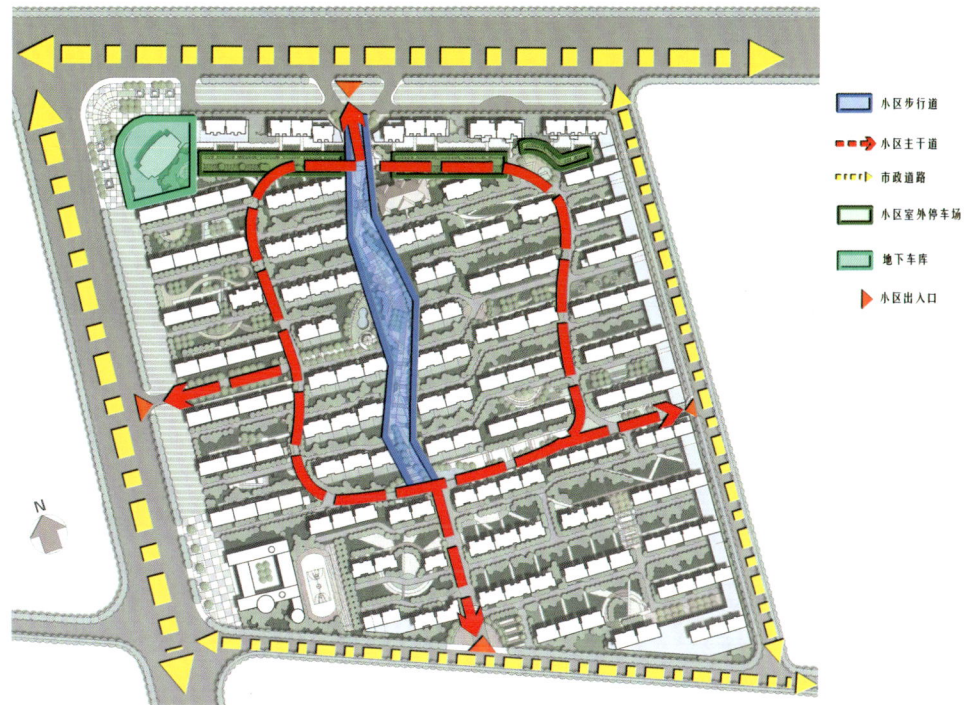

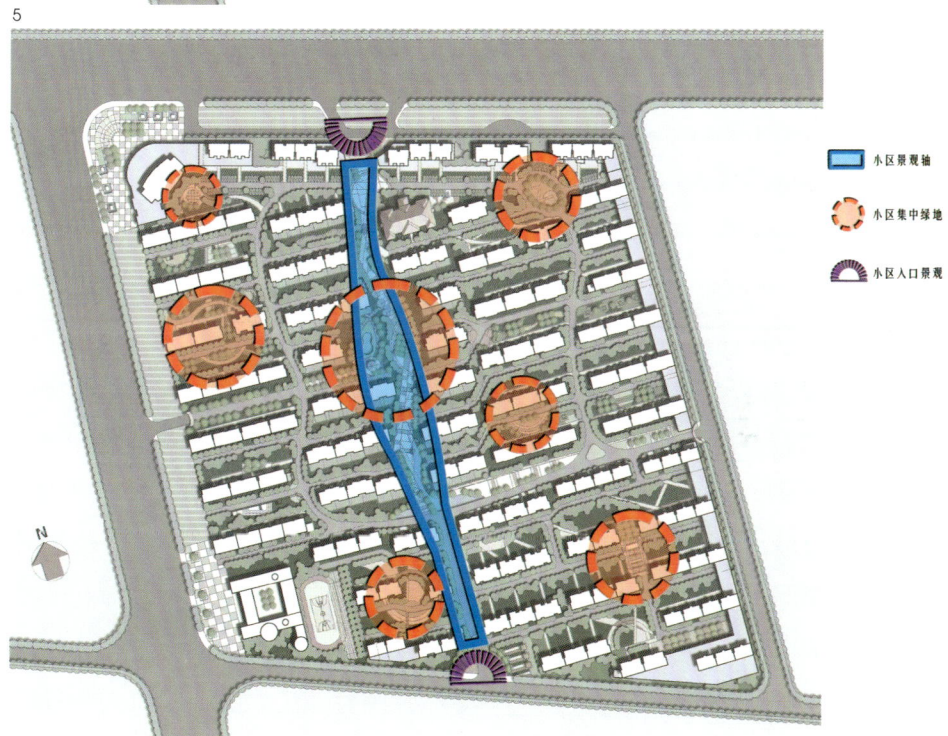

小区步行道
小区主干道
市政道路
小区室外停车场
地下车库
小区出入口

5

6

小区景观轴
小区集中绿地
小区入口景观

4. 总体鸟瞰图
5. 交通分析图
6. 绿化分析图

238

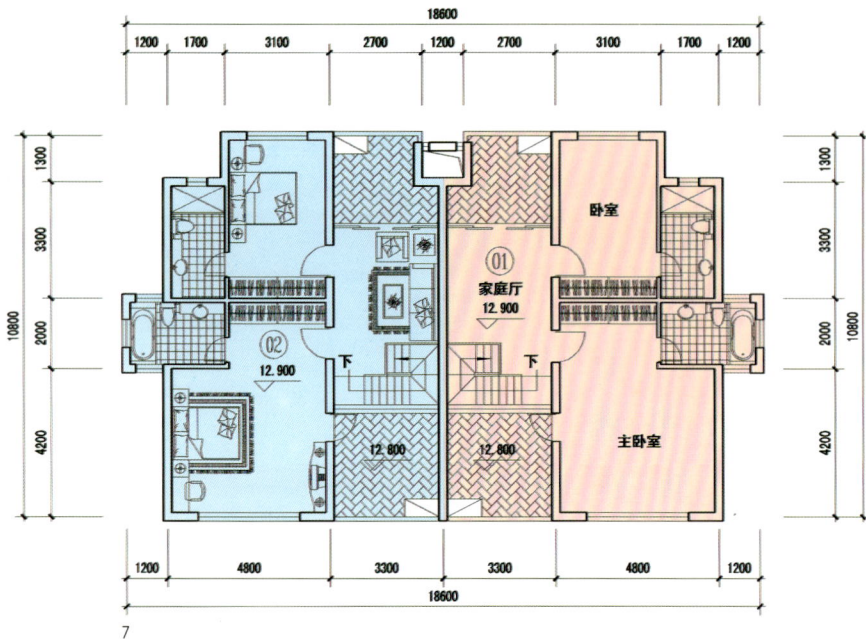

7

8

9

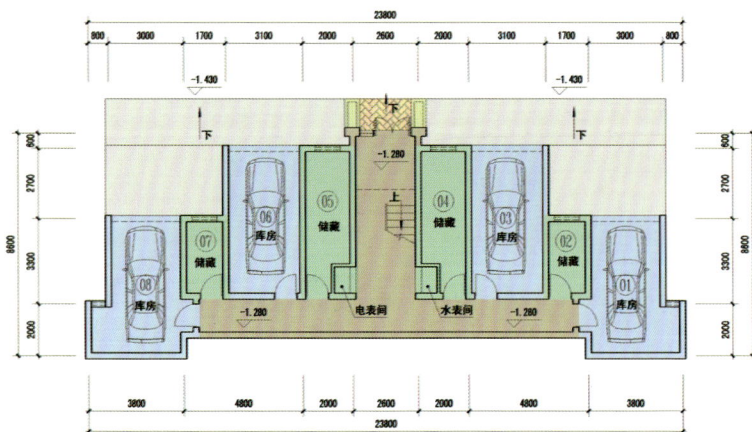

10

Floor area: 218 294 m^2

Building area: 278 448 m^2

Greening rate: 23.7%

Volume rate: 1.28

Texas Heights Century City is located in Hedong Center which lies to the south of Oriental Red Road, the wets of Xingyuan Road, the south of Dongfeng North Street, which plays an important role in the scale of the center. The design focuses on the central landscape axis and at the same time integrates with the five double-layer stack, five-storey green house, six-storey dwellings and eleven-storey small high-rise residents, which makes up the group space in picturesque disorder. All the residents are ensured with good conditions of direction and most of them are guaranteed the nice feature. The designer makes use of different amplexation ways of the single buildings to make up a well-conceived yards in all sizes, so that everyone in the group can have a public garden in front of their houses. Also, several large public lawns are set up in particular to configured with fitness, activity squares, and other sports facilities. It provides the residents with external activity , and creates a good external environment to establish the relatively stable sense of belonging and social relationship, which aims to satisfy their demands to understand make acquaintance with each other, provide a more charming living space.

The western group of the community includes a commercial building, primary school and several kinds of dwellings with the two-room, three-room or green houde. It is made full use of the central landscape near the lawn, which enriches the internal space of the group and its stereoscopic effect. The northeastern group includes a supermarket, and the dwellings with three- bedroom and stack-based. Several groups are separated by the lawns in order to ensure the function of the respective regions and the integrity of useness. The southeastern group includes a terminal market and dwellings with three-room and multi-story. Apart from a good direction of the internal groups, it has the large floor space that intexgrates with the green space and greening in front of the building, which fully embodies the "people-oriented" design conception.

7. A+型阁楼层平面图

8/13. 效果图

9. A+型首层平面图

10. A+型底层平面图

11. A+型二层平面图

12. A+型三层平面图

14

14. K、K+型标准层平面图
15. 会所效果图
16. 商业街效果图

15

Facade selection is the designer uses noo-elassical approach and relative plain colors on the design of the multi-storey dwellings. Besides, it integrates with the brick, stone and paint to make a different expression, which presents the construction style of simply and prudent, young and not frivolous. The design of commercial legislation is applied in geographical style and obsorbes coastal architectural style. Town, galleries, skylights and other elements are combined together.

Greening Design: In accordance with the flat topography of the characteristics of a partial transformation of the black, a green path, green heap mountain, green parking lawn and others are adopted, which makes up the landscape and in pictureque disorder to be a series of comfortable, leisure green space. There are a 25-meter-long and a 30-meter-long green belts in the western and northern of the community, which becomes natural defense against the external noise and upgrade the quality of living. Two or more green space are focused on by the green system. The southeast has a very board field of green space, which integrates perfectly with architecture and garden. The greening of the whole community can be configured various stadiums and leisure centers. Residents can enjoy the leisure of life and watch the landscape at the same time. The primary school of the west is assigned by all kinds of green belts with the combination of the stadium, and integrated with the municipal greening together to create a flower garden modern primary school. The green of the whole community is integrated with main and minor position, dynamic and static, and Structured can upgrade people's spirit in the entertainment.

Developer: Shenzhen longgang real estate holding company
Location: 12th Area ,Longgang Center,Shenzhen City
Design: Shenzhen Dongda Architectural Design Co.Ltd.
The area of the first phase land: 107 700 m²
The area of the first phase construction: 171 200 m²
The ratio of volume: ≤1.683
The number of buildings: 11
The number of dwellings: 641 sets
Parking Space: 470(including 119 on the floor,351 underground)
The area of the second phase land: 37 740 m²
The area of the second phase construction: 171 200 m²

1

东方沁园

开 发 商：深圳市建设控股龙岗房地产公司
项 目 地 址：深圳市龙岗中心城12区
建 筑 设 计：深圳市东大建筑设计有限公司
一期用地面积：107 700 m²
一期建筑面积：171 200 m²
容 积 率：≤1.683
栋 数：11栋
户 数：641套
二期用地面积：37 740 m²
二期建筑面积：97 610 m²

2

1. 园林景观
2. 售楼处
3. 鸟瞰图

东方沁园位于龙岗中心城12区，小区三面环路，自然景观优美，占地面积近10万平方米，总建筑面积17万平方米，由多层、小高层一梯二户建筑形态组成，以居家型的三房、四房户型为主。为提高居住舒适度和小区品质感，东方沁园在龙岗率先引进多层带电梯设计，辅之以2万多平方米的个性化园林绿化，是领龙岗居住潮流之先的精品社区。规划及单体设计力图创造出一种休闲、安逸的生活化空间，赋予更多的人文活动场所，成为本地区的高档居住小区。

Oriental Community is located in 12th area of the center of Longgang ,of which three sides are surrounded by roads .It is decorated by the beautiful natural scenery and it covers about 100,000 square meters in area. Besides, the total area of the construction is 170,000 square meters which consists of multi-storey, small high-rise building that is one storey with two apartments. The style of the apartments is based on three-room or four-room. To improve the living quality of comfort and sense of fashion, Oriental Community introduces the design of multi-storey with lift, supplemented by more than 20,000 square meters of personalized landscape, which is leading the trend of living in such a boutique community in Longgang. The planning and the design of apartments aim to create a casual, cozy living space and supply more activity centers, which becomes the superior quality residential quarter in this area.

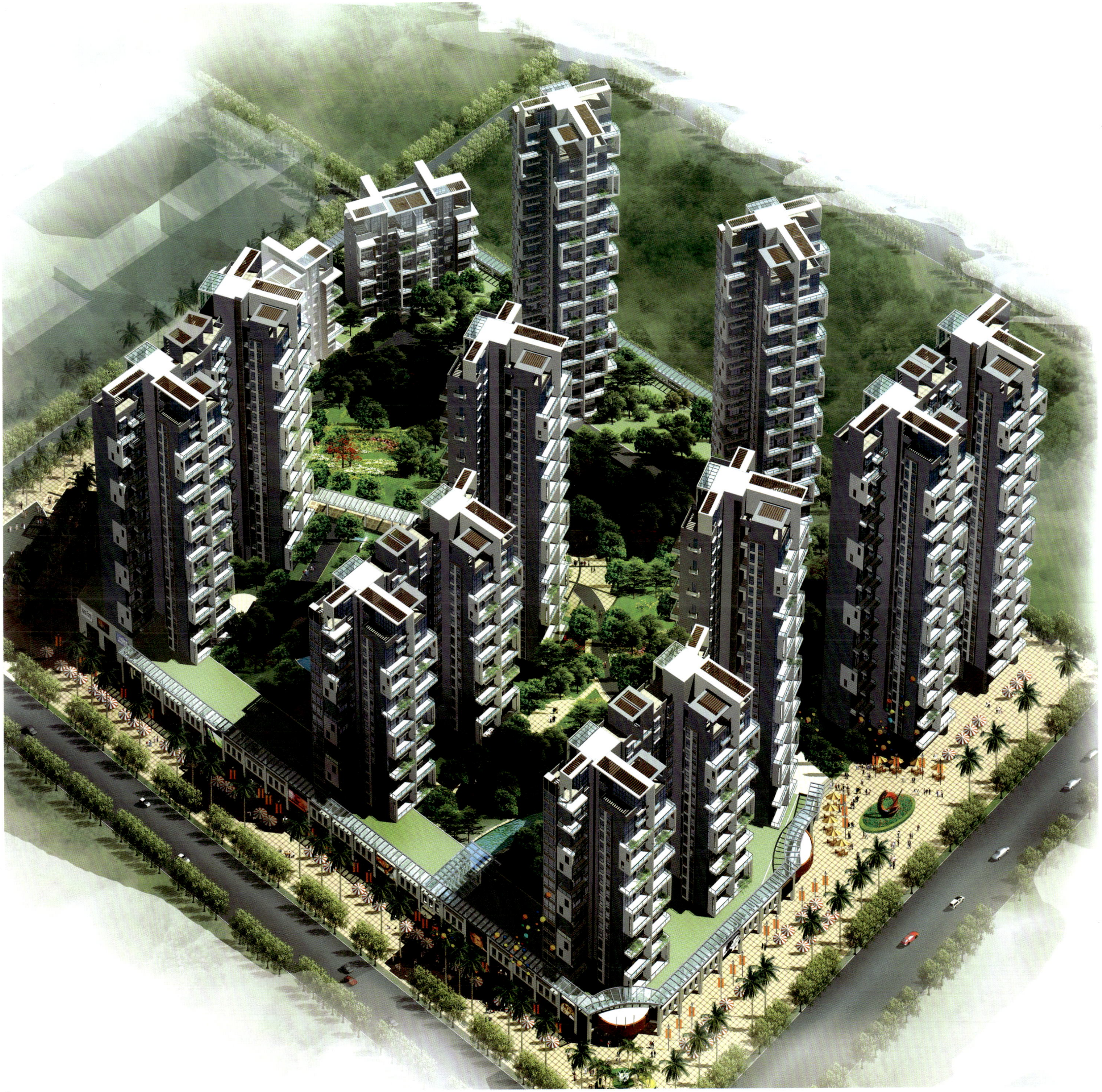

3

4. 总平面图
5/6. 园林景观
7. 建筑效果图

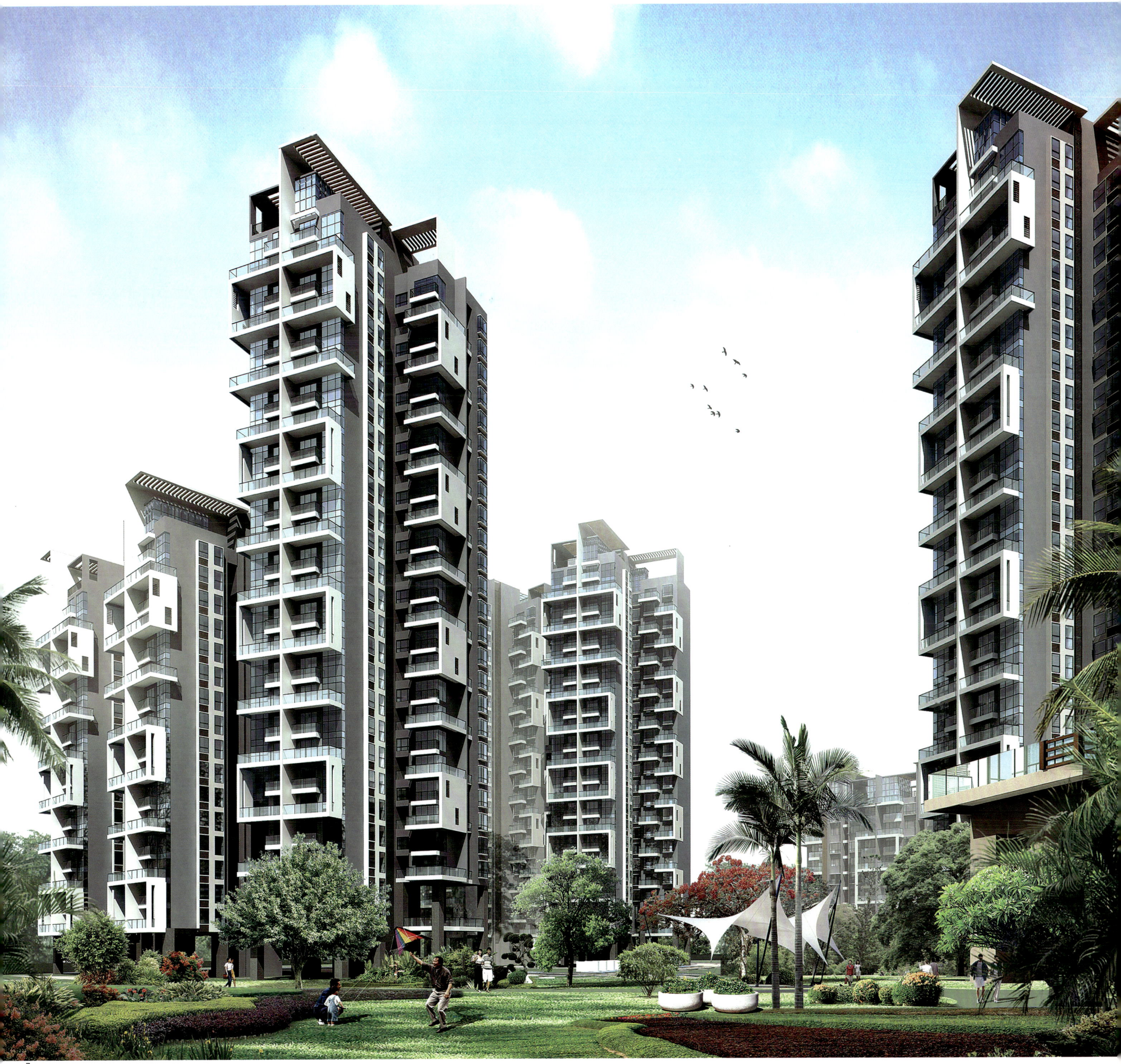

7

8

8. 夜景效果图
9/10/11/12. 园林景观实拍

9

10

11

12

1

万科四季花城

开 发 商：北京万科四季花城房地产开发有限公司
建 筑 设 计：AUNA国际建筑设计事务所
　　　　　　德国（WSP）建筑设计
　　　　　　中科建筑设计研究院
室 内 设 计：北京空间易想艺术设计有限公司
占 地 面 积：约200 000 m^2
建 筑 面 积：约300 000 m^2
容 积 率：1.56

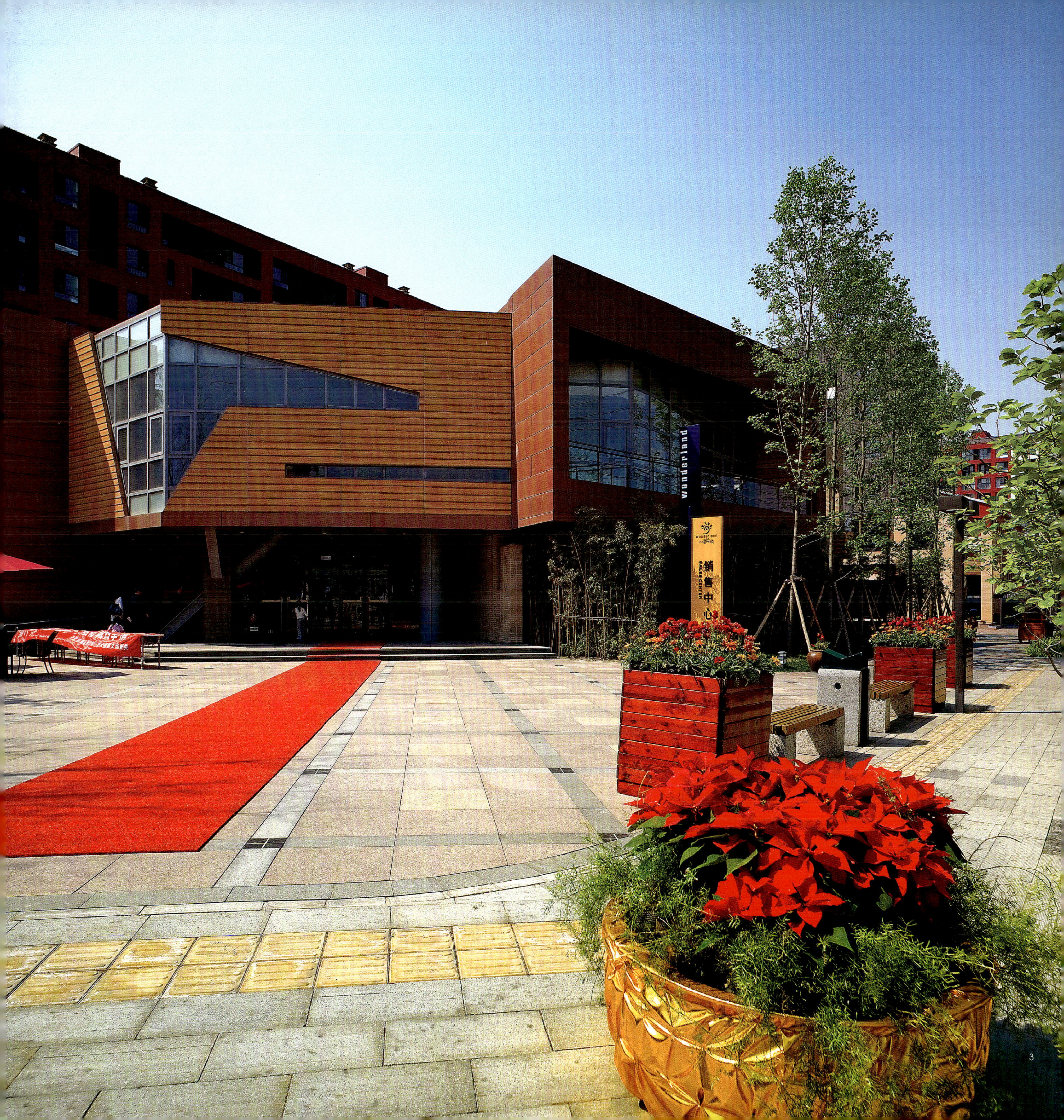
wonderland
wonderland
销售中心
SALES CENTER

4

四季花城总体空间结构以滨河绿带为主干，由顺平路商务空间引入，沿河伸展到地段内部，在与都市核心路交汇处形成城市活动中心，再沿都市核心路东西展开，形成地段城市空间的骨架。

步行流线沿公共道路及绿化空间设立，能够直接与主要公共设施及社区服务设施直接相连，加强居民步行使用城市公共功能的可能性。在步行环境的具体设计上，重视道路中的步行空间，保障步行系统的完整性，树立社区内"步行优先"的发展理念。

景观设计：社区富有生命力的两条动静脉东西向中央景观轴由商业广场、多排林荫树、下沉式广场、水池、休闲区组成十字景观轴线；南北向中央绿化带有大面积中央绿地，形成优美的林荫步道。东西、南北两条轴线形成主体社区十字景观轴线，就像两条富有生命力的动静脉，而新城市洋房则有理、有序的分布其中。中央景观区位于十字景观轴的核心地带，是一个复合型的景观带。由景观塔、中心广场、溪塘以及儿童活动区域、树阵等景观组成。以微地形为区格与连接，辅以果岭上的植被种植，使

各区域相对独立，而又相辅相成，曲径通幽，步移景异。为社区业主提供了一个综合性的活动场所，满足人们亲近自然的休闲需求。在主入口处，一个公共广场以及商业中心与整个城市形成开放共享；十字景观轴线，是社区内的主要休闲、交流场所，属于半开放空间；Block院落内的绿植、小品，共同营造出私密的安静休息场所。四季花城依此将整个社区分为三重生活空间，从而实现景观环境的多功能和可参与性，也让人和人、人和景观之间的沟通更多了些方式，更多了些生动。

251

4. 总体鸟瞰图

5/6/7. 建筑局部

Floor area: about 200 000 m²
Building area: about 300 000 m²
Volume rate: 1.56

Vanke Four Season Flower Town is in Group C in the Wangquan Homeland, close next to the city main road Shunxi Road, only 800m away from Shunyi City Commercial Center,9km from the airport and about 30km from Sanyuan Bridge. It is a residential community of high quality and low denstiy with low-storey and high-storey as its main parts, its total floor area is 300,000m² . Wonderful production, advanced plan, abundant features, comfortable living space, lively commercial plaza and high-quality estate make up such a new-town living modelling block.

General planning rationale: in order to satisfy man's need for social communication, and sense of belonging, we learned new urbanism planning rationale from America, pay attention to public space, city image and symbol to form a multi and healthy community. Meanwhile we adpot concept as German-like cautious planning, by reseaching into the condition and shape of the land and then sorting out a briefest and most effective method to reach the goal of making space structure clear and land reasonably used.

Succeeding the rationale of Four Season Flower Town, Vanke takes full consideration of city skeliton between plate of Wangquan Temple and Shunyi New Town to provide residents with central-city-level life, creating an open, energetic, cultural, safe, and modern civic living area.
General space structure of the town is revolved around a green belt del that is introduced from Shunping Business Space, reaching into the inner road and meeting with main roads of the city. So here comes the city activitiy center an again it spreads from the roads, acts as skeliton of the city space.

8

9

10

253

11

12

13

8. 实景图
9/10. 花园小景
11/12/13. 建筑立面

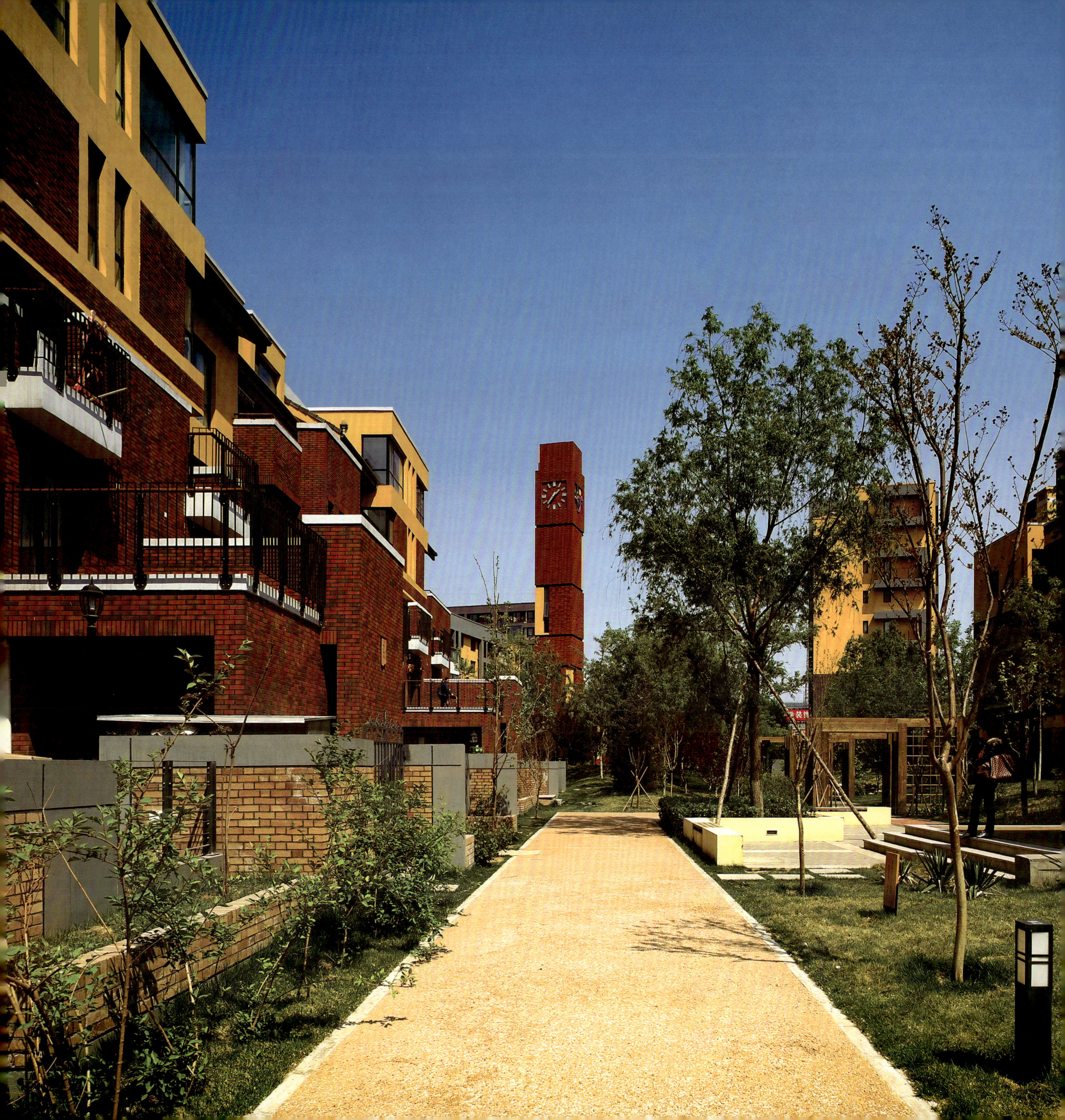

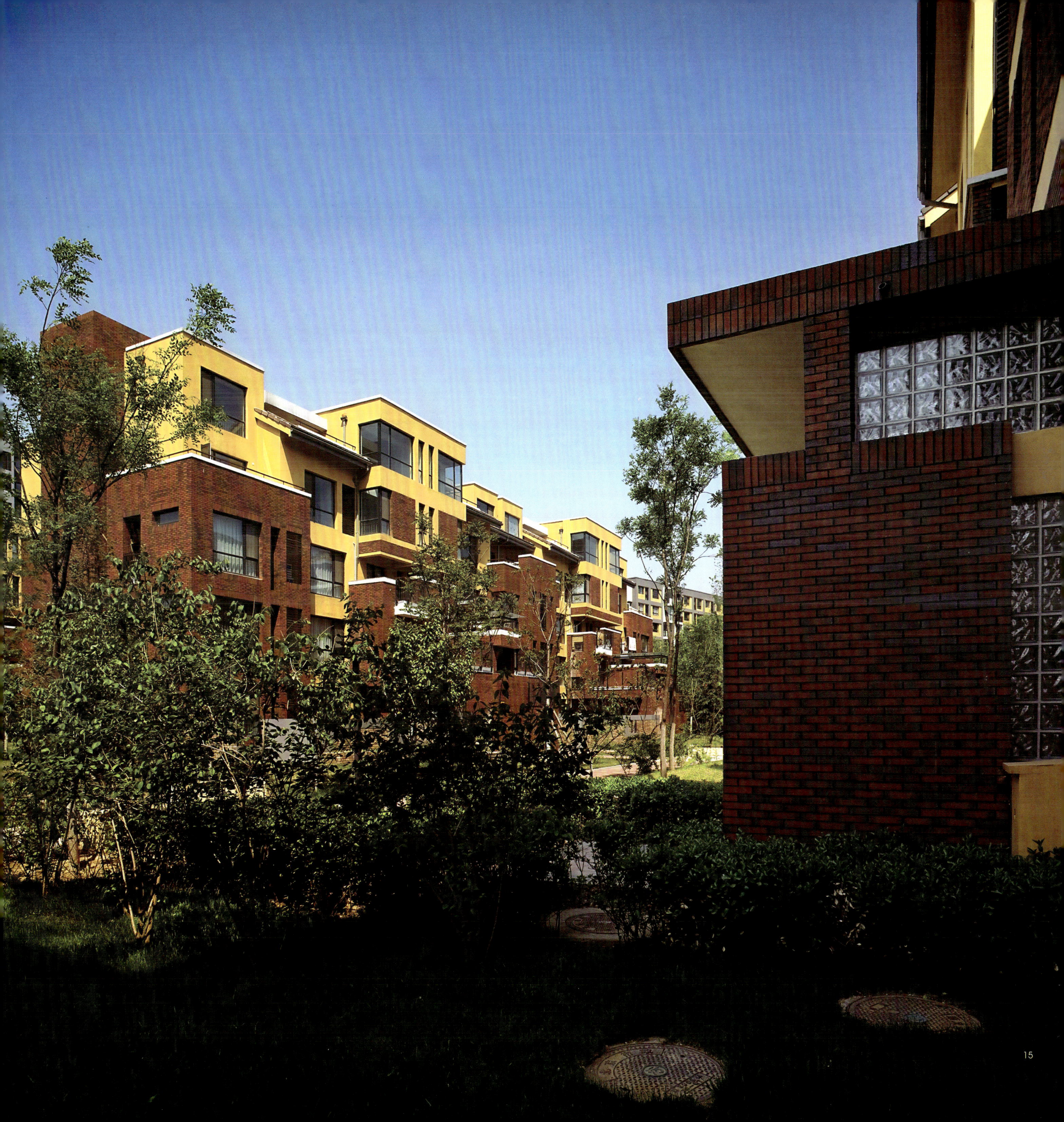

16

The walking lines are set up along the publicroad and the greening space, directly connecting with main public facilities and the community serve facilities, which ensure that residents can use them on their walk. In terms of detailed design for the walking environment, space on the road and the walking system is emphasized, rationale of "Walking First" is call up as a slogan in the community.

Landscape design: 2 central feature axises, one goes eastwards and the other goes wesatowards, like 2 lively arteries of the community are made up of the the commercial plaza, rows of trees, sinking plaza, pools and leisure area, a north-south-towards central greening belt becomes beautiful avenue with large area of central lwan. Tow such axies make the main community a crossing feature, while new city western houses ditributed neatly inside. This mixed feature belt then consists of the feature towe, the central park ,the ponds , the children activity zone and the trees. Plants on the green seperate each zone but connect them too accordig to the slight difference in terrain. You can watch different views as you walk through the recess. At the main entry, a public plaza and commercial center can be shared by the residents, and they can get close to nature. The crossing feature axis is a semi-open space for leaisure and communication. Plants create private and quiet rest place. From this Four Season Flower Town deviedes the community into triple living space, realizing environment's being multipurpose and joinable, also providing more way for communication between people and people, people and the environment.

257

14. 中央景观轴林荫道
15. 中央绿地
16. 水池景观
17. 楼距规划
18. C组团外景
19. 建筑立面

20. 样板房客厅
21. 样板房饭厅
22. 样板房展示区
23. 样板房卧室

中国特色楼盘（II）

广州佳图文化传播有限公司　主编

CHINESE
CHARACTERISTICS ESTATE (II)

深圳出版发行集团
海天出版社

图书再版编目（CIP）数据

中国特色楼盘（II）／广州佳图文化传播有限公司主编.
深圳：海天出版社，2009.2
　（佳图建筑系列）
　ISBN 978-7-80747-474-6

　I．中…　II．广…　III.住宅-建筑设计-中国　IV.TU241

中国版本图书馆 CIP 数据核字（2008）第 198695 号

中国特色楼盘（II）
ZHONGGUO TESE　LOUPAN　（II）

出 品 人：陈锦涛
出版策划：毛世屏
责任编辑：王　颖（0755-83460593　E-mail:6021@sina.com）
责任校对：周　强
责任技编：钟愉琼
策　　划：佳图文化
装帧设计：杨先周

———————————————————————

出版发行　海天出版社
地　　址　深圳市彩田南路海天大厦（518033）
网　　址　www.htph.com.cn
电　　话　0755-83460137（批发）　83460397（邮购）
印　　刷　深圳市彩美印刷有限公司
版　　次　2009 年 2 月第 1 版
印　　次　2009 年 2 月第 1 次印刷
开　　本　889mm×1194mm　1/12
印　　张　22
总 定 价　560.00 元（I、II 两册）

———————————————————————

目录
CONTENTS

高层住宅

High-rise Residence

高层住宅

1

2

万科金域东郡

开 发 商：深圳市万科城市风景房地产开发有限公司
项 目 地 址：深圳市大工业区行政一路北侧
规划建筑设计：深圳市博万建筑设计事务所
景 观 设 计：SITE CONCEPTS INTERNATIONAL

万科金域东郡位于深圳东部新城（大工业区）行政商务区中心公园旁，该区域是规划的行政商务的核心区域。本项目则是该区域内的第一个商品房开发项目。迫于政府开发周期短的时间要求，以及3.0容积率的规划要求，深圳万科房地产公司决定将建筑设计以及建筑材料部品装饰均委托给设计单位进行。

项目地形方正，用地东、南、北侧均为待开发居住用地，西侧为规划中的中心中学。用地地形北高南低，用地南侧与北侧道路约有5米高差。因此解决好高差问题在规划设计中起着关键性的作用。

该项目由4栋16～23层的住宅以及沿街商铺等配套设施组成。由于项目周边地块尚未开发，外部环境存在很大的不确定性，因而在建筑布局上采取了以我为主的设计思路，着重营造小区内部环境。4栋住宅以大围合的形式布置于用地周边，南北各布置一栋板式住宅，中部东西两侧各布置一栋点式住宅，住宅层数依地形从南到北依次递增，分别是16层、22层、23层。这种布局方式可以在用地内部形成很大的内部庭院，并且有利于改善小区内部的通风和日照，改善后排住宅的视线。

在竖向设计上，该项目充分利用了原有地形南北两侧的高差，把小区内庭院整个抬高至与北侧道路相近的标高，充分利用原小区内庭院与南侧道路形成了约5米的高差，车库及商业设施相当合适的充满了内庭院下层的空间。这种竖向设计所带来的优点是极大的减少了建设地下车库所需要开挖的土方量，同时对于强化小区庭院空间与外部城市空间的相对独立性，丰富小区的空间层次感受，达到人车完全分流及方便车库车辆的进出等方面都有很好的效果。

由于本项目执行"国六条"的要求，即"新审批、新开工的商品住房建设，套型建筑面积90平方米以下住房（含经济适用住房）面积所占比重，必须达到开发建设总面积的70%以上。"故项目是以中小户型为主，在户内空间的设计上，如何在有限的建筑面积里提供更多更合理的使用空间就成为了设计必须要考虑的问题。因而，本项目在户内空间（包括厨房和卫生间）大量地使用了落地凸窗，双层高阳台等建筑语汇，并且对每一件家具的尺寸及摆放都作了细致的考虑。另外，在结构设计上充分的考虑了室内空间的可变性，室内的大部分墙体都可以根据需要拆除而重组空间，以适应不同家庭、不同时期的需求。在公共空间的设计上，充分体现了无微不至及以人为本的精神。每栋住宅的首层均设计了层高高达到7.4米的架空层，既改善了小区通风条件，又给住户提供了一个令人愉悦的活动交流空间。

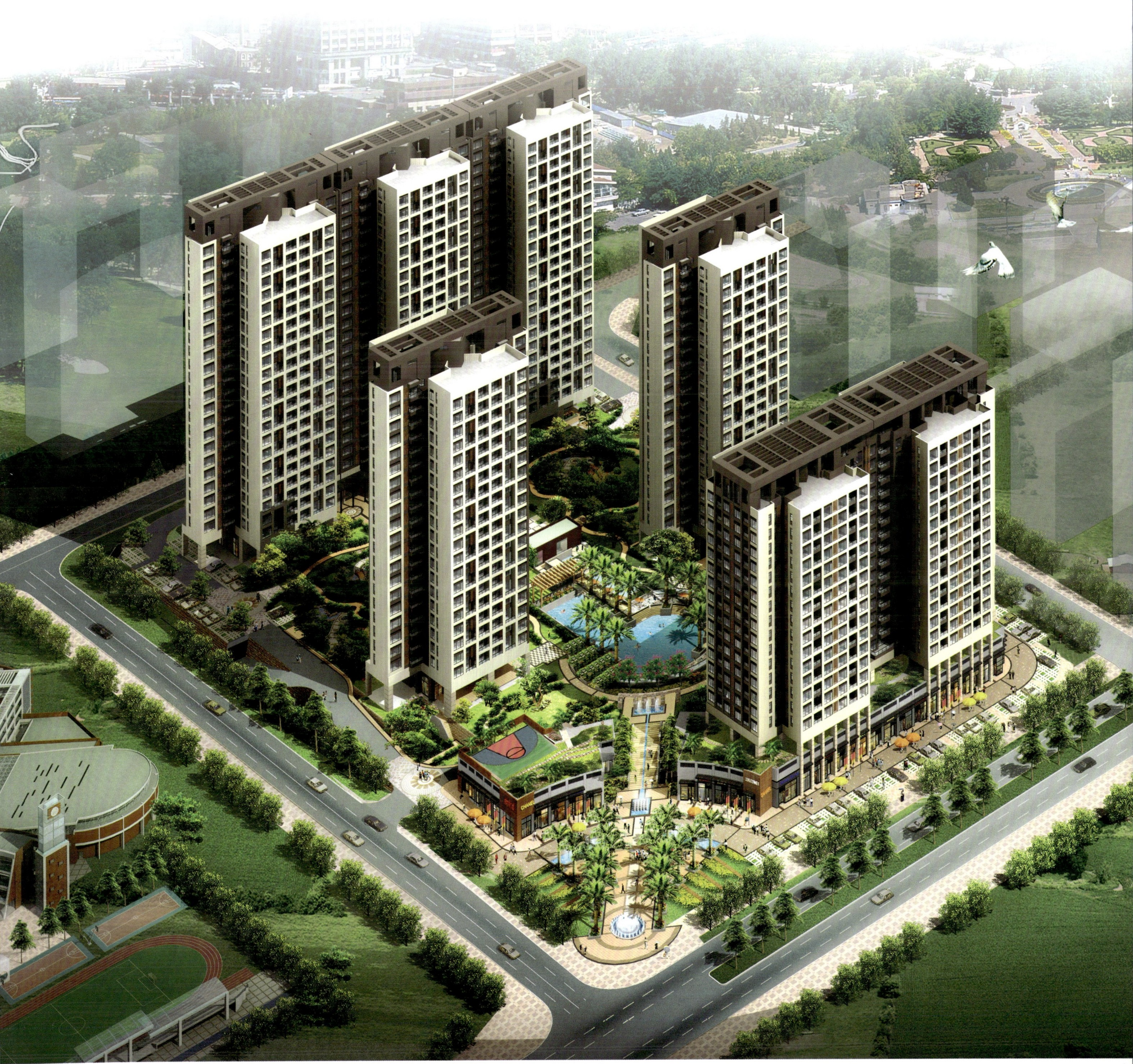

项目开发时间要求短、平、快。经与发展商商榷，建筑选择了简洁的现代风格。设计利用建筑横竖向交错的墙体、双层高的阳台隔构，形成了建筑的外墙骨架。外墙以浅色面砖为主，配以木色百叶分隔，线条流畅、棱角分明，加上下部洞石、面砖及涂料等细部的深入刻划，营造了现代感极强的居住氛围。亭台水榭间，舒适、文明的现代生活已入其中！

商业立面，可以说是小区的点睛之笔。她既展示着楼盘的时代与性格，又要营造好商业氛围。商业要文雅中体现时尚特色。金域东郡的商业模式，是以商铺为主的店面形式。设计在商业立面刻画上，采用型钢、方木等硬朗的建筑材料，考虑好每一间店面的入口、橱窗及广告位置，加强重点部位的刻画。在设计与施工过程中，先后以模型推敲、现场作样、细部深化等工作方式进行推敲，使建筑设计及外墙把握控制都取得了很好的效果，受到了万科和社会的认可。

6

7

8

9

10

11

12

Project location: North Side, NO.1 Xingzheng Road, Dagongye District, Shenzhen

Architectural planning & designing: Shenzhen Bowan Architectural Design Firm

Landscape design: SITE CONCEPTS INTERNATIONAL

Gross floor space: 26 218 m^2

Gross building area: 101 155 m^2

Construction floors: 23 floors on the ground, 2 floors underground

Structure of the main body: Frame-supported Shear Wall

Building coverage: 24.5

Plot ratio: 3

Afforesting rate: 35.5%

Total units: 868

Parking spaces: 797

Vanke Jinyu Eastern County is located beside the central park of the administrative and commercial area in Eastern New town (Dagongye District), Shenzhen. It is the core section and the first commodity houses development project of the area. Compelled by government's short schedule requirement and the one of a ratio of 3.0, Shenzhen Vanke Real Estate Company Limited decided to delegate to the designing company the architectural design and decoration of the architectures' materials.

The project gets a square terrain and the terrain near building where the east, south, north side all are prepared for developing residential areas, even from the west side is designed to a middle school in the central according to the planning. The terrain which worth side is high but south side is quiet low what is more, the road between south and north exists above 5 meters altitude distance. Therefore, the key to the planning is to solve the problem about the distance in height. The project consists of an department on 4 floors from 16-23 stories as well as variety of stores and facilities open at the streets. As the massif approached is undeveloped, it has quiet a great definition in the external environment, so in the layout of the building designers adopt a concept of "basis of oneself", focusing on creating a internal environment for communities.

Such 4 floors arranged in the form of enclosure constructed around the building. From the south and north, each side holds a plate department, in the middle to east and west side it puts on a point-shaped department. The level of residential from south to north, it turns higher and higher by terrain, including 16 stories, 22 stories and 23 stories. The way of layout make the building a large internal yard, taking advantage of improving condition about shinning and fresh air, improving the eyesight of the residential behind.

In the vertical-directed, the project makes fall use of original distance of altitude between south and north side, leading the yard a relevant altitude. With road north, making the best of above 5 meters altitude which formed by inner yard and road in the north. Garage and business facilities filled with space of yard upstairs appropriately. The design's advantage of vertical-directed is in order to decrease areas when building, meanwhile, consolidating yard space and the independence of city space, fulfilling community's sense of space dividing into

13

people and vehicles, in addition, it's very convenient for traffic and make a better effect on it.

The project is carried out at the request of Provisions of The People's Republic of China Concerning the Administration of Achievements in Survey and Drawing, which is, newly approved and constructed commodity houses (including economically affordable residential houses), within the house type area of 90 square meters, must take up over 70% of the gross development building area. So, most of the house types are small and medium sized in this project. Then, how to make reasonable use of the space provided by the limited construction area is what must considered in the project, and floor bay windows, two-story high balcony are used indoors in large amount (including the kitchen and the washroom). Besides, changeable indoor space has been fully considered in terms of the structure design, so most parts of the wall can be removed and rearranged, if needed, to meet different demands of different families in various periods. Design of public space shows careful and humanized concern. First floor of every residence are designed with hollow space of 7.4 meters high, which both improves ventilation condition of the community and provides pleasant communicating space for residents.

The time of developing the project is required to be short, stable, and quick, and its style is decided to be the brief modern one. Outer wall skeleton of the constructions are formed due to the use of the laterally and vertically staggered wall, two-story high balcony duct spacer. Light-colored face bricks with wood color at intervals, smoothly and clearly. Together with that, intensive hole stones, face bricks has created strong modern living atmosphere, and a comfortable and civilized modern life can be enjoyed in such surroundings.

Commercial elevation, which dots the design of the community, not only shows the characteristic of the property and that it standing for our era, but also creates good commercial atmosphere. Business pattern of Jinyu Eastern County is to take shops as main form. Architectural materials such as profile steels and square woods are used, and entry of each shops, show windows and advertisement places are taken into account and strengthened. During the process of design and the construction, we study the models, make experiments on the spot and polish and details. Therefore we get good results through this design and the control of the outer walls, which gain recognition of Vanke and the society.

15

16

17

1

深圳风临域

开 发 商：正中置业集团

项目地址：龙岗区龙岗中心城深惠路与吉祥路交汇处

建 筑 设 计：深圳市城脉建筑设计有限公司

景 观 设 计：深圳市同济人建筑设计有限公司

占地面积：2 877 m²

总建筑面积：58 105 m²

容 积 率：3.35

绿 化 率：18.6%

总 户 数：829

停 车 位：共800

1．效果图
2．建筑局部
3．鸟瞰图

2

四房户型南北通透，除设有入户花园及露台外，还隔层设有空中花园。

2. 第3栋商住楼地上28层，下面有2层地下室用作车库，地下2层兼作平战结合的人防区有2个防护单元，地上1－4层为独立商铺，5层为架空花园，6－28层全为高层公寓，有大一房、小一房和小二房的户型，共计736户。

3. 项目建设与城市公共利益相结合，创造良好的邻里关系与友好的社区，并为四期的旧厂改造预留缺口，形成沿街的连续商业氛围，在一层与二层设计6m及5.60m层高，宽达6m的公共开放空间，24小时开放，其间以自动扶梯相连，并直接通过天桥与吉祥中路相通，空间内设置座椅、绿化景观等设施，并设通高共享中庭，在一、二层均有出入口与四期衔接，目的是为周边社区居民提供便捷与舒适的步行系统，使人们可以更好地使用地铁，方便出行，同时有利商业人流汇集，形成社区与企业共赢的局面。

4. 建筑造型设计挖掘建筑自身错位花园的特征，以交织错落的方壳造型获得玲珑丰富的视觉效果，配合顶部的方型半通透造型，使整体造型鲜明、简洁、清新、柔和。

Developer: Genzon Property Group

Project location: Junction of Shenhui Road and Jixiang Road, Central Town, Longgang District

Architectural design: Shenzhen Citymark Architectural Design Co., Ltd.

Landscape design: Shenzhen Tongji Architects

Floor space: 2 877 m²

Gross building area: 58 105 m²

Plot ratio: 3.35

Afforesting rate: 18.6%

Total units: 829

Parking spaces: 800

Hongji Garden Phase 3 (Fenglinyu) is invested and developed by Genzon Property Group Shenzhen (Group) PLC. It is located in the junction of Shenhui Road and Middle Jixiang Road, Central Town, Longgang District, Shenzhen. Longgang central town has been gradually become a center of politics, economy, culture and business of the district, and it is also one of the eight satellite cities of Shenzhen. The project is made up of four blocks, former three are commercial-residential buildings and the last one is kindergarten. The first two blocks are 17-storyed and parts of them are 18-stoyed (with duplex top stories). There are underground floors are used as garages, facility rooms, and are connected to the buildings. The 2nd to 18th floor are high-rise residential buildings. Block C is 29-storyed and there are two floors underground used for garage and partly for facilities.

5

6

4．全景效果图
5/6．平面图
7/8．入口效果图

7

8

Site planning:

1. To make the space of the courtyard as much as possible and set a kindergarten in, the No.1, No.2 and No.3 Tall Building are built alongside the edge of the building site.

2. The stream of people and traffic is mainly coming from the subway entrance in the First Class Shenhui Road of the Jixiangzhong Road, which enter the internal cell in different way.

3. According to the land layout requirement: In order to create a pleasant community environment, the land has been combined with several kinds of green land, cover beyond 6,000 square meters, in the landscape planning of the courtyard in the three stages to provide a place for the resident to relax and do activities.

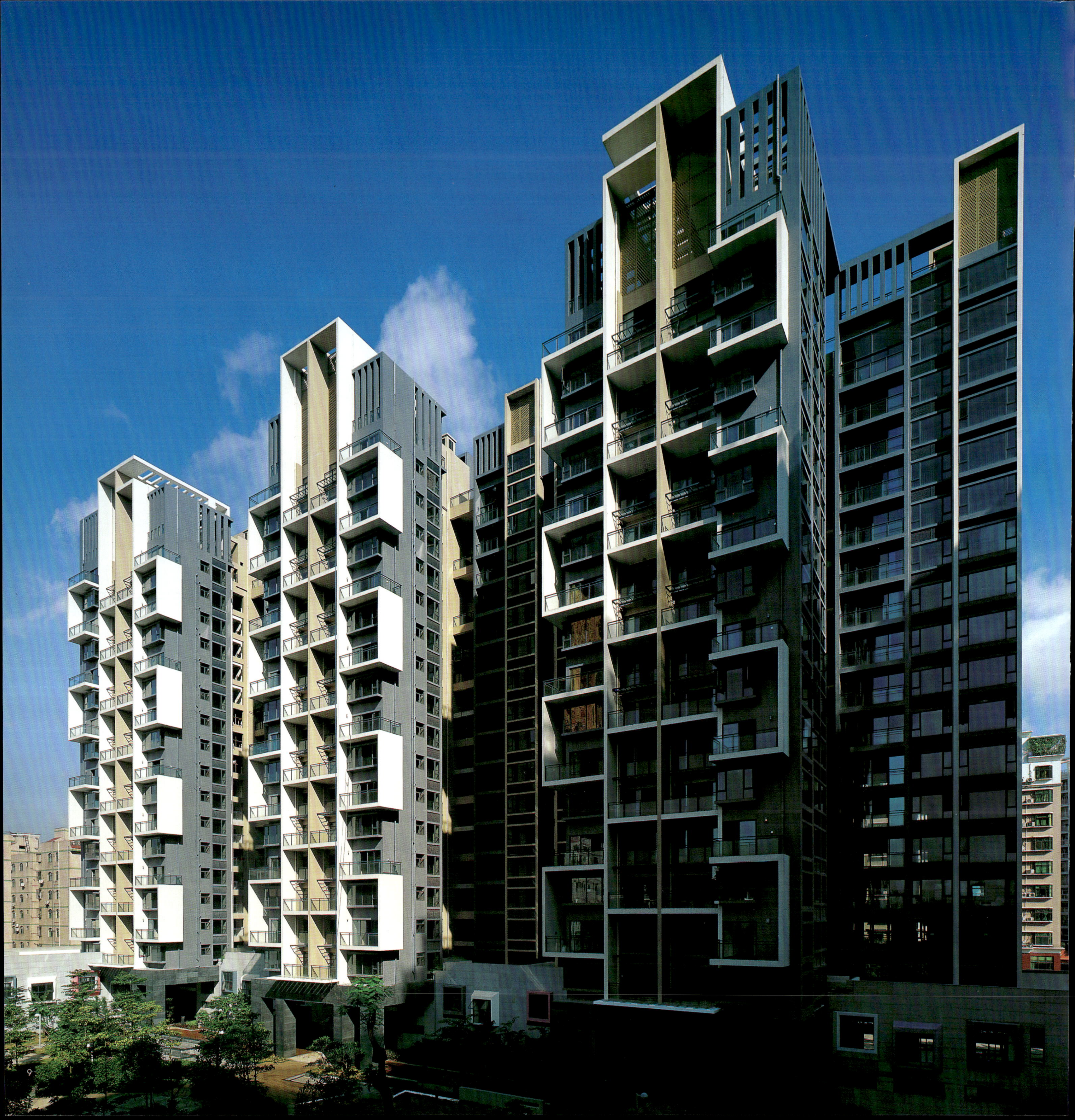
9

Architectural Design:

1. There are 17-storeys in the ground and part of 18-storeys (the top storey is the compound apartment) of the NO.1 and NO.2 Business Complexes. The basement storey is the independent shop. There are high-rise apartments from the second storey to the eighteenth storey, including 320 households with different type of flat such as two-bedroom economic housing, three-bedroom economic housing, four-bedroom comfortable housing, five-bedroom duplex room and so on. The two-bedroom and three-bedroom economic housing have the entrance garden and the large viewing balcony. It alternates so it can be exclusive use, also can be used with the parlor or others. The four-comfortable housing has the hanging garden in the interlayer, besides the entrance garden and balcony, north and sound permeability.

2. There are 28 floors on the ground of block C and there are 2 floors underground used for garages, partly used protection with 2 protecting units. First to fourth floor on the ground are independent shops, the fifth floor are hollow gardens, ad sixth to twenty-eighth floor are high-rise apartments whose house types are one large room, one small room and two small rooms, containing 736 units in total.

3. Combination of the construction and the public interest in the project create good neighborhood relationship and friendly community. It also provides opportunity for the remold of the old

factory, forming continuous commercial atmosphere along the street. Public opening space of 6 and 5.60 meters high and 6 meters wide on the first and second floors opens for 24 hours. They are connected by auto escalator and directly connect with Middle Jixiang Road through the overpass. Inside the space there are facilities like benches and greening features. Besides, shared atrium is built, which gets gates to the Phase 4 on both the first and second floor, so as to provide convenient and comfortable walking system for the residents. In the meantime, the community and the enterprises share benefits from the gathering flow of the commercial crowds.

4. With the advantage of being a disordered garden, the architecture shows a visual effect of delicate and abundant in the shape of square case along with the half spacious square on the tops, turning on a distinctive, brief, fresh and soft overall shape.

9/10．建筑立面富于表现力

11/12．建成后的效果

1

1. 园林效果
2. 建筑效果
3. 从商业街透视

2

中海璟晖华庭

开 发 商：中海发展（广州）有限公司
项目地址：广州市天河区珠江新城金穗路与猎德路交界处
占地面积：20 000 m²
总建筑面积：200 000 m²
容 积 率：7
绿 化 率：30%
总 户 数：600
停 车 位：500

中海璟晖华庭位于珠江新城金穗路与猎德路交界处，广州珠江新城CBD的核心居住区，定位为CBD街区高档景观住宅，是目前广州珠江新城CBD内稀缺公园景观高端物业，占地2.2万平方米，总建筑面积约20万平方米。项目共分两期开发，其中一期A1-A5栋已经于2008年6月全部交付业主使用，二期A6-A10栋在2008年10月25日面市，率先推出的单位为A9-A10栋，户型涵盖了89平方米的南向精致两房，121-139平方米的南向和望园林舒适三房，以及163-164平方米的南北对流经典四房。在寸土寸金高楼林立的珠江新城，项目向东即可俯瞰珠江公园近800米的纵深景观，向内可欣赏110米的超宽小区园林。由于珠江新城住宅容积率都非常高，一般住宅小区园林面积都非常小，地块与地块之间

除了拥有280000平方米优越公园景观视野外，中海璟晖华庭项目整体规划为"U"型围合结构，从而充分利用空间，形成了高达12000平方米超大私属园林，同时又构筑出封闭独立的小区园林，有效规避了外部的噪音干扰，在繁华的CBD中营造出安静、私密居住空间。目前珠江新城在售住宅项目大部分园林面积均只有几千平方米，有的甚至只有1000平方米左右。中海璟晖华庭项目在整个珠江新城CBD区域内属于罕见大面积园林社区。精致、富有层次感的园林设计、小区泳池形成了珠江新城内少有的小区景观和丰富生活情趣。

中海璟晖华庭项目除拥有珠江公园稀缺城市景观，向内有超

线猎德站只需五分钟，交通配套极为便利；同时社区内外"幼儿园+小学+中学"一条龙式完善的教育配套。珠江新城商业配套：南侧兴盛路路宽45米，总长度约1公里，西连珠江大道东路、东连珠江公园西门。在景观上联系城市新中轴线中心广场和珠江公园，客观上形成珠江新城东西向商业活动轴线。中海璟晖华庭项目在珠江公园正西侧，可东向正望珠江公园这一稀缺城市公园景观。兴盛路两侧建成具有岭南特色的"双层骑楼"结构，根据珠江新城规划，兴盛路将成为珠江新城唯一一条商业步行街，商业规划前景非常看好。

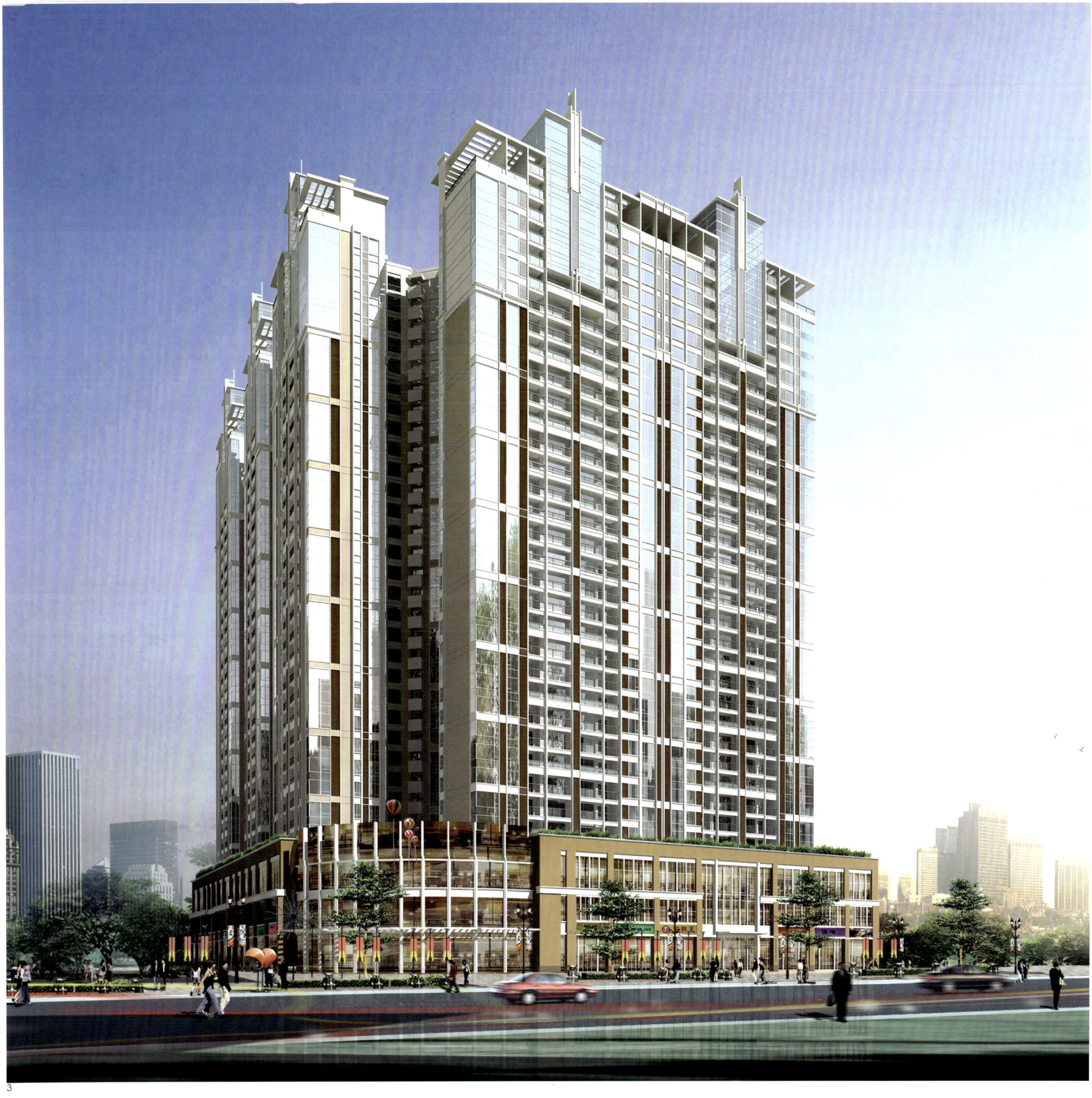

3

Project location: Intersection of Jinshui Road and Liede Road, Zhujiang New Town, Tianhe District, Guangzhou

Developer: China Overseas Development Co., Ltd. (Guangzhou)

Floor area: 20 000 m^2

Gross building area: 200 000 m^2

Plot ratio: 7

Afforesting rate: 30%

Total units: 600

Parking spaces: 500

1. 社区主入口
2. 鸣泉水景广场
3. 浅溪石景
4. 廊桥小筑
5. 叠翠庭院
6. 林荫碧波泳池
7. 泳池配套词
8. 逸水岛吧
9. 幼儿园
10. 亲子童趣广场
11. 掬水石景
12. 沉思广场
13. 荫林庭院
14. 荷塘莲影
15. 弯啼水榭
16. 人行次入口

4. 总平面图
5. 泳池意境
6. 庭院俯瞰
7. 泳池休息区

8/9/10. 园林实景

China Overseas Park Royal, a core CBD residential area in Zhujiang New Town, is located in the intersection of Jinshui Road and Liede Road. It is oriented as a high-class CBD block landscape residence which is at present a rare high-end park landscape property, covering an area of 22,000 square meters and gross building area of about 200,000 square meters. The project is developed in two phases. A1-A5 of the former one have been delivered to the owners on June, 2008 and A6-A10 of the latter one have been introduced to the market on 25th, Oct, 2008. A9 and A10, which have been firstly promoted, cover the house types of delicate southern-aspect house of 89 square meters, 3-comfortable-room with southern-aspect garden of 121 to 139 square meters and classic four-room of advective southern and

northern aspect of 163 to 164 square meters. Here in Zhujiang New Town, of high value and with tall buildings, you can overlook out of the landscape of Zhujiang Park of near 800 meters from the east of the project and enjoy extremely wide garden of 110 meters inside the project. Extremely wide distance between buildings of 110 meters of China Overseas Park Royal is rather rare in Zhujiang New Town, because plot ratios of most residences in Zhujiang New Town are very high, garden areas of ordinary residential quarters are small and distance between sections are too near.

Besides excellent park landscape of 280,000 square meters, Project China Overseas Park Royal is generally designed is in the structure

of "U", making full use of space to for an extremely large private garden which is as high as 12,000 square meters. Meanwhile, it constructs a closed and independent residential quarter garden, effectively avoiding noise from the outside and creating a quiet and private living space in prosperous CBD. So far, areas of most part of garden of residential project on sale in Zhujiang Ne Town are only several thousand square meters, some are even only around 1,000 square meters. This project is a rare large-area garden community within the CBD area in the whole Zhujiang New Town, with its delicate garden design in abundant layers and swimming pools.

The project has rare city landscape of Zhujiang Park and extremely large private gardens, wide distance between buildings of 110 meters and good lighting and ventilation. It is also right next to the sole commercial streets in the town -----Xingsheng commercial pedestrian streets which is only 5 minutes' walk from Metro Line5 Liede Station. In the meantime, there are a perfect one-line educational system including kindergarten, primary school and middle school inside and outside the community. Necessary business system in the town: Xingsheng Road of 45 meters' wide and about 1 kilometers' long in the south, East Zhujiang Avenue in the west and West Gate of Zhujiang Park in the east. The landscape is connected to the city's New Middle Axis Central Plaza and Zhujiang Park, objectively forming Zhujiang New Town's west-eastern business axis. Constructions on the both sides of Xingsheng Road are built in the Linnan-featured structure of "double verandah". The road is said to have good business prospect according to the plan of Zhujiang New Town and to its' being the only commercial pedestrian streets here.

11

12

13

11．样板房卧室

12．样板房客厅

13．样板房窗外

1

中海珠海海逸山庄

开　发　商：中海地产（珠海）有限公司

项 目 地 址：珠海市香洲银坑

建筑设计公司：城脉建筑设计（深圳）有限公司

总用地面积：107 800 m²

总建筑面积：360 000m²

建筑容积率：2.5

楼　　　　层：≤49层

1．全景夜景图图
2．海面透视效果
3．鸟瞰图

2

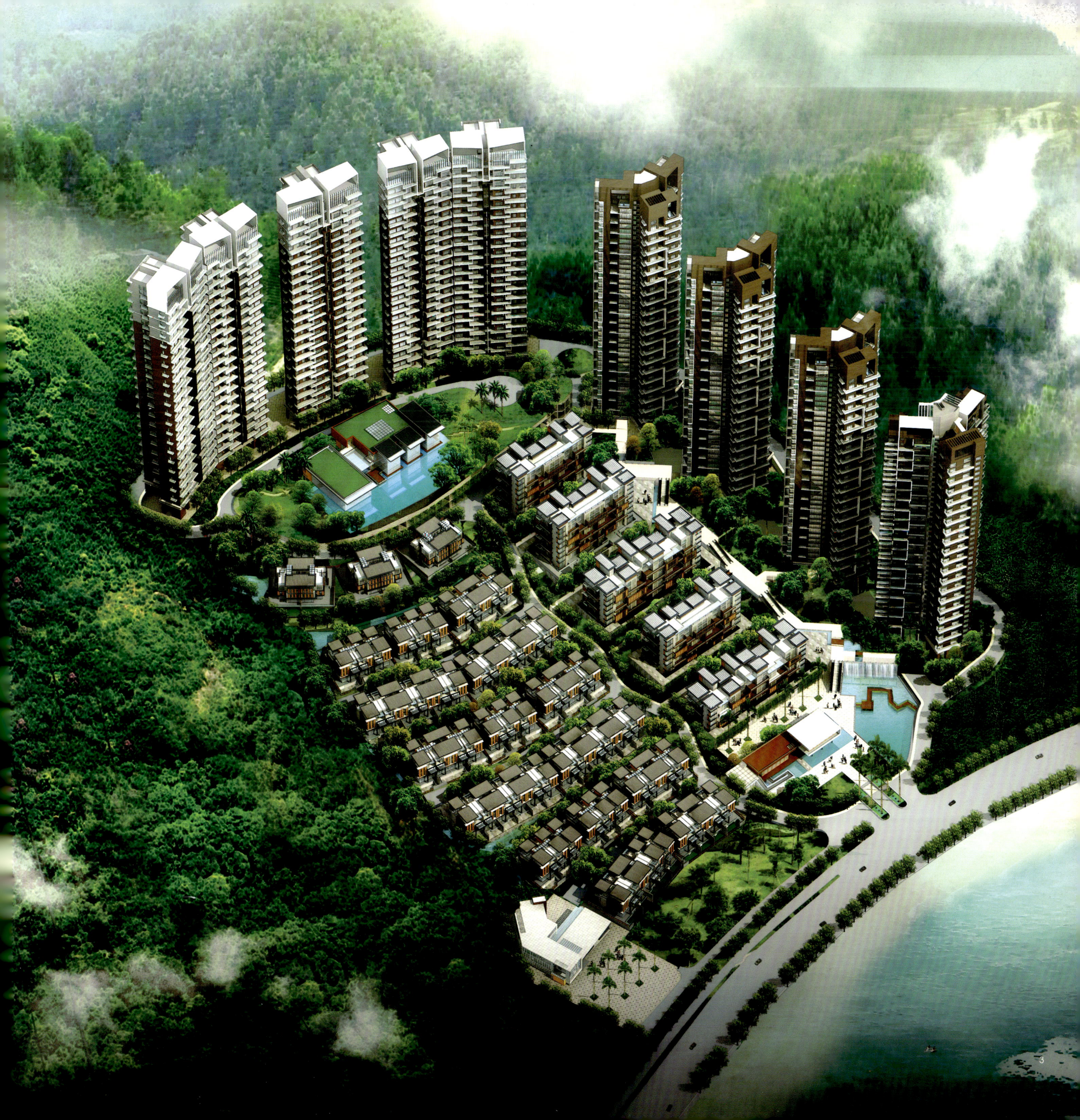

小区的道路系统以为人及车提供高效、安全和便捷的交通为主旨。

小区竖向设计利用多层地下室逐级抬高场地，减少了土石方量和对周边山体的破坏，解决了现状复杂的场地高差问题，为小区争取更多的海景视线。

依赖项目优越的地理条件，借助中心城区进一步完善和集聚的功能，开发商将致力把珠海银坑项目打造为一个具有创新户型的和优美小区环境的高档山海景观资源的居住小区，使其成为珠海市港湾大道上标志性建筑物和城市的新名片。

4．总平面图
5．绿化分析图
6．规划结构分析图
7．高层效果图
8．全景效果图

Project address: Yinkeng, Zhuhai, Zhuhai

Developer: China Overseas Holdings Limited (Zhuhai)

Architectural Design Company: Shenzhen-based CityMark Architects and Engineers

Total land–using area: 10.78 m^2

Gross floor area: 360,000 m^2

Construction volume rate: 2.5

Floors: ≤49

Zhuhai is a garden-pattern seaside holiday city, a shining pearl on the side of South China Sea and is honored as City of Islands. Project Zhuhai Yinkeng of China Overseas Real Estate is located in the northeast of precinct center of Hongkong in Zhuhai City, facing the ocean in its back with unique ocean landscape and lying against the foot of the continuous Phoenix Mountain. As necessary facilities of the future central area further improved, the value of the land will be raised to a great extent, providing excellent living necessary as well as cultural feature resources in the future residential quarter.

Mixed-living pattern is adopted in the scheme that is matching mode of high-rise plus low- and multi-layer. Such is a way of effective land using appeal to the limit of living density and land saving. Therefore, resources such as landscape, land and necessary can be fully made use of and complement each other, meanwhile the middle class in different growing periods can share common living space which is good for formation of community culture and cultural atmosphere.

In the planning design, extremely high-rise standard point plane takes the layout of one stair three or one stair two, and the designing method of elevation down to zero, spreading gradually along the root in the northeast direction with our expectation to reduce oppression from the city space because of the extremely high-rise construction and to gain maximum vision sight of the ocean landscape for the residents, finally integrating the construction with the mountain body. You can enjoy the most delighting cultural feature on the city axis in the east Gangwan Avenue as the skyline which is rich in space layers rising from the coastline.

By introducing the water feature, low-density residence becomes a neighborhood in the layout of the form of islands. And the low-rise houses are built with modern material, exquisite details. Neat process method of the roof represents oriental, classical and elegant architectural image.

Low-density and high-density areas are relatively independent by two climbing parkways throughout the whole layout. So with the entry commercial plaza as its beginning, the main community walkways is between the two parkways, up along the space between extreme high-rise and multi-layers and ends in from the chambers on the top. Such range of space constitutes a amin open feature axis of the community.

9

10

11

9．西立面图
10．立面图
11．港湾大道视点效果图
12/13．效果图

12

13

Goal of the community transportation system is provide effectiveness, safety and convenience.

Community elevation planning takes advantages of multi-layout of the basements to rise up the field, reducing earthwork and the damage to surrounding mountain body. Vision of the ocean landscape is therefore gained by the solution to solve the complicated distance problem of the field at present.

Relying on the superior geological condition and appealing to the function of the further improvement and concentration of the central area, the developer devote to molding Project Zhuhai Yinkeng a new residential quarter with beautiful environment and high-class ocean landscape resource. Then it will become a symbolic construction on the Gangwan Avenue in Zhuhai City and a new name card for the city.

14

15

16

17

18

1

1. 全景图　　　　7. 31-40层平面
2. 模型图　　　　8. 3-30层平面
3. 剖面图　　　　9. 景观分析图
4. 效果图　　　　10. 一层平面图
5. 总平面图　　　11. 二层平面图
6. 41-50层平面

青岛东海路九号超高层住宅

开 发 商：青岛阳光新地置业有限公司
建 筑 设 计：澳洲U/A设计国际集团
　　　　　　 江苏省建筑设计研究院有限公司
室 内 设 计：加拿大FK设计集团
景 观 设 计：贝尔高林香港有限公司
总用地面积：10 800 m²
总建筑面积：85 057 m²
住宅面积：76 844 m²
共建面积：8 213 m²
容 积 率：7.88

2

青岛东海路九号位于青岛市东海路及延安三路交界处北角的黄金地段，地块南向大海，距海岸线约二百米，北向青岛市中心，青岛市著名景区太平山、八大关、五四广场等都近在咫尺。同时，地块临近青岛市CBD中心区，交通十分便利。

该地块的规划共由两座50层的塔楼及两层裙楼组成，塔楼整体平面呈前尖后缓的船形，两片玻璃墙体如同海帆或羽翼，面向大海，扬帆乘风破浪而去，简洁有力的造型为青岛市增添了充满喻意的天际轮廓。建筑外立面以低辐射透明玻璃幕墙覆盖，晶莹通透，务求这两个高耸的塔楼可以融入辽阔的大海及宽广的蓝天之中，以其透明的效果减少建筑体量对周围城市建筑的压迫感，并增添建筑的高科技含量。楼体立面的弧面反映了海洋的柔美，白云的飘逸。弧形立面由下向上，由前向后的不断收分，以弧形玻璃幕墙在30，40，50层多次退台，形成立面上的波涛效果，有如晶莹通透的海面被海风吹皱，形成了一波波的浪花。塔楼顶部的处理为画龙点睛重要部分，为了突显塔楼的高耸，50层以上的玻璃幕墙不断的向上收分，形成一双巨大弧形的玻璃帆，在海风之中鼓足了风，似要带动整个青岛市破浪而去，充满活力。位于50层以上，玻璃帆的包围中有一中庭，设置了住客观景台，酒廊、空中花园等设施。在夜间点亮，于150米高处璀璨生挥，犹如海上的灯塔一般，以至大海中央的船只也能欣赏到它的壮丽美景。

住宅塔楼底部的群楼为玻璃幕墙，利于阳光透射及使用者观景。形体概念源自行驶中船下激起的浪花，浪尖向两边扬起，围绕着船身。立面上裙楼的屋顶妨如海面上展翅欲飞的海鸥，翅膀张开，带动着海水，尽显有机的动感。它流动的曲面外墙最大限度的展开了观景采光面，展开的双翅也预示着青岛一飞冲天的潜力及鹏程万里的美好未来。

本案户型内布局的设计考虑到现代生活的节奏以及家庭模式的变化，创造了满足大活动空间、大流动空间的新式户型。该户型以大面积的开窗，灵活的封闭式观景阳台来加大客厅的面积，同时强化个性私人空间的生活模式。反传统的厨房设计是户型设计中另一特色，厨柜、早餐桌在大部分户型中与客厅、餐厅等生活空间融为一体，不再是在厨房中孤立的家具。这样不仅加大了空间的灵活性，也进一步体现了现代生活的家居布置。

为了使空间更具弹性，令户室中的私人以及公共生活空间富于变化，方案中采取了推拉式隔墙的概念，用以划分观景阳台、卧室、客厅以及生活室。这样不仅打破了传统固定隔墙的局限，令空间可以相互组合，同时又保证了卧室空间的隐私性。在沿外墙面上设置的隔墙，打开时更增大了各空间的观景面积，使户内形成一个一气呵成的观海走廊。

住宅塔楼为住户设置了前后入口大堂，住户可以由正面海鸥形通廊或塔楼北部的次大堂到达私人电梯厅。两个大堂均为两层通高的大空间设计，给人以通透，宏伟的感受。住户会所均以最高档次设计，供住户享受的设施有室内大型游泳池、健身房、桑拿、商务会馆、多功能室、棋牌室等。沿着住宅塔楼间的玻璃走廊，在首层裙楼内设有酒吧、西餐厅、咖啡店、小型超市、药店、小型银行等便利店铺，以利于住户使用。

3

4

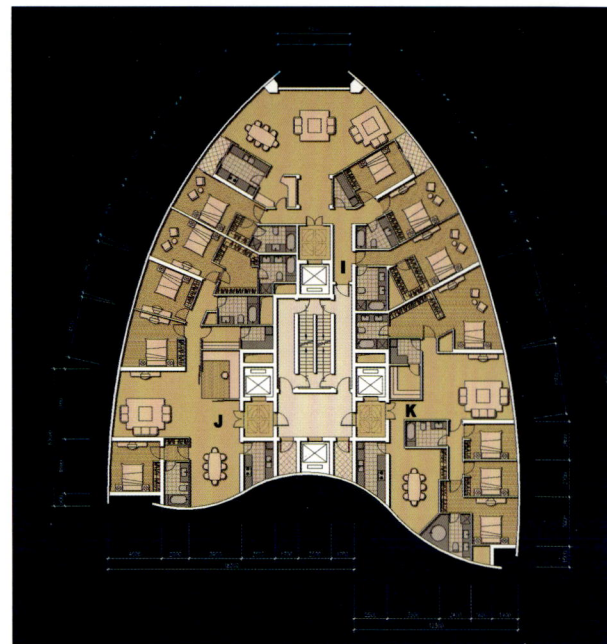

6

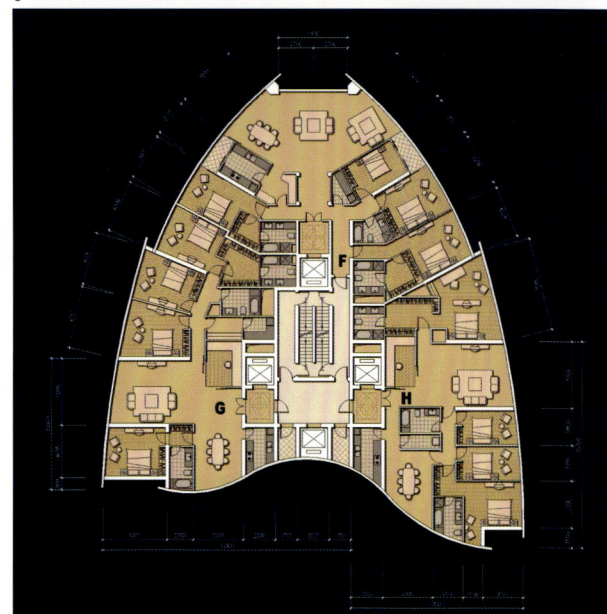

7

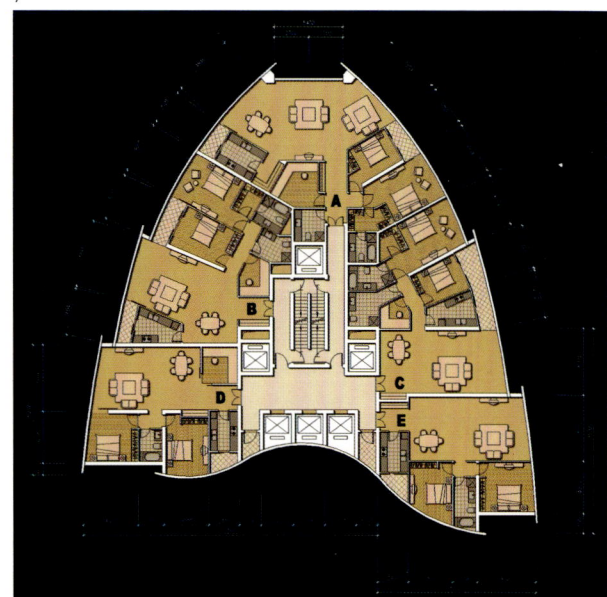

8

5

Floor area: 10 800 m²

Building area: 85 057 m²

Volume rate: 7.88

Qingdan No.9 Donghai Road is located in Qingdao City's inter section point of Donghai Road and the north corner of No.3 Yanan Road's glod land sector, the land is in the south to the sea, away coastline about 200m, north to Qingdao Center. There are famous scenic spots such as Victoria Peak, Ba Da Guan, Five four Square, all located near here. At the same time ,the land approaches Qingdao City CBD center areas, which transportation is quite convenient.

The planning of the land consists of two fifty-storey tower as well as two-storey buildings, the total plane of the tower builds like a boat-shaped which is sharp in the front and plain in the backside, the wall with two pieces of glass as sea sail or wing, facing to the sea, sailing through the wind and splashing the waves, this simply but powerful styly increase an outline which is filled with figurantiue. Elevation is covered with lower radiation glass wall, which is clear and see-through, to inform the two high tower blend in large sea and wide sky, in this clear effect, decreasing the sense of pressure about architecture to surround architecture in the city, at the same time, adding to architecture's quality of high technology. The curve surface of building reflects softy of the ocean, the graceful cloud. Curve surface from top to the bottom, from front to the backside depress generally, by he arc glass wall 30,40,50 stories, informing the wave effect, which looks like the clear ocean was broke by wind ,turning to several waves.

As an important part of dealing with top tower, to show the height of tower, over 50 stories glass wall depresses to the top, informing a par of large arc sail, filled with strengthfrom the wind, filling with energy which looks moving all the Qingdao City by the waves. Located over 50 stories, between the glass sail there's a hall, playing some facilities such as scene spots and pub and park for guests. Lighting at night, it shines in the level of 150 meters,liked a tower on the sea, so that ships sail on the central sea just can watch the beautiful and grant scenery.

The bootom buildings from towers are glass wall, which is good to the use of light and watch scene. Body concept is linked to waves aroused by ships, waves leaped by two sides, surrouding the ship. Three-dimensional top of buildings as sea gull shich will wing on the wae, opening the wings, moving sea water, showering energetic movement. The fluid and curved walls show on the light side to be a large extent, the opening wings also predicts Qingdao's bright future.

In this case, the design from department's layout considers the pave of modern life and the change of family modds,creating a new department which satisfies large space for activities. The department based on wide size to make windows, unflexible and closed balcony add to sitting room's areas, meanwhile strengthen personnel space of life model. Another feature is the design of broken-tradtional kitchen, in most department, cupboadrd, table blend in sitting room, dining room, it doesn't to be an isolated furniture in the kitchen.

On the whole , to do like this not only strengthen stable of space, but also present decoration of modern life in the way.

To make the space elastiaty, making personal and public life more changeable, he case adopts the concepts of party wall, using it to devide into balcony, bedroom, sitting room. In this way not only breaks limited of traditional wall, combining with space, but also garanteeing private space in bedroom. Along outer wall parition, when opens it increases areas of each space, informing a corridor to watch wae scenery.

Tower Building offer the front and backside big hall for owners, owners can walk through secondary hall to private life hall by sea gull liked corridor or northern tower building. Two hall are designed large space, giving great feelings. The chambers both designed by top level, the facilities provided for owners such as large swimming pool, workout, bath, Business institute,multi-functional office and chessroom and so on. Along glass cooridor between tower buildings, on the first buildings there are pub, canteen, cafe, market, drugstore, small bank,etc. Some convenient store, which is good for owners to use.

9

A座30层以下景观范围

B座30层以下景观范围

注：A，B座30层以上均为开阔景观

11

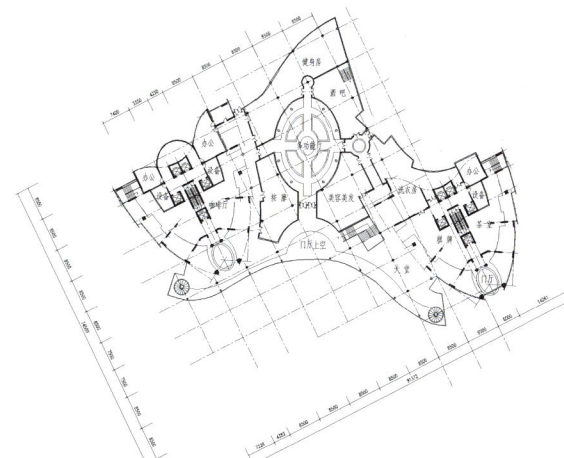

12

1. 夜景图
2. 正面透视图

上海海上财智中心

开　发　商：上海新耀房地产开发有限公司
项目地址：澳杨浦区关山路139号
建筑设计：都市元素
占地面积：20 604 m²
总建筑面积：80 000 m²
总 户 数：85

上海海上财智中心即东上海中心是一商务居住混合型社区，号称全功能MINI城市综合体，东上海城市商务酒店（四星级RAMADA酒店）、东上海企业中心、东上海国际公寓、东上海商业广场四个部分组成。总建筑面积82000平方米，位于黄兴路国顺东路，五角场繁华一线。集高尚公寓、办公和高功能商业于一体，临黄兴路的为商用，关山路为居住。所以定位于酒店式主题管理，号称星级式优质服务，住户为富裕白领，品位商务人士。就商铺前景看，本楼盘属于商业街的经典旺铺。除了受五角场商圈辐射影响，能够吸引四面八方的五角场消费群，同时还面向黄兴公园周边诸多高档居住小区，紧靠城市"绿肺"黄兴绿地，致使该区域已经发展成为当前杨浦区最大、最主要的生态居住集中区域，加上其本身办公楼和住宅楼的业主及酒店所带来的充沛客源，使得其自身的人气更为有保证。

五角场中心前沿，全功能MINI城市综合体。本项目位于黄兴路国顺东路，五角场繁华一线，向北笔直300米，即刻畅游都市繁华盛景，东上海中心，集高尚公寓、高品质办公和高功能商业于一体，全面满足高品质生活、商务、投资、购物需求。酒店式主题管理，星级式优质服务，东上海中心已经为都市精英一一准备。无法复制的地段见证东上海中心的荣耀，品质与品位的构筑创造东上海中心的价值，8万平米高品质功能型城市综合特区，先进的建筑规划和管理经营模式，全新的精英领地，与五角场中心繁华共享，一齐闪耀，共同拥有属于自己的上海东门户荣耀版图。东上海中心的房型为65平方米的一房，101平方米的二房。

Developer: Shanghai Xinyao Real Estate Development Co., Ltd

Project Address: NO.139 ,Guanshan Road ,Yangpu District.

Building Design: Metro elements

Area Covered: 20 604 m^2

Total Building Coverage: 80 000 m^2

Number of Households: 85

The Center of Nautical Wealth and Wisdom in Shanghai , a community combine commercial and residential,which is called MINI urban equation in full function, consists of East Shanghai Business Hotel(Four-star RAMADA Hotel),the corperation centre of East Shanghai,the International Apartment of East Shanghai and the Commercial Squares of East Shanghai.The total area of it is 82000 square meters and it is located in Huangxin Road and Guoshundong Road ,where is called the flourishing line of Wujiao Court. It is a combination of gracious apartment, offices and high-function commercial. The Huangxin Road is used for commercial while the Guanshan Road is used for residential.So management of the center is hotel-orientated topic and it is called the star-quality service.The households there are rich white collars and the businessmen. For the commercial prospect, this mansion is reside in the classical prosperous store of the shopping mall.Aside from the circle radiation influence by the commerce in Wujiao Court ,which attracts the consumers from all quarters,the center is facing a lot of high-grade residential areas around the Huangxing Park and next to the Huangxing Green Space,the green lung of the city.

The MINI urban equation in full function ,the forefront center of Wujiao Court. This project is located in Huangxing Road and Shundong Road, the so-called flourishing line of Wujiao Court. When go northward straightly by 300 meters ,you can enjoy the East Shanghai Center, the busy grain view of metro,which combines the gracious apartment, high-gradeoffices and high-function commercial .It meets your needs of high quality life , business, investment and shopping demand.Management of hotel-orientated topic and the star-quality service are provided for the metro essence by the East Shanghai Center.The unique location witness the honor of the East Shanghai Center,and the construction of quality and taste create the value of it . high quality functional integrative district of 80000m^2 the advanced building planning and management model and the bran-new manor of essence share the prosperity,shining and own the honor artwork of east Shanghai together with the Wujiao Court Center .The types of apartment include one-room apartment of 65m^2 and the two-room apartment of 101m^2

3．总平面图
4．鸟瞰图

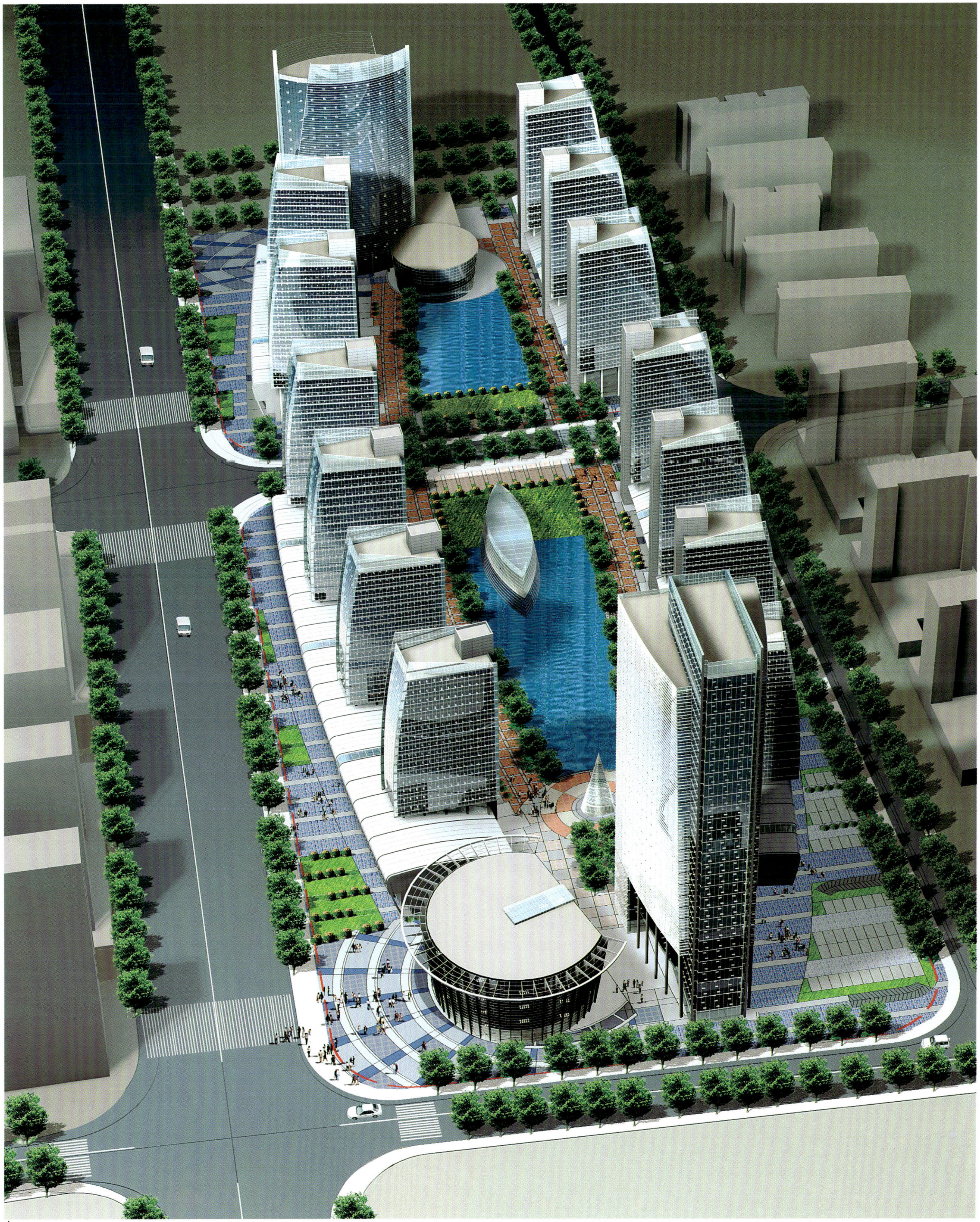

4

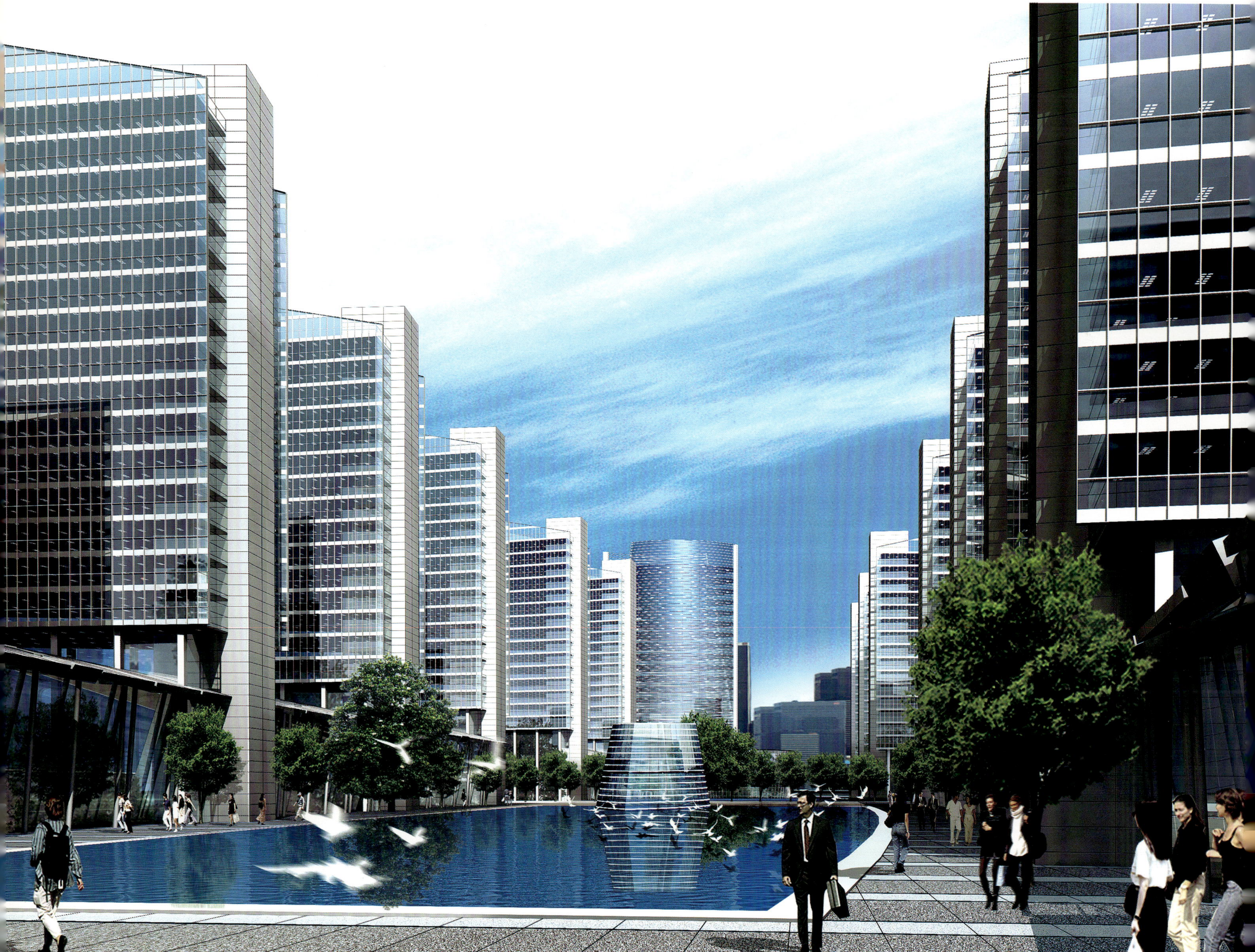

5

5. 内庭效果图
6/7. 透视图效果

1

望京国际商业中心

开　发　商：北京正鹏房地产开发有限公司
建筑设计：华通设计顾问工程有限公司
占地面积：170 000 m²
建筑面积：180 000 m²

2

　　望京地区位于北京市区东北部的四环路与五环路之间，东南侧与京顺路及首都机场高速路相邻，西南侧临近阜荣街，西北侧临近望京地区内环路阜通大街(东)，东北侧临近望京街。隶属朝阳区管辖，是市区中相对独立的，具有工作、居住、娱乐等功能，并设有扩大地区级生活服务设施和市级大型公建的综合性新区。按照该地区最新的城市规划功能划分，以北小河为界，分成南北两大块。北小河南侧统称为"望京新城"，其功能以望京中心区公共设施为中心，以居住为主，兼文化、教育、办公等功能的综

合区；北小河北侧确定为"望京科技产业园区"，其功能以高科技产业及研发为主，兼有居住、商服等相应配套服务功能的综合区。

　　望京地区总用地规模为17.8平方公里(其中西侧绿化隔离带用地约为1平方公里)，规划总居住人口为33万人，现有常住人口近15万人。总体布局由地区功能链、自然生态链、景观绿化轴及公共交通通道等主导系统构成，其中包括：

(1)1个地区级主商业中心、5个次商业中心、8个社区商业中心和6个文化娱乐中心；

(2)8个区间公园、38个社区绿地和3条区间绿色通道；

(3)广顺南北大街为地区功能性景观主轴，北小河为水景主轴，望京轴绿化主轴。

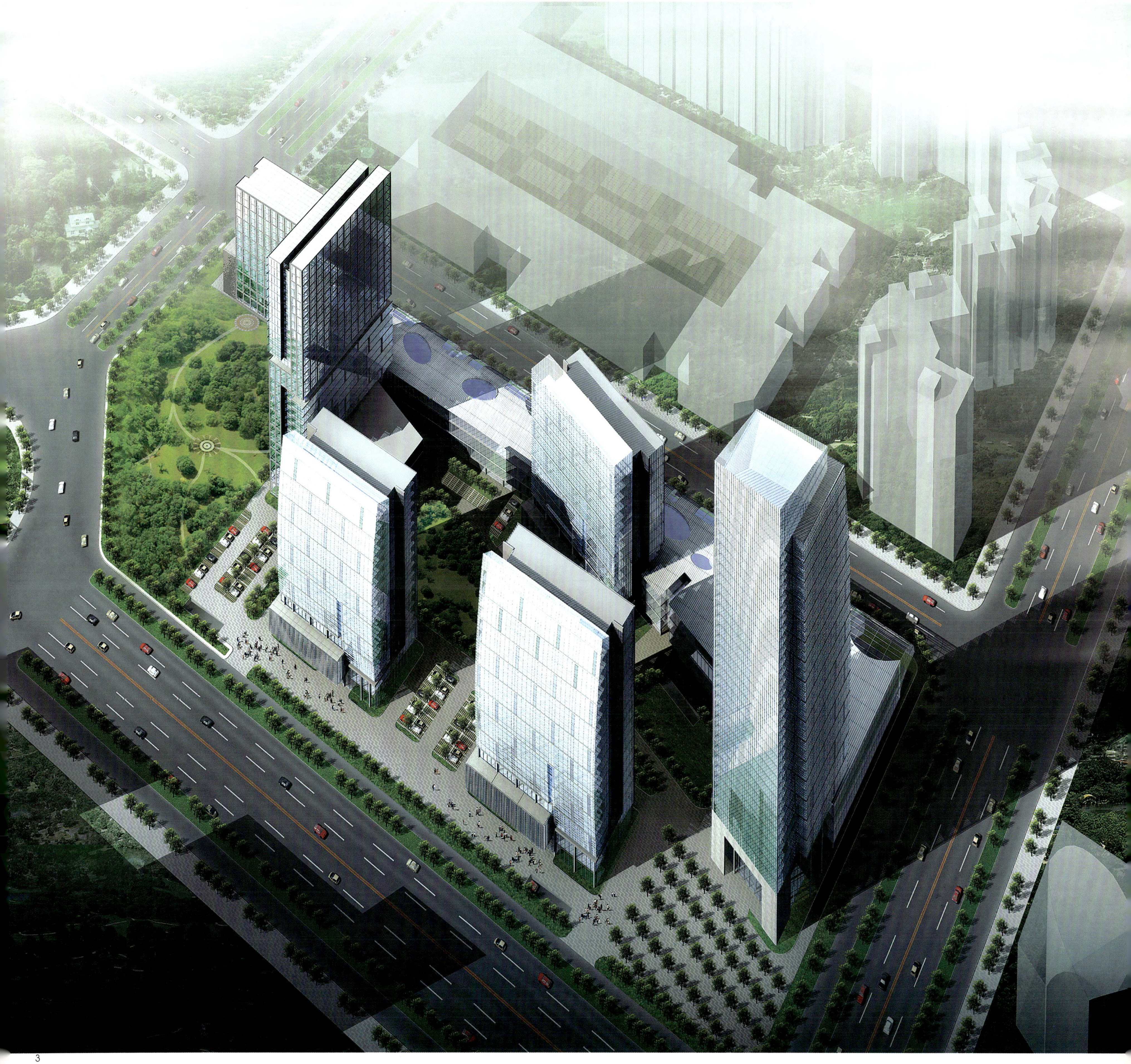

3

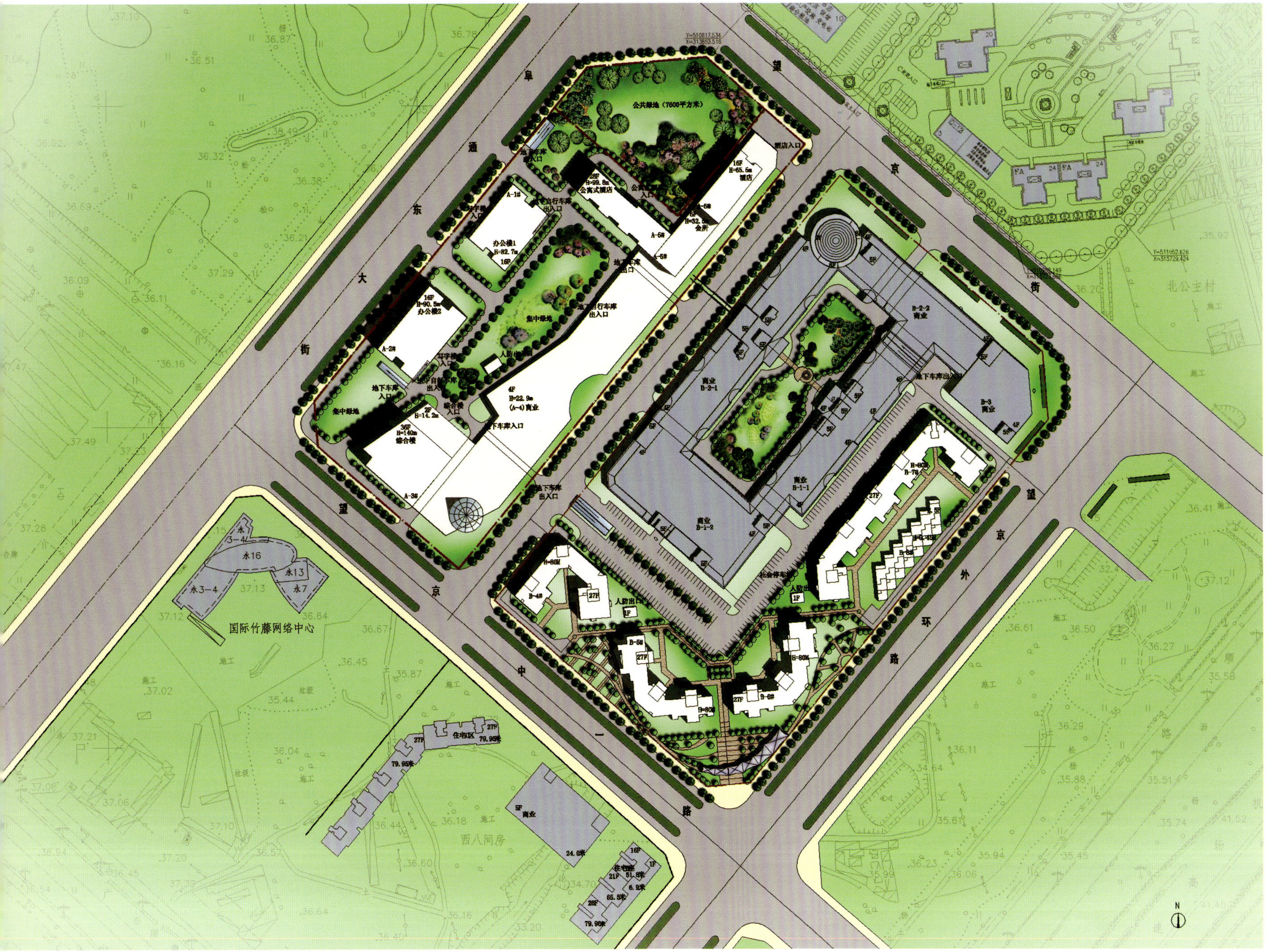

4

项目特点

该项目是集“商业、高级住宅、写字楼”三大物业形态于一身的建筑综合体。该项目总规模超过55万平方米，其中18万平方米为精装公寓，另有12万平方米中央休憩式STREET MALL，更有25万平方米中央OFFICE集群。平层、跃层近40种精装户型，高层塔、板结合，2.9米层高，2梯4户，1户1位。

社区配套

12万平方米全业态休闲商业配套Street Mall，集购物、餐饮、娱乐、休闲、社交、商务等功能于一体，包含大型百货和超市大卖场及各类专卖店、游乐设施、文化广场、休闲餐饮，为业主提供一站式休闲服务。

4．总平面图
5．交通分析图
6．景观分析图
7．人流分析图
8．建筑效果图

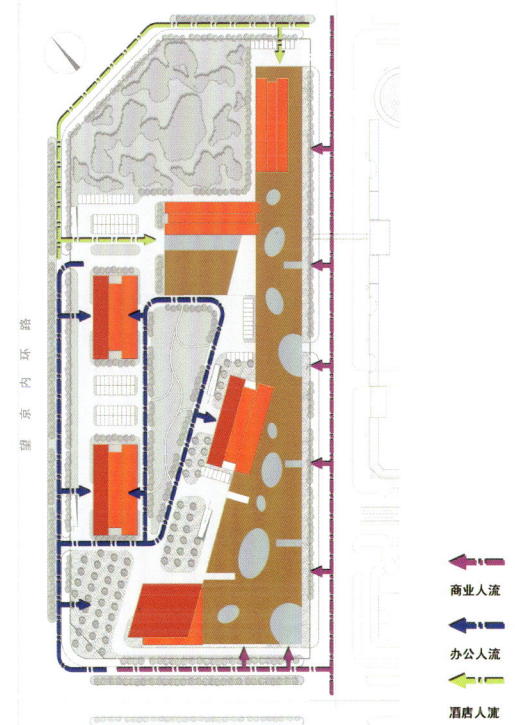

望京内环路

酒店停车

商业停车

办公停车

车行主要道路

人行道路

5

望京内环路

主要景观点

景观空间分析

景观视线分析

6

望京内环路

商业人流

办公人流

酒店人流

7

8

9

to as "Wangjing New City". The function focus on the public facilities of Wangjing Center, based on residential, is mixed use area which offers lots of functions such as culture, education and office. From the North River is " Wangjing Technology Industries Park", it's based on high-level technology industries, it's a comprehensive district with residential, business service and such accordingly functions.

Wangjing area for the total land size of 17.8 square kilometers (west side of Green isolation belt which is about one square km land), planning for the total population is about 330,000, the existing resident population is near 150,000. The general layout features linked by the region, the ecological chain, green landscape shaft and public transport access-a system, which including:

(1) a regional level the main commercial center, five of the second commercial center, eight community centers and six commercial culture and entertainment centers;

(2)8 interval parks, 38community green space and three green channel interval;

(3)Guangshun North and South Main Street area for the main function of the landscape, the North River for the Waterscape spindle, Wangjing Green spindle axis.

Characteristics of the project

The project is set " commercial , senior housing ,office space," three major forms of property in a building complex. The total size is over 550,000 square meters, of which 180,000 square meters for the hardcover apartment, and another 120,000 square meters for Central Open-STREET MALL, even more 250,000 square meter office central cluster.-Layer, the spring layer near 40 hardcover apartments high-rise towers, with boards, 2.9m – high, 2 ladder 2, a1.

10

Developer: Beijing Zhengpeng Real Estate Co., Ltd.
Architectural design: Walton Design & Consulting Engineering Co., Ltd
Land–using area: 170 000 m²
Floor area: 180 000 m²

Wangjing district is located in the Northeast of Beijing, between the fourth Road and the fifth Road, southeast approaches to Jingshun Road and the Capital Airport railway, southwest to Burong Street, northeast approaches to Wangjing Street. It belongs to Chaoyang District, it is such relatively independent areas, which provides many function for people's work, residential conditions and even entertainment. In addition, with expanded regional living or service facilities and king sized comprehensive new district which build by government. According to the city's planning, for the sector north of the river, divided into two blocks about south and north. On the south of North River collecting referred

Community support

All 120,000 square meters of commercial with leisure form supporting Street Mall, shopping, dining, entertainment, leisure, social, business, and other functions in one whole, which including large department stores and supermarkets various specialty shops, recreational facilities, cultural square, leisure food to provide one-stop leisure services for owners.

9. 实景图
10. 入口局部
11. 商业楼一层平面
12. 商业楼二层平面
13. 商业楼三层平面
14. 商业楼四层平面
15. 实景局部
16. 俯瞰效果图

11

12

13

14

15

16

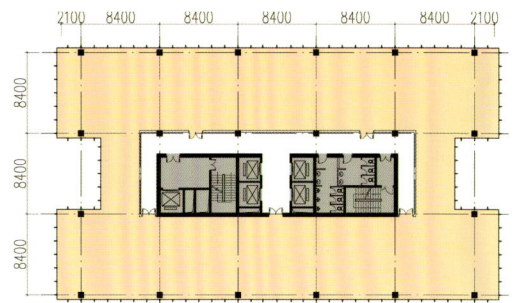

19

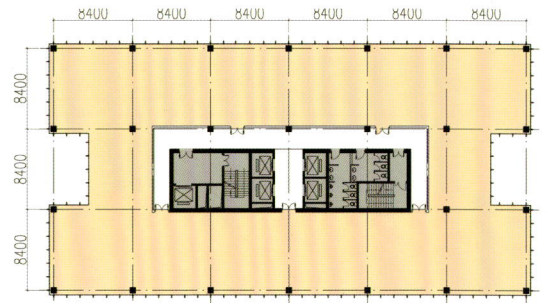

20

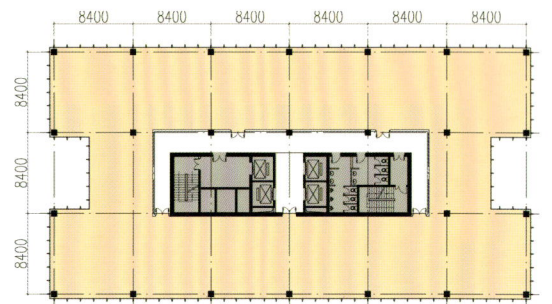

21

22

17

18

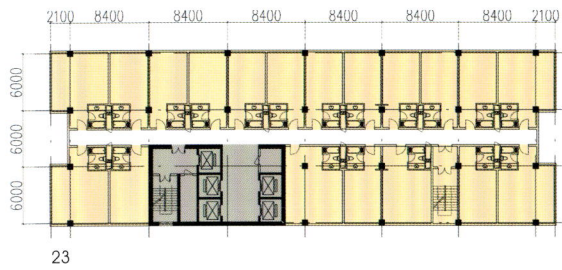
23

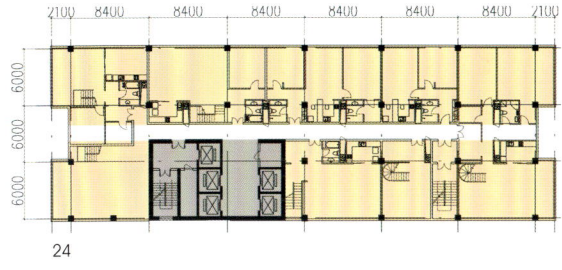
24

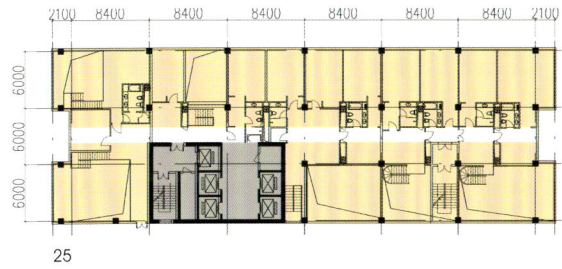
25

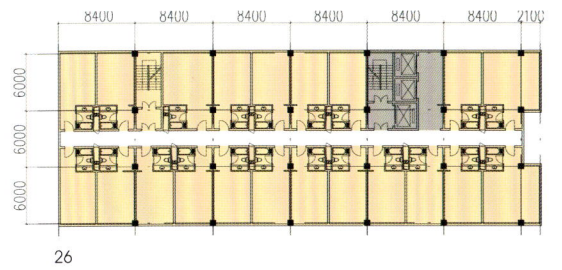
26

27

28

17/18．办公区实景图　　23．公寓式酒店标准层平面

19．办公室1标准层平面图　24．公寓式酒店复式一层平面

20．办公室2标准层平面图　25．公寓式酒店复式二层平面

21．办公室3标准层平面图　26．商务酒店标准层平面

22．剖面示意图　　　　　27/28．酒店实景图

48

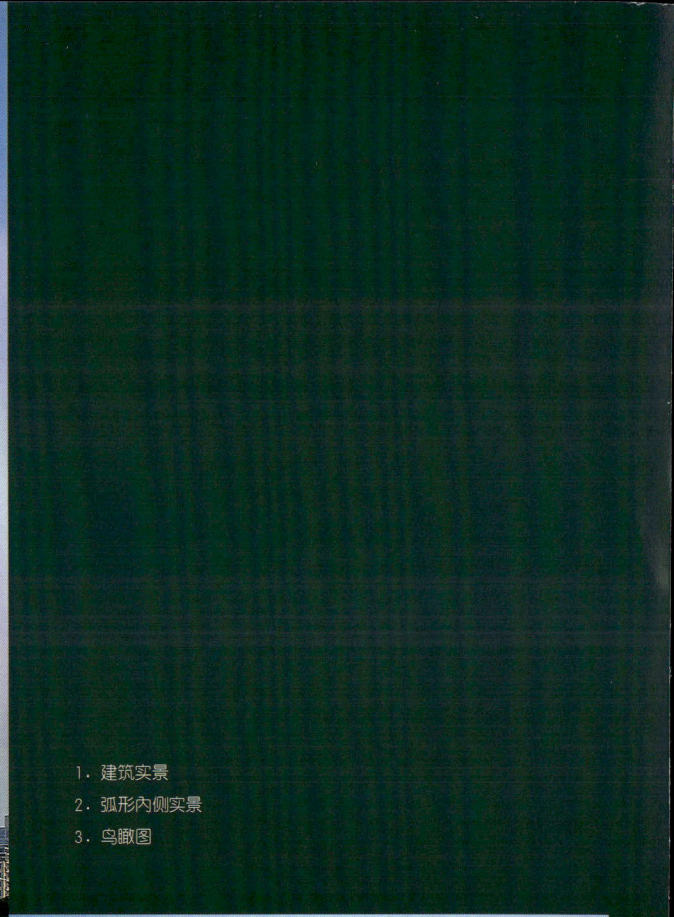

1. 建筑实景
2. 弧形内侧实景
3. 鸟瞰图

中浙太阳国际公寓

开 发 商: 浙江中浙房地产开发有限公司
项目地址: 杭州滨江区之江二号路以东
建筑设计: 华森建筑工程设计顾问有限公司
总占地面积: 50 000 m²
总建筑面积: 180 000 m²
建筑密度: 12.6%
容 积 率: 2.98
绿 化 率: 55%

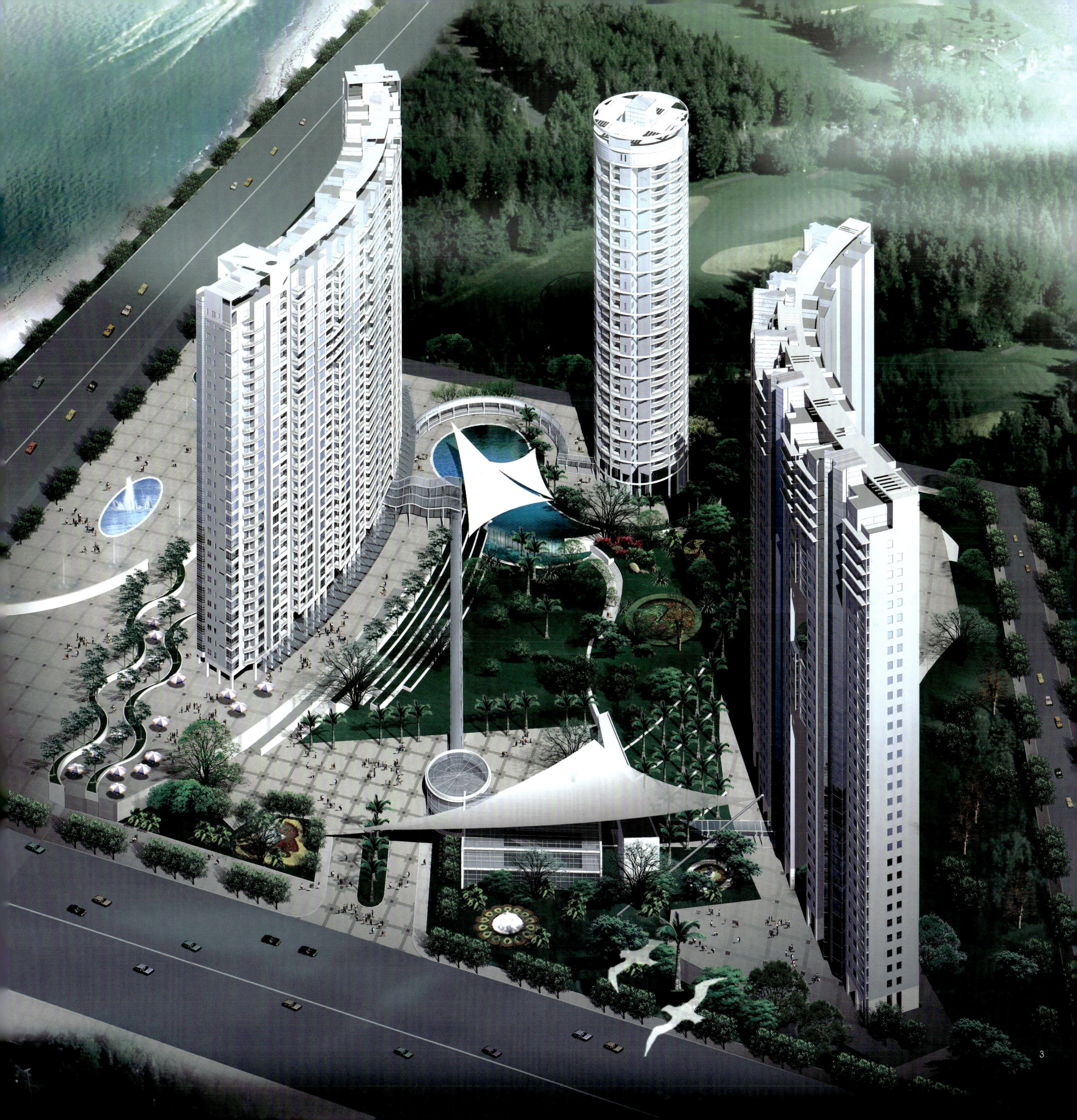

4

Developer: Zhejiang zhongzhe Real Estate Development Co., Ltd
Building Design: Huasen Architectural & Engineering Designing
Consultant Ltd.
Floor area: 50 000 m^2
Building area: 180 000 m^2
Greening rate: 55%
Volume rate: 2.98

Tie-young International Apartments adopts plate form construction with reverse arc, which in a largest scale merge the river feature into community life, meanwhile maximizing feature room within the community. It is an equal and open community. One of the most important principles of its plan is that every residents can enjoy as much of the community's feature resources as possible. Three fully open courtyards are therefore set in the plan to better avoid oppressiveness that usually existed in high-storey residential blocks.

The design allows residential of the block to reach all areas through the semi-overt corridor in the air after they enter the entry hall, where they have easy access of elevator to the corridors. Passage along the river platform also adopts such design. Commercial facilities is in the lower layer and parallel to the road, not only making it convenient for running a business, but avoid disturbance for residence.

As a residenial block, constuction of modern plate form is its main part, the scheme of such plan can well resolve living problem of population density area. In terms of construction density, idea of high-storey construction of plate form has expanded living space feature, meanwhile determined the minimum area of floor areas by its shape. Therefore, construction density in the plan only takes up 12.6%,

gaining quite large space for greening environment, making the 1st and 2nd buildings get a minimum distant of more than 115m and a maximum one of over 145m, such community environment is so valuable in such city with a dense population. There are lobby hall of 1500 m^2, chamer of 3,000 m^2, community life function area of 6,000 m^2, underground garage of 24,000 m^2 and central courtyard of 40,000 m^2 for you, and differen housing types with areas ranging from 55m^2 to 280 m^2 from which you can choose like single apartment, common housing apartment and luxurous buildings.

With people's local living habits and the use of the reiver feature both planned already, we have promoted quality of indoor and outdoor to the largest extent and reasonably made up for the distant between the basement and river bank. A design of 24,000 m^2 –large garage realized pedestrians and vehicles seperation. And a outdoor greening space of 40,000 m^2 is created under this low-density construction arrangement. Finally, a North-South-towards communicating space in the low storeys has successfully solved how high-storey residential house have made people feel oppressive.

Aims of creating a unique and symbolic construction feature for the North and South bank of the Qiantang River is in Tie-young International Apartment's plan. Breakthrough in arragement of traditional high-storey residential house has been made and inside living space has been reasonably solved, we have create a living park of public construction.

4. 连廊
5. 侧面

6. 裙楼
7. 局部
8. 连廊
9. 底层透视

7

8

9

1

青岛新世界数码港

建筑设计：澳洲U&A设计国际集团

　　　　　北京澳亚中元建筑设计咨询有限公司

主设计师：潘泰、潘洋、邓黔、熊艺

规划用地：210 000 m²

总占地面积：149 600 m²

容积率：5.96

绿化系数：35%

停车位：地下283，地上78

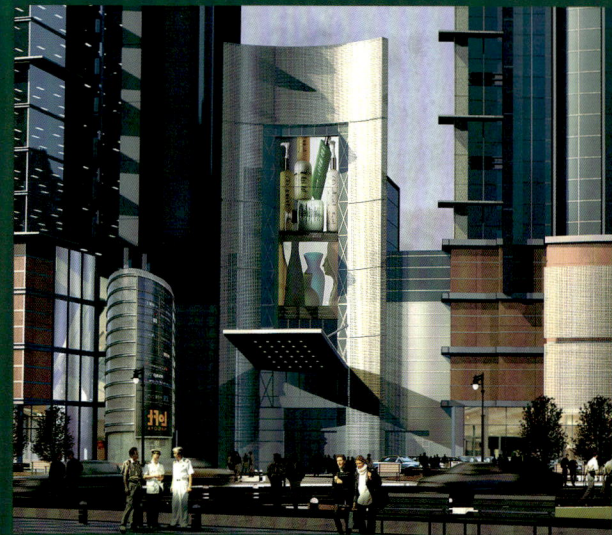

1. 总平面图
2. 建筑局部
3. 鸟瞰图
4. 人视效果
5. 顶部
6. 二层平面图
7. 一层平面图

2

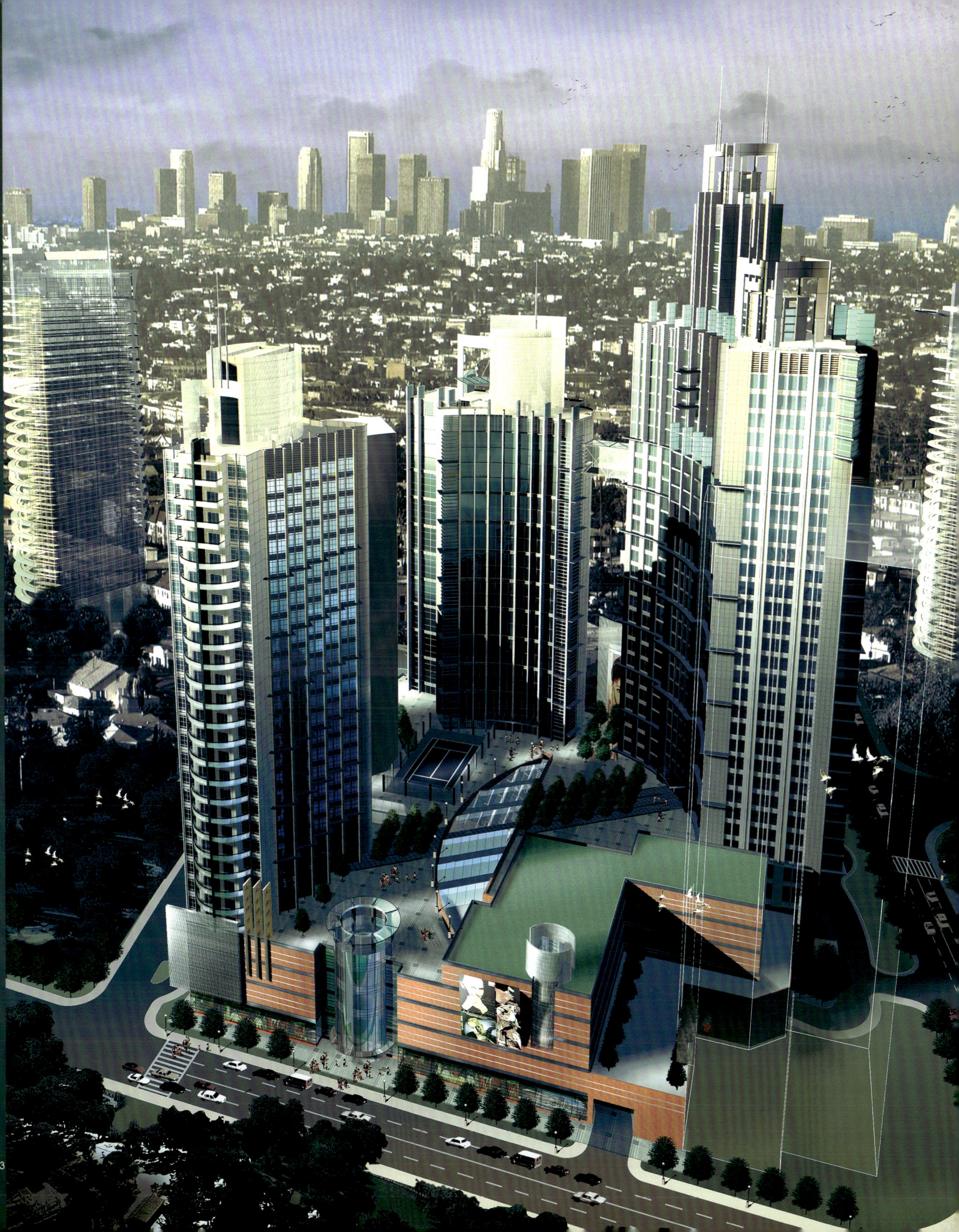

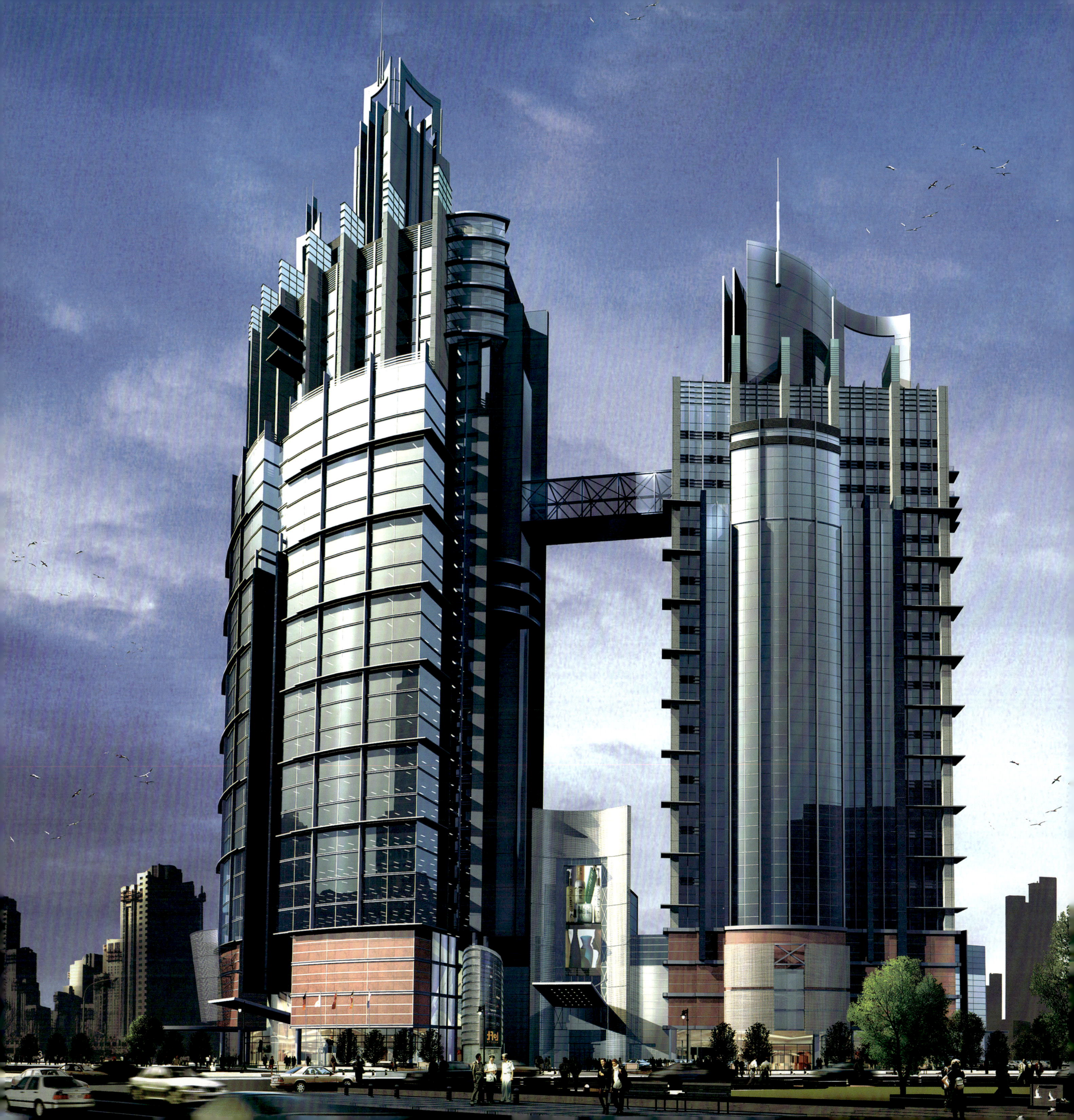

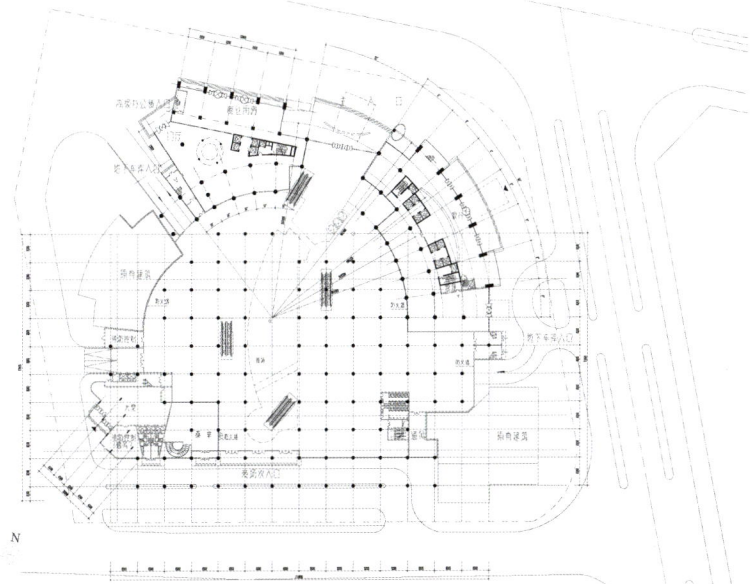

5

The principles of the monomer programme is considered carefully to make it possible to bring the living, entertainment and leisure into effect, under the promise of meeting the needs of modern business and conditions of office. The offices' building areas cover 33929.00 square metres. And there are 36-floors and two- storeys underground. There is a five-storeys commercial area, and the office is from the sixth floor with accepting intelligent building design it build the environment of the digital office in the information age, which provied us with the highly efficient, fast and comfortable, working condition. There are 24172.90 square meters of the hotels with 36-floors. The business area is under the fourth floor and from the fifth floor are the departments with hotel's service(SOHO Office). The design is intended to make full use of the advantage of the morden communications for creating a barrier-free environment of office. So on vacation people also can catch the commercial opportunities anytime to make the continuous office come true. And the area of youth apartment covers 19,700 square meters. There are 28-floor , the commercial area is under the fourthfloor. And the standard hostel is from the fifth environment for the single senior staff. All of the three main high-storey apartments take cast-in-place reinforced concrete and sheat wall structure,houses and underground of which are frame shear walls. The Facade and model embody the high technology, information, and the sense of the era.

Architectural Design: Universal Atelier International Group
Beijing Australian Zhongyuan Architectural Design Consulting Ltd.
Chief Designer: Pantai, Panyang, Dengqian, Xiongyi
Planning for: 2.1 hectares
Floor area: 149 600 m^2
Volume: 5.96
Green factor: 35%
Parking space: 283 underground, 78 on the ground

The project of Qingdao New World Of Cyberport is situated in the easten of the Qingddao city. The cyberport is in the Nanjing Road in the east, links with Hongkong Road in the north,and next to Shenzhen Road. The planning area is about 2.1 hectares. There are one 36-storeys office, one 28-storeys department (SOHO Office) where the service is the same with hotel, and a 28-storeys Youth's Department in the plan. The project is pitched at the modern comprehensive area of commerce and resident, for foreign clients and senior staff of foreign invested enterprises.

The design is in view of existing situation, combining the existing buildings with the planning well to form an organic whole. It gets the maximum comprehensive utllization of space and the old and new architecture complement each other. A unique idea, flowing lines and rich presentation molded body, create a powerful and outdoor space environment that full of modern flavor. The design lay stress on solvig a few contracitions between static and dynamic, high and low, living and business, stream of people and traffic flow, and new construction and old buildings.

The purpose of the project's green design is the multi-level and continuous three dimensional green. The design have made more efficient in the flat roof and the space of atrian, besides the area greening and road greening to afforest in mid-air. It makes the link between inside and outside, up and down. The Harbour waterscape Falls in the central area link three diffenrent buildings to form a outdoor landscape with local characteristics, combining the green waterscape with a short to create the pleasant atmosphere.

N

6

N

7

Floor area: 88 941 m^2
Building area: 458 510 m^2
Greening rate: 42%
Volume rate: 3.8

1．建筑局部
2．二期D1、D2栋住宅立面
3．鸟瞰图

1

东湖春树里

建 筑 设 计：加拿大汉克（H.N.K）国际建筑设计
　　　　　　广州东略汉克建筑设计顾问有限公司
合 作 设 计：武汉城开建筑设计有限公司
项 目 地 址：武昌水果湖街徐东路50号
总用地面积：88 941m^2
总建筑面积：458 510 m^2
绿 化 率：42%
容 积 率：3.8

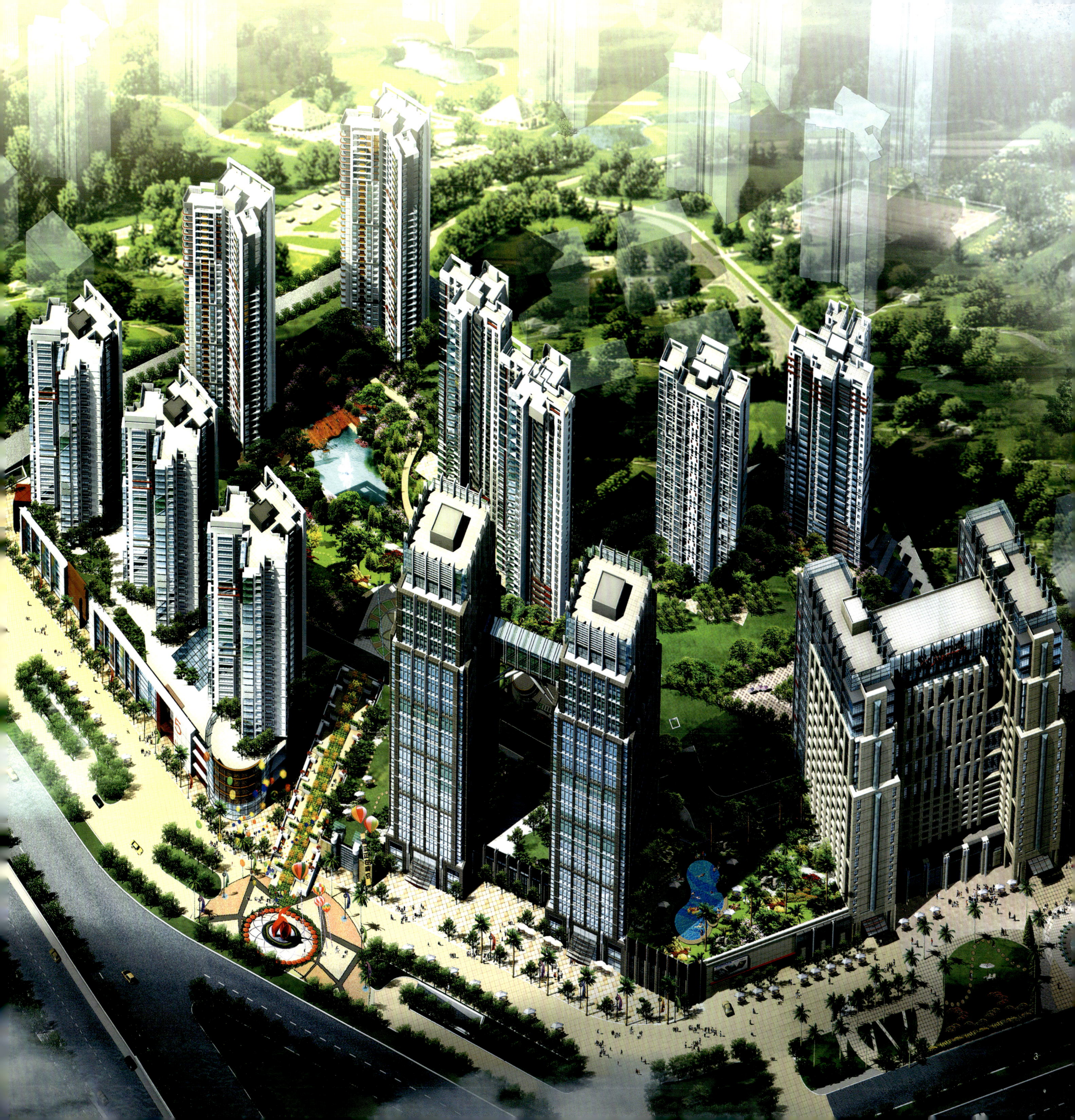

4

小区景观轴线 景观轴线

小区主景观节点 小区主景观节点

小区园景视线 小区园景视线

小区街景视线 小区街景视线

小区望东湖景观视线 小区望东湖景观视线

5

6

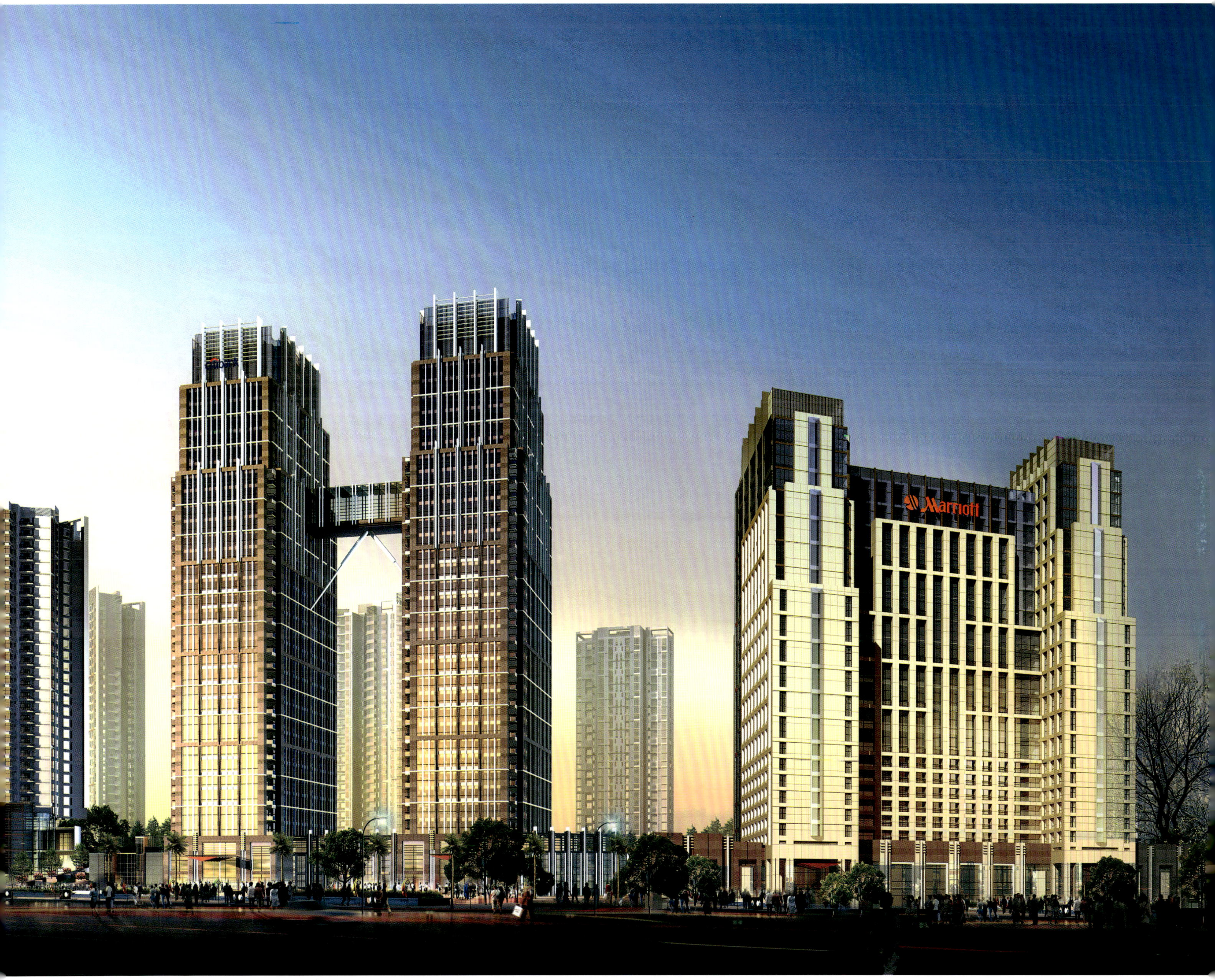

4. 总平面图
5. 景观分析图
6. 建筑效果图

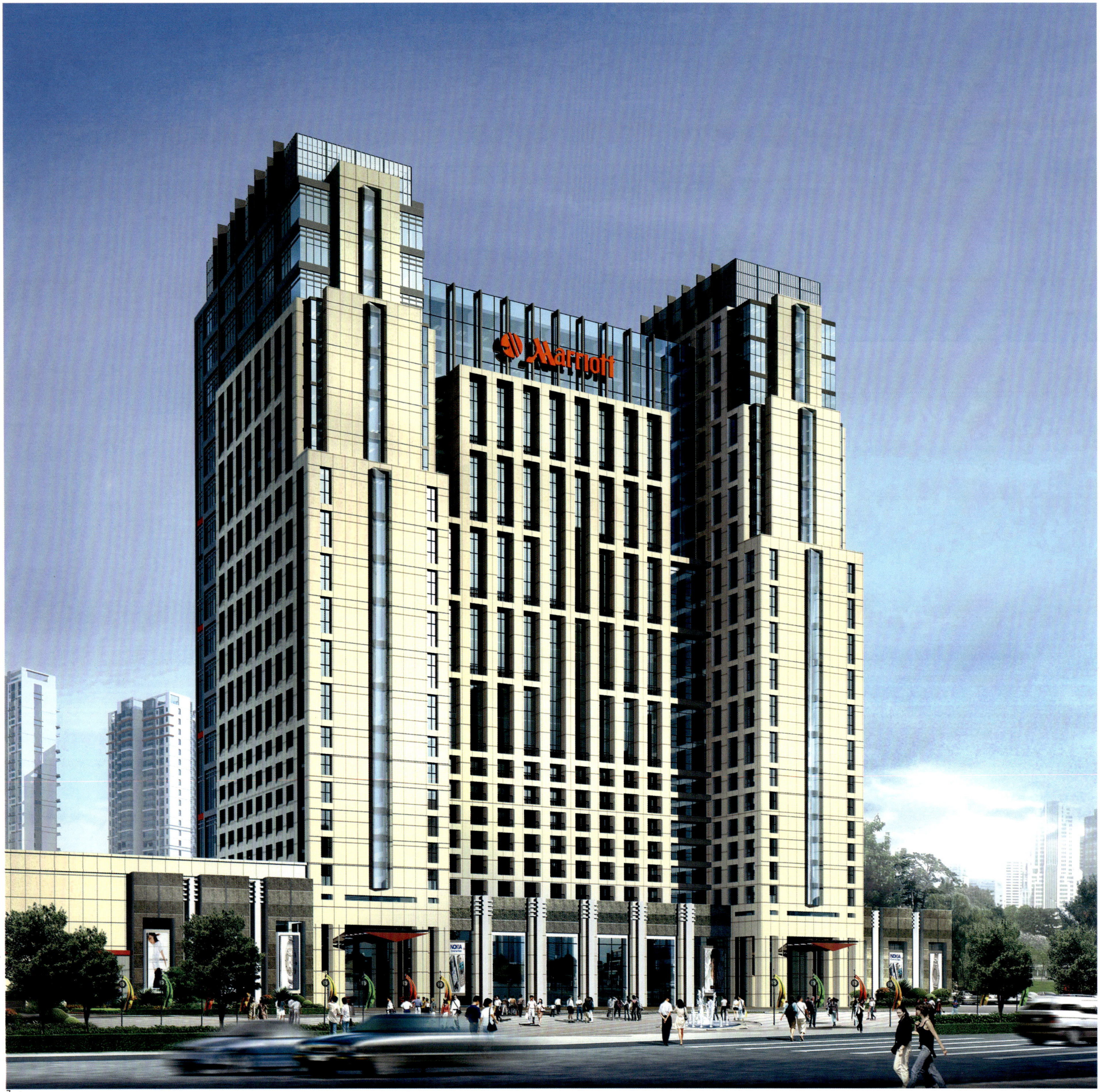

7

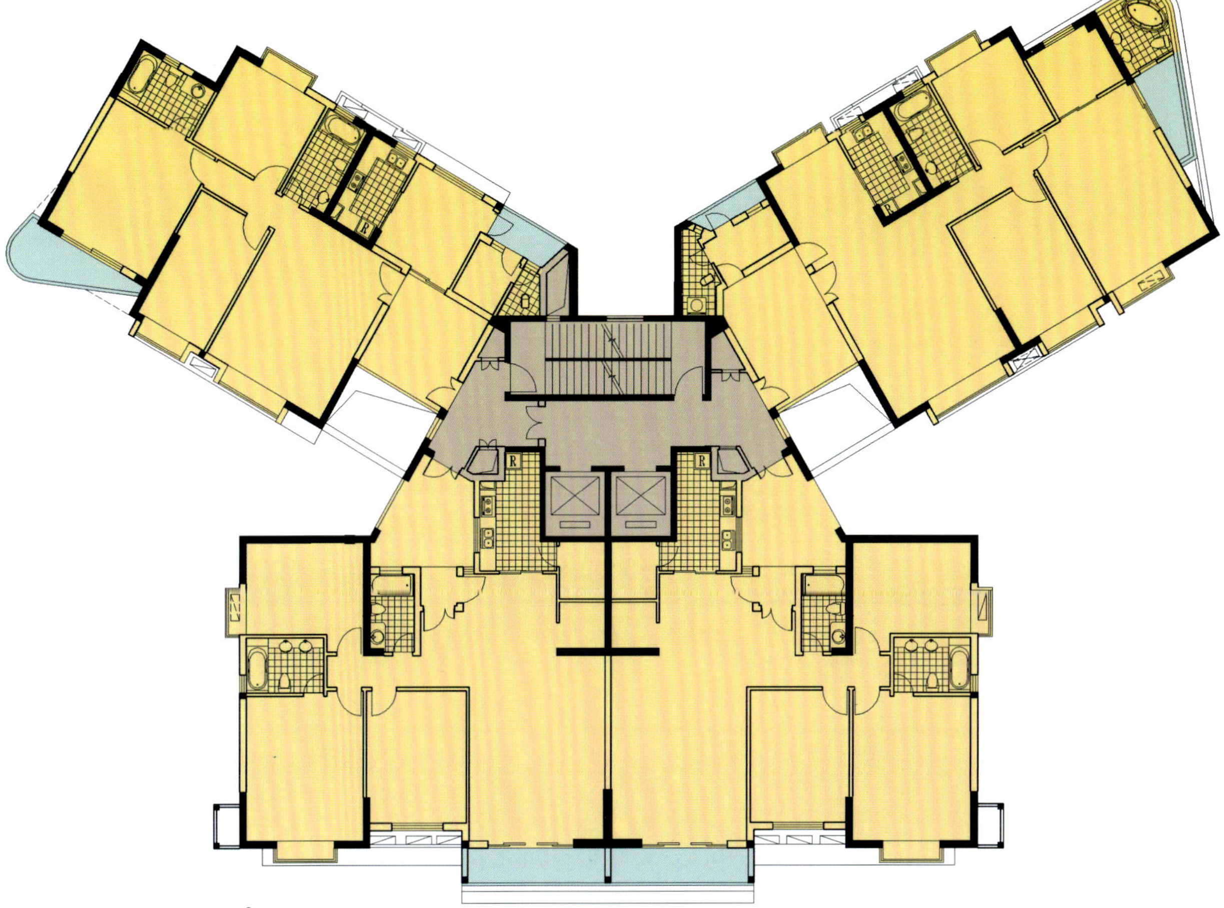

8

9

7．二期商务酒店
8．B栋平面图
9．C栋平面图

10. 超高层办公楼立面－由玻璃、金属和石材肌理构成
11. 小区内景
12. 临徐东路透视效果图

10

11

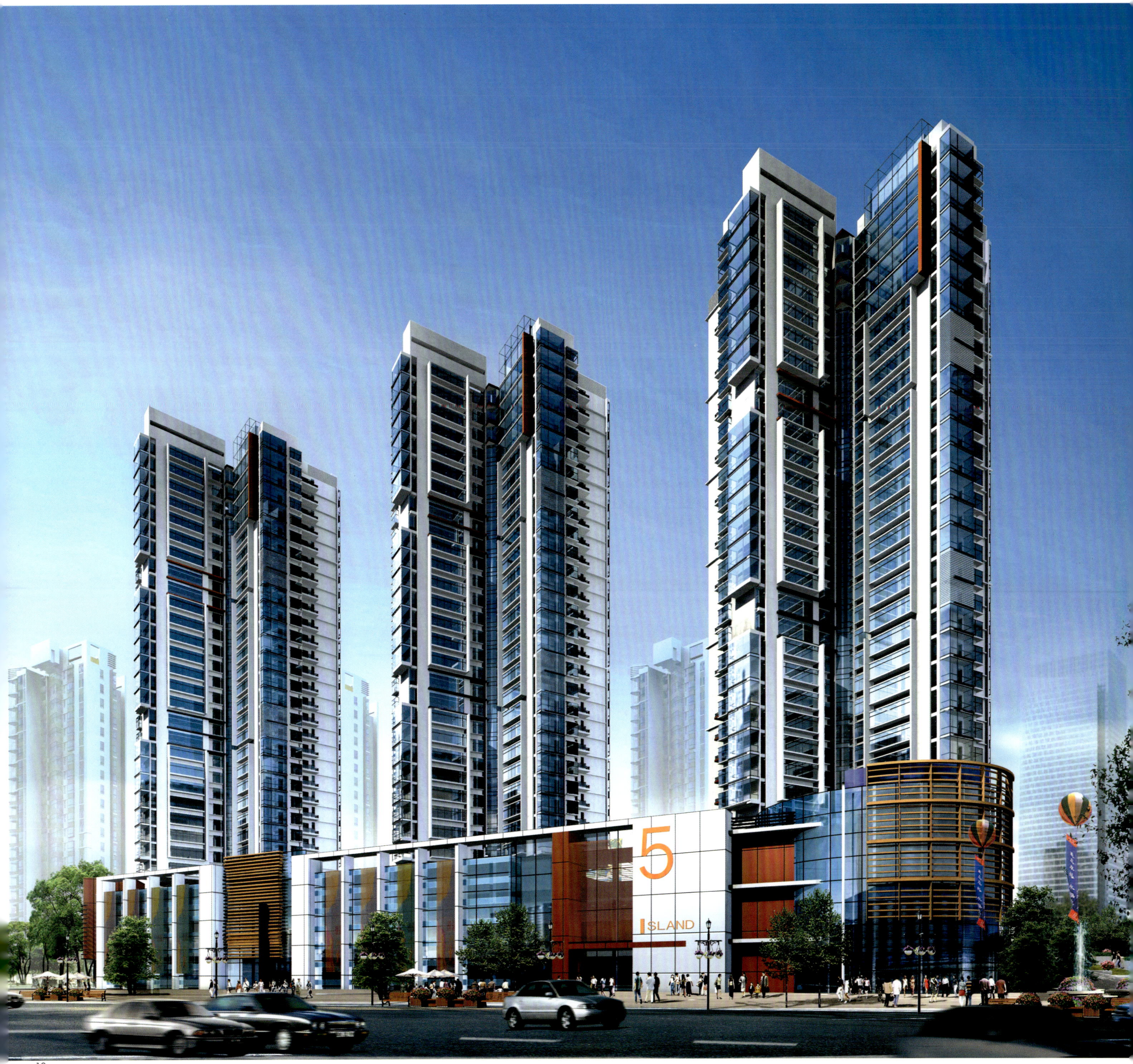

12

十一号路 3.600

3.000 3.800 3.800 4.400 4.000 4.000 4.000 4.800
3.800 3.800 11.000 10.400 4.200 4.200
3.750 10.400 10.400 10.400 10.400 4.000
10.400 10.400 10.400 11.000
10.400 10.400 10.400 3.750
9.800 9.800 10.400
4.150 4.350
4.150 9.800 9.800 9.800 4.350
4.150 9.800 4.350
3.00 7.300 7.300
7.40 7.300 7.300 8.600 8.600 11.300
7.300 7.350 7.300
7.300 7.500 7.400 3.75 会所 CLUB
7.300 7.350 6.30
幼儿园 6.900
KINDERGARTEN 7.300 7.90
5.35
4.06
6.00

1

広 东 深 圳 沙 头 角 蓝 郡 广 场

广东深圳沙头角蓝郡广场

发 展 商：深圳市冠懋房地产开发有限公司

项 目 地 点：深圳市沙头角

建 筑 设 计：深圳市东大建筑设计有限公司

结 构 设 计：深圳市筑博设计有限公司

用 地 面 积：36 273 m²

总建筑面积：170 000 m²

总 户 数：592

停 车 位：800

1．竖向设计图
2．建筑效果图正面
3．建筑效果图侧面

2

广东深圳沙头角蓝郡广场位于盐田区海景路与金融路交汇处，明思克航母旁边，毗邻盐田CBD城市客厅景观和海滨栈道，周边建筑密度极低。以明思克航母作为视觉中心，山海相连的碧海蓝湾作为舞台背景，一条长达8公里的木制海滨栈道，为海与人之间搭起了最好的沟通桥梁。蓝郡广场地上部分为两栋34层高层住宅，地下二层为停车库兼六级人防地下室，地下一层为停车库，半地下室为商业及车库。地上一层为架空层。2—30层为复式住宅，屋顶为带海水泳池的空中别墅。

的适应性；能持续发展；能适应市场需要；能适应家庭结构的变化。蓝郡独有的空中叠墅，更多的是以别墅的标准打造全复式高层洋房。较大面积的国际豪宅人居空间划分，160—300平方米的纯大宅产品，全南北向纯复式设计，高速直入户电梯、挑高阔绰入户花园、私家空中泳池、270°独享海湾景致等等，均是以世界顶级豪宅的礼遇致臻打造，力求将"叠墅"稀缺的品质自如挥洒。规划设计从住户居住舒适性、私密性出发，将更科学、更具美感、更讲求自然充分融合的理念融入，打造城市独有的上品生

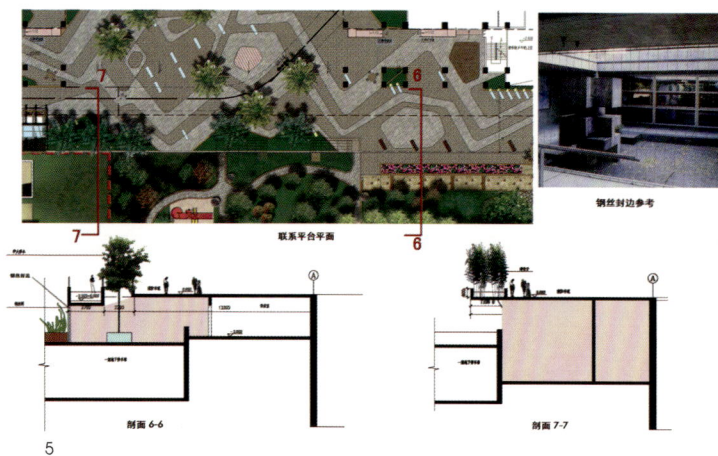

4

5

6

Floor area: 36 273 m^2

Building area: 170 000 m^2

The number of dwellings: 592

Guangdong Shenzhen Shataujiao Blue County Plaza in Yantian District is located at the interchange of Haijing Road and Jinrong Road. It's next to the aircraft carrier Ming Sike and is adjacent to the landscape of Yantian CBD city living room and the Waterfront Plank Road,where the surround building density is very low there.With the aircraft carrier Ming Sike as its visual core and the blue sea and bay, which is linked together with the mountains,as the background of stage, an 8-kilometre long woodern waterfront plank road became the best bridge between human and the sea.The overground part of the Blue County Plaza is two 34- storey high-rise dwelling building.The second storey underground doubled as parking garage and 6-level air raid shelter and the first layer underground is parking garage. A semibasement was built for commercial and carbarn.. The ground floor was built as an overhead tier.There are duplex apartments from the 20th storey to the 30th storey of the high-rise dwelling building and there is air villas on the top of the building.

The maximal features of this engineering design are the creation of air stacked villa , the arrival of family elevator at every household, the extended balconies, Large home gardens and the conception of presenter area which is for the requirement for the market.Every building has a neat and

4．商业街标准段剖面

5．外挑平台设计意向

6．商业街效果图

7．商业东入口手绘图

8．室内大厅效果图

7

8

clean exterior look which represents the scientificaalness and sense of times of Living technology.The creation of the adaptability of living space can be continually developed and adapt to the demand of the market.It can also adapt to the change of the family construction.Most of the air stacked villas,which are peculiar to the Blue County Plaza,are high-rise luxurious compound apartments as the standard to villas.

The space division of the large-area international luxurious residence ,the product of 160-300-squaremetre pure mansion,the design of meridional compound apartments,the High-speed family elevators,the ostentatious home gardens,the private air swimming pool and the 270 °exclusive view of gulf are built as the top-grade mansion of the world. The conceptual planning develops a particular elegant life of city,which is based on the comfort and the privacy of the the enants living, with the Ideas of being more scientific ,more aesthertic and requiring the full integration of nature.

In 2008, Guangdong Shenzhen Shataujiao Blue County Plaza won the prize "United Nations global environmental and ecological communities Habitat Award" at the Third Global Forum on Human Settlements.

1

2

天鹅湾

发 展 商：城启集团

建筑设计：澳大利亚高臣建筑事务所
　　　　　广东新豪斯建筑设计有限公司

占地面积：53 000 m²

建筑面积：150 000 m²

容 积 率：2.83

绿 化 率：40.6%

停 车 位：500

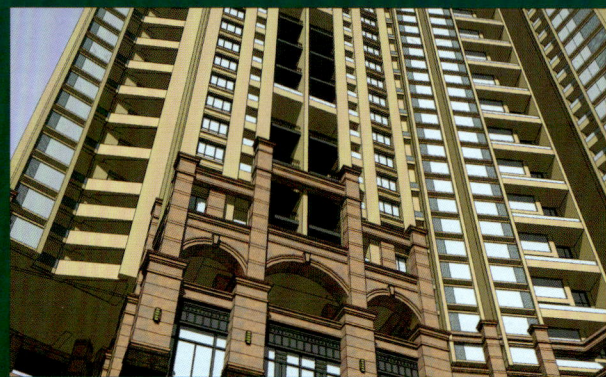

天鹅湾项目的立面造型采用了新古典主义的设计手法.在现代建筑中融入古典主义的元素并以古典建筑美学理念来重新组织建筑的各个元素,使建筑作品体现出古典建筑的高贵、典雅的不凡气质。这是一种对建筑文化传统的新探索,是对传统美学的回归。

天鹅湾项目临江而立,平面呈优美的S形,占据一线江景。整个建筑总高31层,亭亭玉立,宛若天鹅出水,俯瞰美丽的珠江,向西远眺著名的白鹅潭,与白天鹅饭店遥遥相望。整个楼盘定位为一线江景的高尚地产项目,所以采用有贵族气息的新古典主义的风格是一种合适的选择。整个S形平面事实上是由五个独立但相接的塔楼组成,在设计中将五个塔楼的突出部位作重点造型处理,弱化相接部分。设计时充分考虑了建筑已经施工完成的现状,在尽量不伤

害原有功能的前提下来进行造型处理。重点保留了沿江立面可以遥望珠江的卖点,造型时尽量避免影响原有的大窗和大阳台的开阔视野,但在细部的刻画中巧妙地融入了古典主义的厚重,端庄感。

在下部裙房和顶部做了重点的造型处理,特别是入口门厅及架空层部分采用了比较纯正的古典建筑细部,但又不是简单地模仿与抄袭,而是取其精华,进行提炼,融入现代的材料与色调,如金属、深灰色等,使整个建筑古而不旧,贵而不俗。

建筑的主体颜色为淡黄色,裙房为深棕色,少量深灰色糅合,营造出高雅、豪华又与众不同的品味,符合高尚住宅的要求。

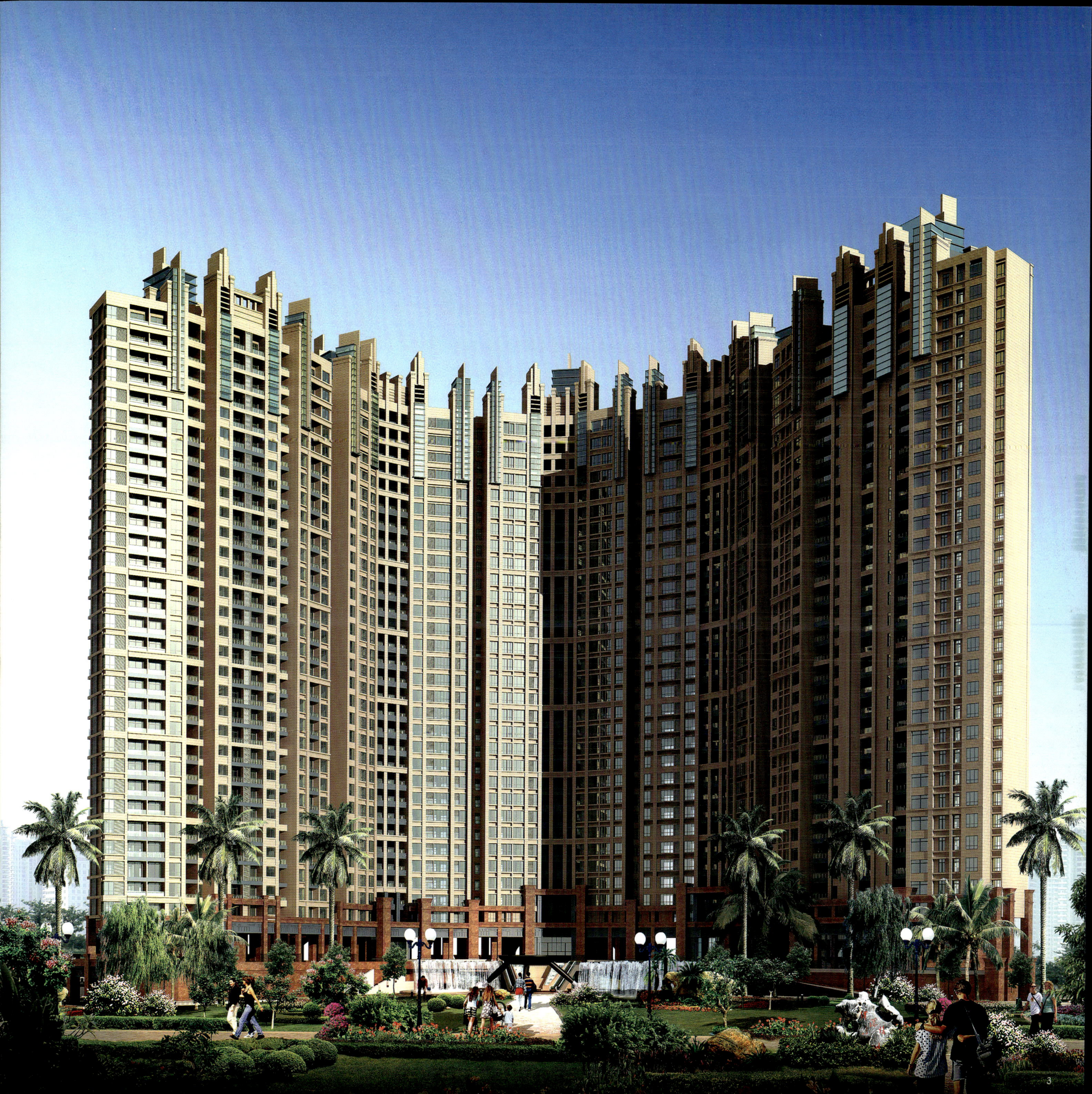

Developer: City Key Group

Architectural design: Gordon Chen Architect

Guangdong Xinhaosi architectural Design Co., Ltd.

Area of the land: 53 000 m^2

Area of construction: 150 000 m^2

Volume rate: 2.83

Greening rate: 40.6%

Parking spaces: 500

The elevation image of Swan Bay project adopted neoclassical design ploy. Blended with classical elements and classical architectural aesthetic ideas, the modern buildings reflect the classical nobility, the elegant and extraordinary temperament. This technique of expression is a new exploration of cultural tradition and the return of traditional aesthetic.

The Swan Bay project is located by the river, which presents the graceful "S"-shape and is a beautiful landscape. The whole building is 31 floors in total which is tall and erect as like a swan from water overlooks glamorous Pearl River and looks at the famous the Baie tan from west. The location of the entire building is considered as the noble real estate project of a whole line of high river view, so that we sure the noble and neoclassical style we have adopted is an appropriate choice. The whole "S"-shaped surface consists of five separate but conterminous towers, and we focused on dealing with the modeling of the obtrusive part of the five towers and weakened conterminous part while designing. We have taken into full account of the completed status quo of the construction and dealt with the modeling but not damage maintaining the original function. The selling point that we have retained especially is people can view the Pearl River in the distance along the river. We tried our best to avoid affecting the original large windows and balconies of the wide field of vision, and blended in classical dignity skillfully in detail.

In the lower part of the podium and the top, the designers have emphasized to deal with the modeling, in particularly the entrance of the hall and the part built on stilts which have adopted the pure classical architectural details, but they are not simply copying and imitation and they are refined to use the modern materials and colors, such as metals, dark grey, etc., which makes the building venerable but not old, noble but not common. The main color of the construction is light yellow, and podium is dark brown with little dark grey which creates an atmosphere of elegance and a taste of splendor and distinctiveness, with the magnificent dwellings demands.

1

国美明天第一城

发 展 商 ：北京明天房地产开发有限公司
规 划 设 计 ：华通设计顾问工程有限公司
总建筑面积： 600 000 m²

2

国美明天第一城项目位于北京亚北立水桥畔，占地含代征用地72万平方米。基于用地位置北临北五环毗临国家森林公园，用地内还有22万平米的代征绿化用地，设计提出了"BLOCK街区"规划理念+绿色生态栖居理念。BLOCK街区理念的注入将穿越的几条规划道路根据不同特性设计成充满活力的商业街区，变消极为积极因素。塑造了商业景观街，临河休闲景观大道，通过对各种体量和不同业态的商业配套的科学规划，将生活便利和舒适提高到新的水平。明天第一城具有居住区的静谧，却有别于一般的商业集散地。它以满足人们现代都市生活多样性的需求为前提，是居住、商业、商务、购物、休闲、娱乐交融的复合型社区形态，是更高层次的整合。

规划强调住宅园区和商业街的融洽共处，强调友好而保持私密的邻里关系，强调与城市肌理的和谐共生，强调公共区域和私

密区域的协调，强调社区的安全性。这样一个社区，不仅为房地产开发引进一种全新的开发思路，而且为当地消费者带来了一种时尚优雅的生活方式。此规划建设了一个集住宅生活与休闲配套设施于一体的新型都市居住宅，而不仅仅是建造一批住宅。以新都市主义的精神偏向于环境的设计及其社会后果的一面，促进地区的发展活力，使之成为理想居住区，提倡社区中的公共空间和由此导致的人际交往。引入复合性开发理念，将居民的SOHO办公居住、娱乐、购物、休闲有机结合，形成高质量的生活体系。注重消除城市与人、人与人、人与自然之间的隔阂，主要表现为功能复合，建筑复合，商业复合，阶层复合，文化复合。

规划中同时引入绿色生态设计，强调健康运动。借周边大面积的城市绿化带，在各个功能住宅区内部建立与之相联系的整体生态循环系统，将新鲜氧气与绿色源源不断输送到各区块内部，

营造舒适惬意的生态栖居环境。沿着水系、结合建筑的精心设计，形成以水系景观与陆系商业街区相结合为主线的休闲活动路线，将社区中部连续活力公共服务街区的活力引入，并进行功能互补。

由于在规划上科学地分离了商与住，街区的公建和商业非但没有扰乱街区的平静秩序，而且令街区和城市相互汲取了活力。街区的居民也因此享有了更为便利的城市生活，增加了人际交往，同时也令街道上进行的公共活动更为安全。

3

4. 总平面图

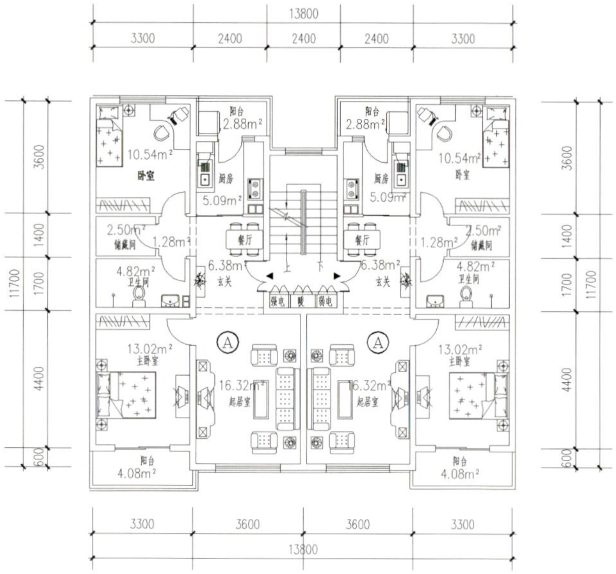

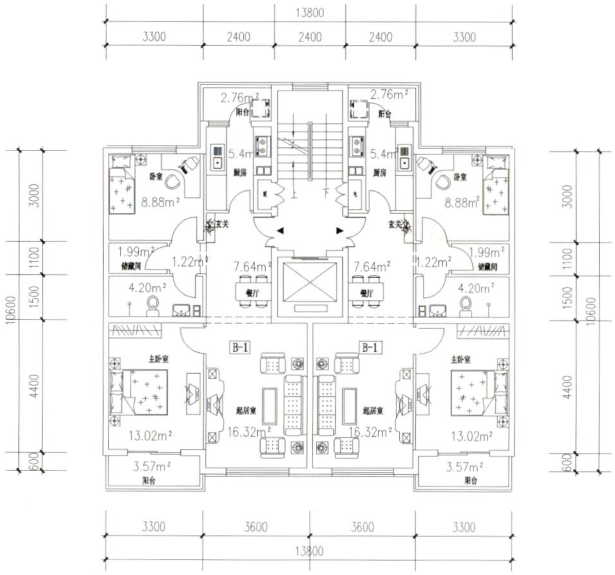

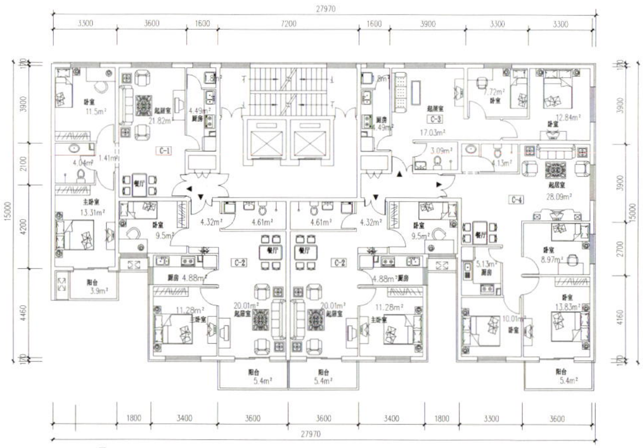

5. A户型标准层平面图

6. B户型标准层平面图

7. C户型标准层平面图

8. 7栋组合标准层平面图

9. 社区景观

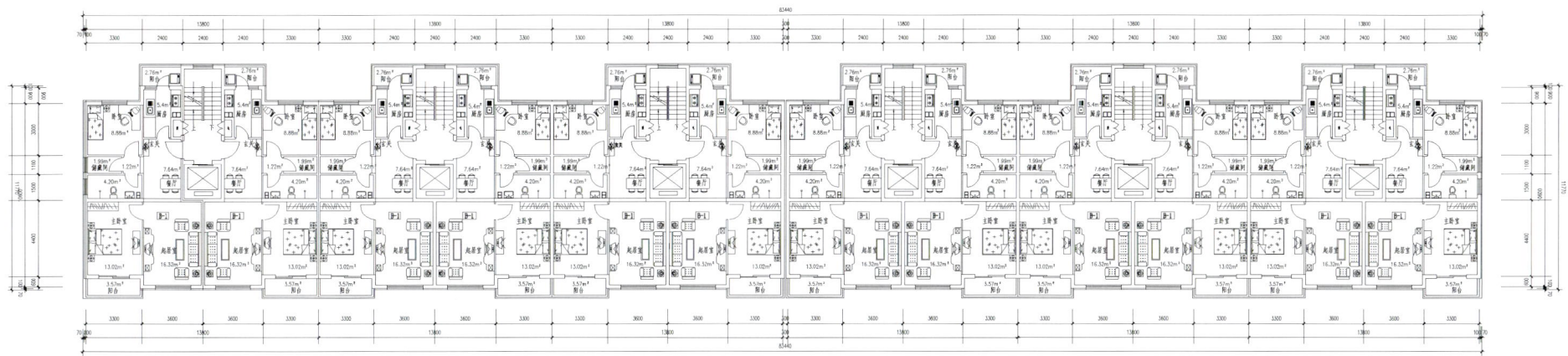

8

9

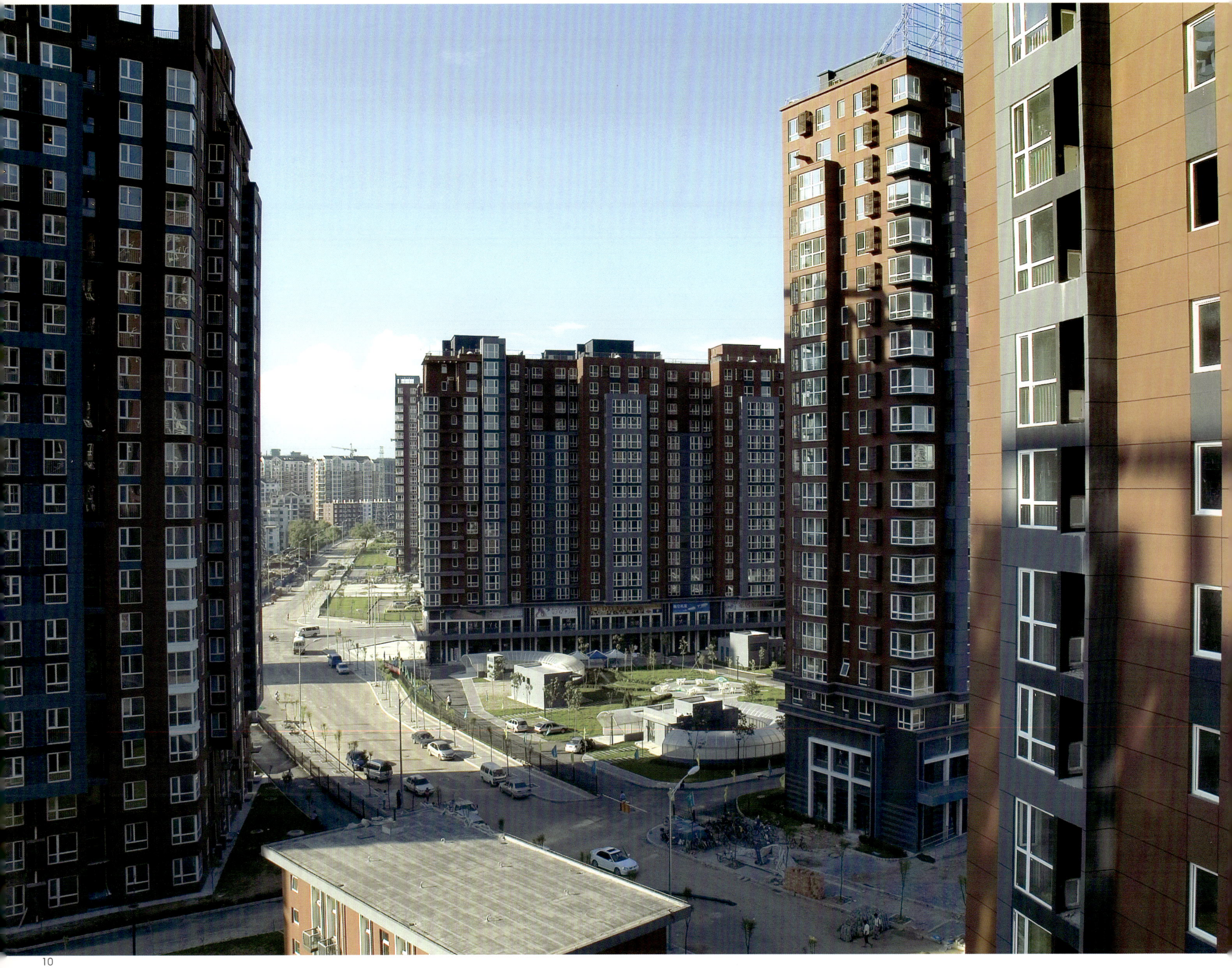

10

11

12

13

10. 建筑实景
11/12. 立面构成
13. 内庭实景

14

Developer: Beijing Tomorrow's property development&Co.,Ltd.
Planner: Huatong Engineering of Design adviser& Co.,Ltd.
Total Area of the construction: 600,00 m²

The No.1 Burg of Guomei Tomorrow is located next to the Beijingyabeili Bridge, and covers an area of 720,000sq.m. including the acting requisition of land. Based on the location of the town, facing to the North five ring and next to the National Forest Park, there are 220,000sq.m. of the acting requisition of green lands belonging to the area of the burg. And the design is made with the BLOCK designed belife and the Green Ecological Housing designed belife. The planning roads which are crossed with the use of the BLOCK, are design into a trading area full of energy, according to their different characters. It transforms negative into position. Building the Commercial Landscape Street and the Lounge Bordered on the River Landscape Road, planning various measures and different commerces scientifically, these make lives have a new advances in convenience and comfort. The No.1 burg is quiet enough just like the uptown , but different from the commercial distribution centre. It is preequisite to meet the demand of the varions modern city lives, and combines commerce, business affairs,shopping,leisure, entertainment with residence to build a multiple community for a higher integration.

15

The plan emphasizes the housing area getting on well with the commercial street,the friendly but keeping privy neighbourhood, the texture of the city life with harmonious coexistence, the public spare harmonizing with the privy space , and the security of the community. The community not only stimulate the property development bringing a new thread of development in , but also bing a fashion and elegant mode of life for the local costomer. The plan is designed to build a new uptwn with the life of housing and forming a complete set of leisurely equipment, but not just a group of house. One aspect of the spirit of the new city principle leans to the environmental design and the result of the the society to be an ideal place, and calls for the public space in the community and the communication resulting from it. With bringing the multiple development belife in, the system of high quality life comes into being. The system makes the SOHO office-housing, entertainment, shopping and leisure well in combination. It attaches importance to eliminate gulf between city and people, people and people, and human and nuture. And the multiple function, the multiple construction, the multiple class, and the multiple culture are the main forms.

Meanwhile, the plan has brought the green ecology in, emphasizing the healty sports.

16

17

There is a system of the whole ecological circulation being built ın which ıs connected with functional uptown by uslng a big area of city green belt in the perimeter. The system can transport the fresh oxygen and green steadily to inside of every area to build a comfortable, pleased life in the green eco;ogical environment. Along the Hydrographic net, there is a perfect design that conbines with the construction , forming into a leisure activity course which takes Landscape of the Hydrographic net and trading area of the land line together into the major line. In this foundation, the community lead in the energy from the Public Service Block which is active in the middle part of the community in order to complement the shortage. Thanks to the scientific separation between business and dwell, the construction of the block and business make the block and city bring the energy from each other, but not disturb the quiet order of the block. It is more convenient for the dweller to live a city life. And you can make friends with each other. Meanwhile, it is safer to hold the public activity in the street.

14. 裙楼
15. 外景
16. 内院
17. 建筑立面

1

2

水岸江南小户型高层住宅

发 展 商：天津松江集团
项目地址：天津市梅江南生态社区
建筑设计：ZPLUS 普瑞思建筑规划设计咨询有限公司
总用地面积：95 400 m²
总建筑面积：103 000 m²
容 积 率：1.09
绿 化 率：50.76%

　　水岸江南位于天津梅江南居住区最南端的11号地，南侧为外环南路，此路与友谊南路及解放南路相连，外部交通便利，周围无遮挡，宜形成南低北高的总体布置，最大限度利用南向。该项目共设置7座高层建筑，中间3栋32层百米高层，顺势排开，与相邻基地上的天鹅湖项目相协调，构成了展翅欲飞的天鹅的另一单翅。在这一排百米高层的南北两侧各有两栋18层54米高层，楼距舒展，交错排列，在采光、观景方面都很优越，达到了占地不多但环境却很优美的效果。

　　该项目以"全功能小户型"为亮点，使中小套型具有高舒适

度的"精密设计"，引领着最新的住宅设计趋势。中小套型的设计难度，应该说要比大套型难得多，首先就是要控制"公摊"，研究最佳梯户比，优化"核心筒"和室外交通走廊的布局，将公摊面积减至最小，因为中小套型住宅内的1平方米套内面积，绝对不是普通意义上的1平方米，它对厅、室的舒适性、尺寸的优化、流线的改善都有作用，需要精打细算，挤出无效空间，尽量提高套内有效使用面积。

　　在设计中，户户南北通透是布局的第一要素。另外，将走道并入客厅和餐厅，客厅与餐厅连通，既节省空间，又使人感觉到开

阔的视野；户户送露台、衣橱，并设飘窗、凸窗，能有效延伸室内有限的空间，而精致、新颖的外观效果，由交错露台导致的特色立面肌理、图案为城市和社区的景观都有所贡献，并使居住者拥有一时尚、清新的优雅环境，在实现功能满足的同时，带来更高的精神层面的享受，体现理性和浪漫的结合。设计外檐时对于中国传统的格窗和现代构成图案都进行了深入的研究，将两者有机结合，体现了案名"水岸江南"的东方传统韵味和现代风尚，对于集合住宅的设计作出了有益的尝试和创新。该项目荣获了"2007年度最具期待的明星楼盘"奖。

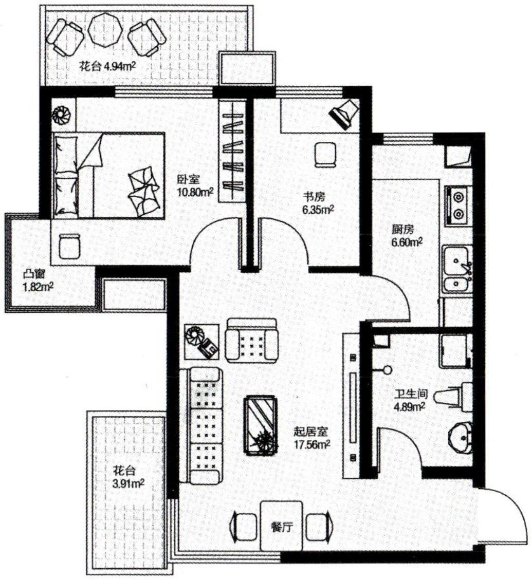

花台 4.94m²
卧室 10.80m²
书房 6.35m²
厨房 6.60m²
凸窗 1.82m²
卫生间 4.89m²
起居室 17.56m²
花台 3.91m²
餐厅
3

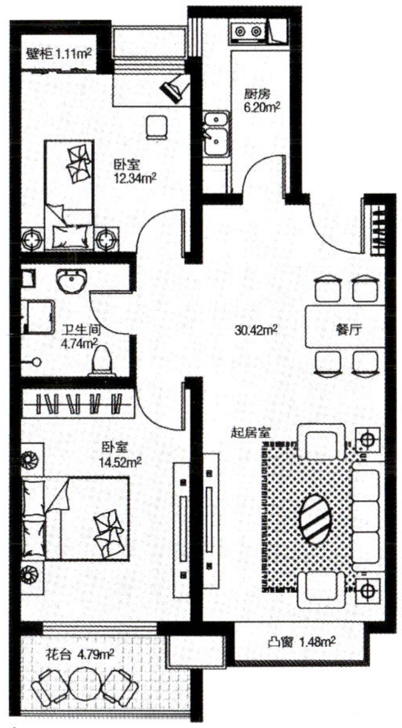

壁柜 1.11m²
厨房 6.20m²
卧室 12.34m²
卫生间 4.74m²
餐厅
30.42m²
卧室 14.52m²
起居室
花台 4.79m²
凸窗 1.48m²
4

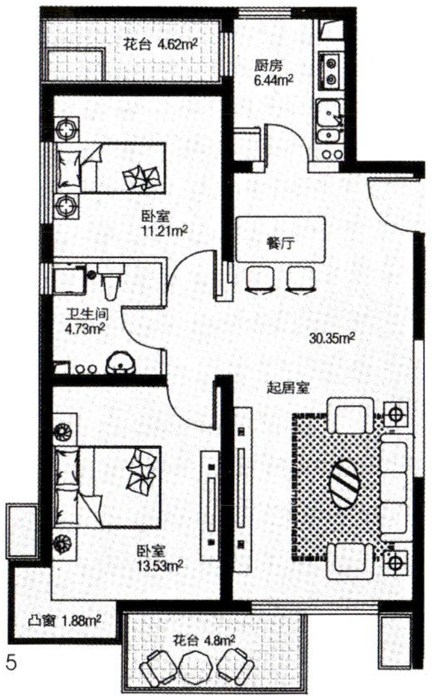

花台 4.62m²
厨房 6.44m²
卧室 11.21m²
卫生间 4.73m²
餐厅
30.35m²
起居室
卧室 13.53m²
凸窗 1.88m²
花台 4.8m²
5

6

3F
3F
3F
3F
3F
3F
3F
3F
3F
3F
3F
3F
3F
3F
3F
3F
9
3F
3F

自行车库出入口
18F
7号楼
变电室 1F
水
自行车库出入口
18F
6号楼
地下车库出入口
地下室边界线
地下车库出
自行车库出入口
32F
8号楼
32F
2号楼
32F
1号楼
变电室 1F
配套公建 2F
自行车库
18F
5号楼
18F
4号楼
地下车库出口
3F
3F
2F
2F
3F

环 岛 西 路
用地红线
25m

84

Developer: Ciity,Tianjin

Project Address: Jiangnan Ecological Community, Tianjin

Aechitectural Design: ZPLUS

Total Land Area: 95 400 m^2

Total Construction Area: 103 000 m^2

Volume Rate: 1.09

Green: 50.76%

Jiangnan waterfront is located in the southern Meijiangnan, Tianjin, district of residence on the 11th. To the south, there is Outer Huannan Road which connect with South Friendship Road and South Liberation Road. It is convenient for the external transport. There is no shelter around, so it is good for forming the North-South low overall layout with the maximum utilization of the south. The project is set up with a total of seven high-rise building. Three 32-storeys about one hundred meters of high- rise resident apartments in the middle, take advantage of physiognomy and arrange in order. The three buildings harmonize with the Lake of Swan of the adjacent base, constituting another sigle wing of the swan which is spreading its wings and wants to fly. There are two 18-storeys buildings about fifty-four meters tall in th two sides, row of the hundred meters building. The distance of the North and South buildings is stretched and arranges with interlacement. It is preeminent in the light and viewing. Tough the area is not large, it has an elegant effect on environment.

The bright spot of the project is "full-featured small apartment". These make the small and medium-sized have a high degree of comfort of "sophisticated design", leading the latest trends in residential design. It should be said that it is more diffcult to design a small and medium-sized than a big one. At first, it is demanded to control the "pool", and then investigates the best rate of the ladder residence. The "core tube" and the layout of outdoor corridors should be optimized to make sure that the area of "pool" is reduced to a minimum. The reason the all these things should be done is that one inner square meter area of the small and medium-sized apartment is not the same with one square meter in the common sense. It has an effect on advancing the comfort of the hall and room, optimizing the size and improving the stream line. So it should be designed exactly to squeeze the useless space out, for maximizing the inner efficient use of space.

In the design, the first element of the layout is the link of the north ands south in every dwelling. Besides, the walking will be incorporated into the living-room and dining-room. The living-room and dining-room are connected up so that not only does it save the space, but also people can feel the broad field of vision. The balcony, clothing cabinet is free for the dwell. And there are drifting windows and convex windows being set up, which can extend the limited indoor space effectively. Moreover, the new refined look have a contribution to the landscape of cities and communities, because of the characteristic elevation of texture and the patterns, led by the

7

8

9

10

wrong fit for balcony. So the residents can have a stylish, clean and elegant environment from it. With meeting the need of functions, in the same time, a higher level of spiritual enjoyment will for you, raflecting the combination of senses and romantic. It have been taken an in-depth study about the patterns combined by the traditional window of grid in China and the modern one while designing the outdoor cornice. And these have beneficial to attempt and innoration for the design of resident collection. The project won the award of "the Most Expecting Star Building 2007".

7. 透视图
8. 功能分析图
9. 景观分析图
10. 绿化分析图
11. 3号楼首层平面图
12. 4号楼首层平面图

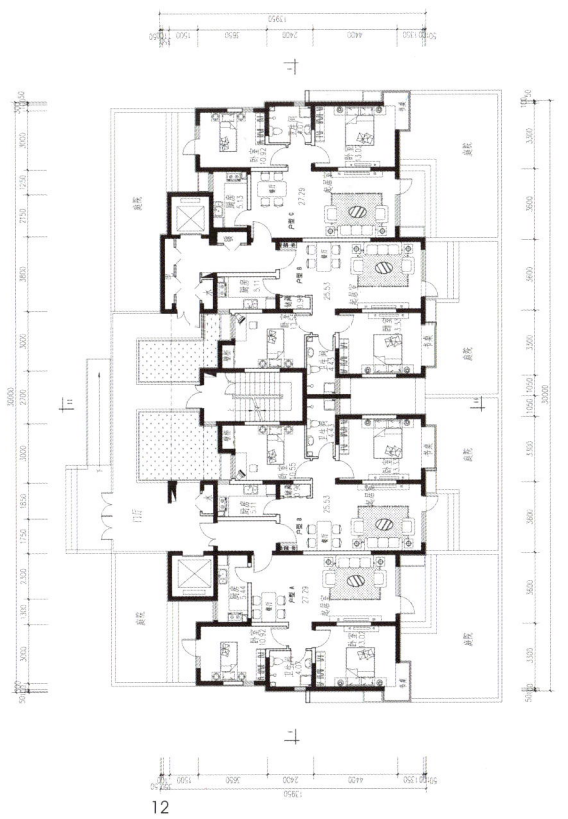

11

12

1

郑州圣菲城二、三期

发 展 商：河南太极置业有限公司

项 目 地 址：郑州金水区鑫苑西路6号

规 划 设 计：加拿大汉克（H.N.K）国际建筑设计
广州东略汉克建筑设计顾问有限公司

建 筑 设 计：河南徐辉建筑工程设计事务所

总占地面积：88 361 m²

总建筑面积：309 631 m²

容 积 率：3.5

绿 化 率：36%

2

郑州圣菲城二、三期项目地块位于郑州市文化路以东，北环路以南，西望东风渠，拥有浓郁的自然水景，东接金融管理学院，占地约11.5万平方米。周边学校、学院林立，生活文化气氛浓厚，是十分理想的人居环境。但地块较为狭长，南北长约800米，东西最窄不足100米，给地块规划带来了较大的难度。

小区规划注重周围环境的有机结构，充分利用周围环境的有利因素。以景观轴线为纽带，结合基地的特点，争取住宅的南向，在二期和三期规划上形成通透的景观长廊，通过徐寨路上二期和三期的入口广场将两期巧妙结合，形成宽阔的休闲商业空间，使得两期各自规划相互独立而在空间上又紧密相连。

由于基地形状的不规则，根据景观及商业价值的不同，以空间形态和景观利用最大化为原则，追求空间布局的变化与有序。

打破空间的均质化。二期点式、板式结合，通透轻灵；三期的围合布局，开敞大气的规划思路都充分展现了本案追求空间通畅、互动的人居空间，打破空间的匀质化，营造更贴近自然的宜人景色，再次体现"浪漫、经典、时尚、个性"的居住风格。

二期：视野最大化的板式结构加上畅通无阻的点式结构，在尽享小区内外水韵、绿意的同时使空间保持跳跃的节奏感。

三期：居住区围合空间可以满足居民的安全，交往，追求良好居住生活质量的需求，形成中心花园，兼顾了内庭院景观和远眺的需要，给人一种开敞、安全、整齐的视觉美感。

小区环境设计中每个细节上都充分体现出对人的关爱。从色彩、材质、结构、尺度以及体量大小，到点点滴滴的细节，通过对物的设计和价值提升达到对居住者在精神上、情趣上尊重，期待能营造出一种盛放每个人自我心灵的空间，使居家的日子与个人因情趣与品位而变得不同。主线贯通的空间骨架。由二期的步行主入口的中心广场起，一道通透的空间景观长廊从二期向南穿过二、三期的入口广场直达三期围合的中心庭院景观，开畅又紧密地将各组团景观串连，与小区外东风渠及两岸的70米宽市改绿化带形成两道内外平行的"水绿"交融的景观带，将本案"浪漫、经典、时尚、个性"的居住风格推向高潮，营造一个充满"现代、优雅、舒适、浪漫"又富于创新的新型社区。

建筑立面创造简洁、现代、时尚、高档的滨水住宅形象。强调体型上的对比及元素的组合手法，尽量减少纯装饰性的构件出现。充分展现平面设计中的特点，用色上以高雅、时尚的米白色搭配暗紫罗兰为主。

经济技术指标

	一期	二期	总用地
用地面积	37008.5 M²	52352.0 M²	89360.5 M²
总建筑面积	118071 M²	201898 M²	319969 M²
地上建筑面积	100071 M²	165938 M²	266009 M²
住宅建筑面积	88569	155485	244054
商业建筑面积	3690 M²	6853 M²	10543 M²
办公建筑面积	4810 M²		4810 M²
会所建筑面积	3539 M²		3539 M²
其中 地上面积	1344 M²		
地下面积	2195 M²		
幼儿园建筑面积		1600 M²	1600 M²
架空层	1658 M²	2000 M²	3658 M²
地下建筑面积	18000 M²	35960 M²	53960 M²
其中人防及车库面积	15154 M²	31600 M²	46754 M²
建筑占地建筑面积	9650 M²	11622 M²	21272 M²
容积率	2.73	3.17	2.97
建筑密度	26.07%	22.20%	23.60%
绿化率	33.26%	34.20%	33.79%
总户数	601 户	1296 户	1897 户
停车位	451	830	1251
居住人数	2103	4147	6250
人均公共绿地	2.90	2.30	2.50

滨水带 圣菲城电脑科技综合园区二期总平面图

1. 二期13号楼实景
2. 二期大堂实景
3. 总平面图

3

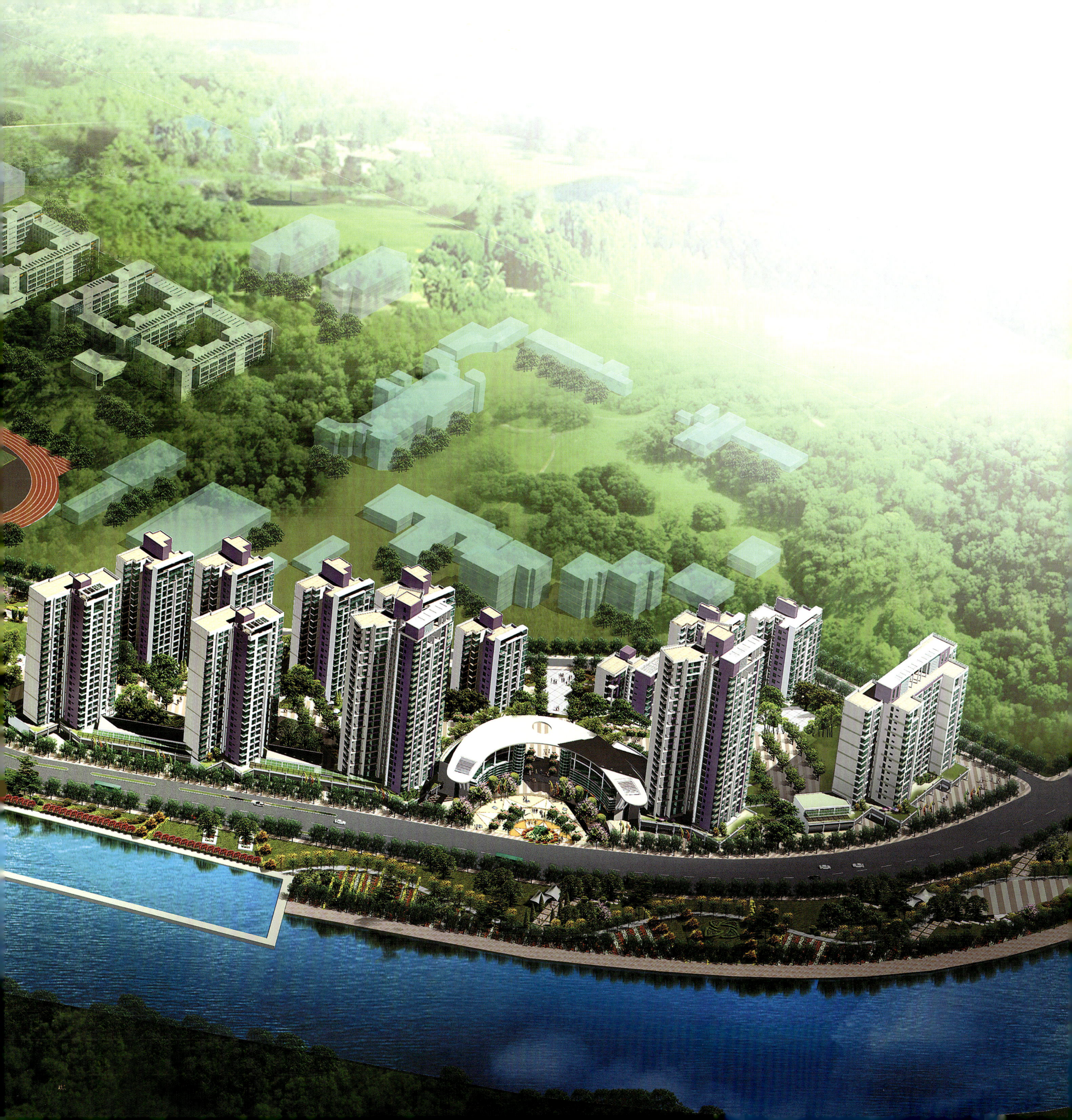

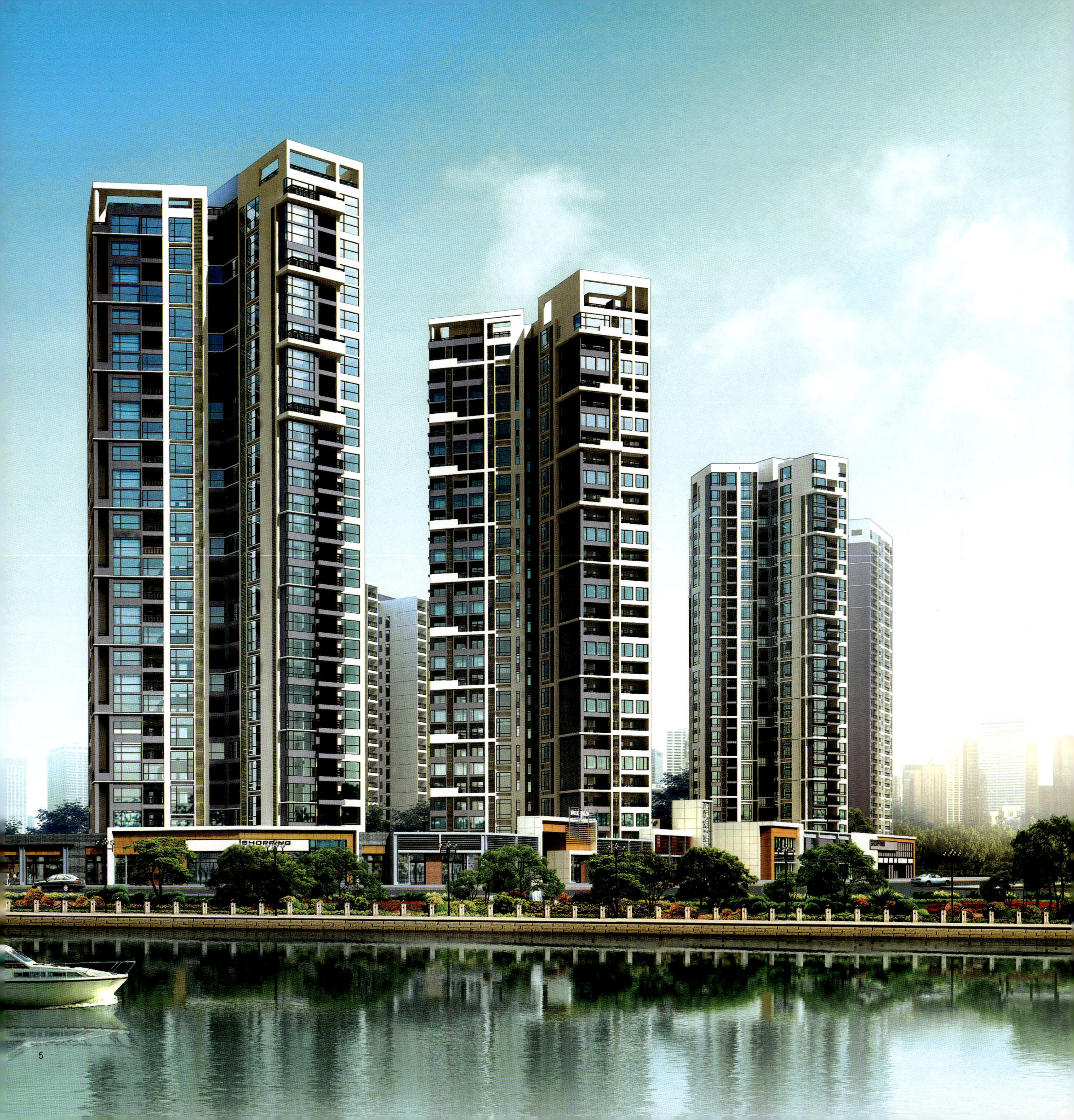
SHOPPING

Floor area: 88 361 m^2
Building area: 309 631 m^2
Greening rate: 36%
Volume rate: 3.5

Plots of 2nd and 3rd phases of Zhengzhou Shengfei Town Project is located to the East of Wenhua Road in Zhengzhou and South of Beihuan Road. It is on the East of Financial College of Management, and faces Dongfeng Qu in the West, possessing rich natural water features. All this takes up an area of 11,500 m^2. Strong atmosphere of cultural life with schools bounding aroud, makes it an ideal environment enough to live in. Nevertheless, this long and norrow land with its length of 800m in the Northsouth direction and narrowest less than 100m in the Westeast has made it quite difficult to carry out the plan for the land.

Plan of the residential block concentrates on the organic structure in its surroundings, advantage of which is made full use of. And we make the axis of the view as a link and combine the features of the basement to approach our goal of making houses south-towards. A scenic corridor included in the plan for the 2nd and 3rd therefore comes into shape, and space for leisure and commerce is allowed due to the plaza in the Xuzhai Road designed ingeniously as both phases' entry. The both therefore are planed respectively but close to each other in terms of space.

5．三期沿东风渠透视图
6．三期22号楼透视图
7．三期24号楼透视图
8．三期17号楼及商铺透视图
9．三期19号楼及商铺透视图

6

7

8

9

10

11

12

13

14

Because of the irregulance of base's shape, we chased the unique change and order of the overall space arrangement according to the variation of the view and its commercial value. The 2nd phase is spacious combination of the point-pattern and the plate pattern. And idea of enclosure in the 3rd phase fully embodies its fluence in space and interactive living environment, creating more natural charming view. That again coincides with the living style of "Romantic, Classical, Fashional and Individual".

2nd phase: Plate-pattern structure with maximum view plus smooth point-pattern structure keeps space its active rhythmn while you are enjoying the water feature and green view inside and outside.

3rd phase: Enclosure provides residents with safty, social communication, and demand for better living quality. The central garden gives consideration on to both feature inside the courtyards and distant view, it also presents sense of broadness, safly and tidiness.

Every details in the block's design shows our careness. We are expecting to create an environment for every resident to release his soul, making it special for each person, by ranging from colors, materials ,structure, scales and the cubic content to promote the design and its value . Space framework is linked up by the axis. Started from the main walking entry of the central plaza, a ventilated scenic corridor passes the mutual plaza toward south, right up to the central scenic yard of the 3rd phase. That closely ties up all parts, becoming a parallel green line along with Dongfeng Qu and the 70-metre green belt on its bank. Style of "Romantic, Classical, Fashional and Individual" is therefore embodied.

The building turn on an image of succinct modern, fasihonal and high-level apartment. Emphasis on the shape's camparison and element is used to reduce component with full decoration. And using the features in the graphic design, an elegant and fashional tone of beige accompanied by pansy is very wonderful.

15

10. 二期10号楼实景图
11. 二期12号楼西立面实景图
12. 二期实景照片
13. 二期沿东风渠透视图
14. 三错层客厅
15. 儿童游戏室

综合

Comprehensive

综合

1

占 地 面 积：340 000 m²
总建筑面积：750 000 m²
容 积 率：2.3
绿 化 率：40%
总 户 数：396
停 车 位：200

1. 小高层板楼效果图
2. 多层洋房效果图
3. 鸟瞰图

2

重庆财富中心

发 展 商 ：重庆香江高科地产发展有限公司
项 目 地 址 ：重庆渝北区北部新区新牌坊新光大道16号
建 筑 设 计 ：德国AXIS设计集团
中国建筑西南设计研究院
重庆市设计院
中国机械工业第三设计研究院
规 划 设 计 ：英国BENOY
澳洲WOODHEAD
香港Husband Retail Consulting商业顾问公司
景 观 设 计 ：日清国际（澳洲）有限公司

重庆财富中心项目地处重庆北部新区核心地段。总规模80万平方米，是集地标性国际甲级写字楼、5A生态写字楼、星级酒店群、国际购物中心、科技创业园、国际公寓等于一体的超大规模建筑综合体。荣膺"2007年重庆最具投资价值的地产"及"2007年购房最满意楼盘"。

重庆财富中心由香江国际与国际顶级专业团队德国AXIS GROUP、香港HRC、英国BENOY、澳大利亚WOODHEAD联袂打造，在建筑综合体之内全面涵盖现代商务办公、时尚购物、餐饮娱乐、居家休闲等多种功能，营造24小时魅力都市精英生活，将城市昼夜运转的轨道完美衔接。

财富中心扼守北部新区黄金门户，占据北部最成熟的区域核心，周边分布财政局、地税局、发改委等100多家市级机关单位，200万平方米的高新产业楼宇，近1000万平方米重庆高尚居住区，5家涉外酒店以及城市中心公园等。得天独厚的地段和外围环境烘托，使财富中心成为北部新区当之无愧的商务核心。

由五星级酒店、三星级酒店、国际甲级写字楼、5A生态写字楼、大型购物中心、服务式公寓、多功能餐饮休闲娱乐中心以及科技创业园筑成的商务组团，为商务精英提供办公、居住、购物、休闲、娱乐、社交等各种需求，轻松分享一流商务配套服务，将生活角色与场所转换的时间成本缩短，让同时拥有高效工作和高品质的生活成为可能，高效商务圈为您完美呈现。

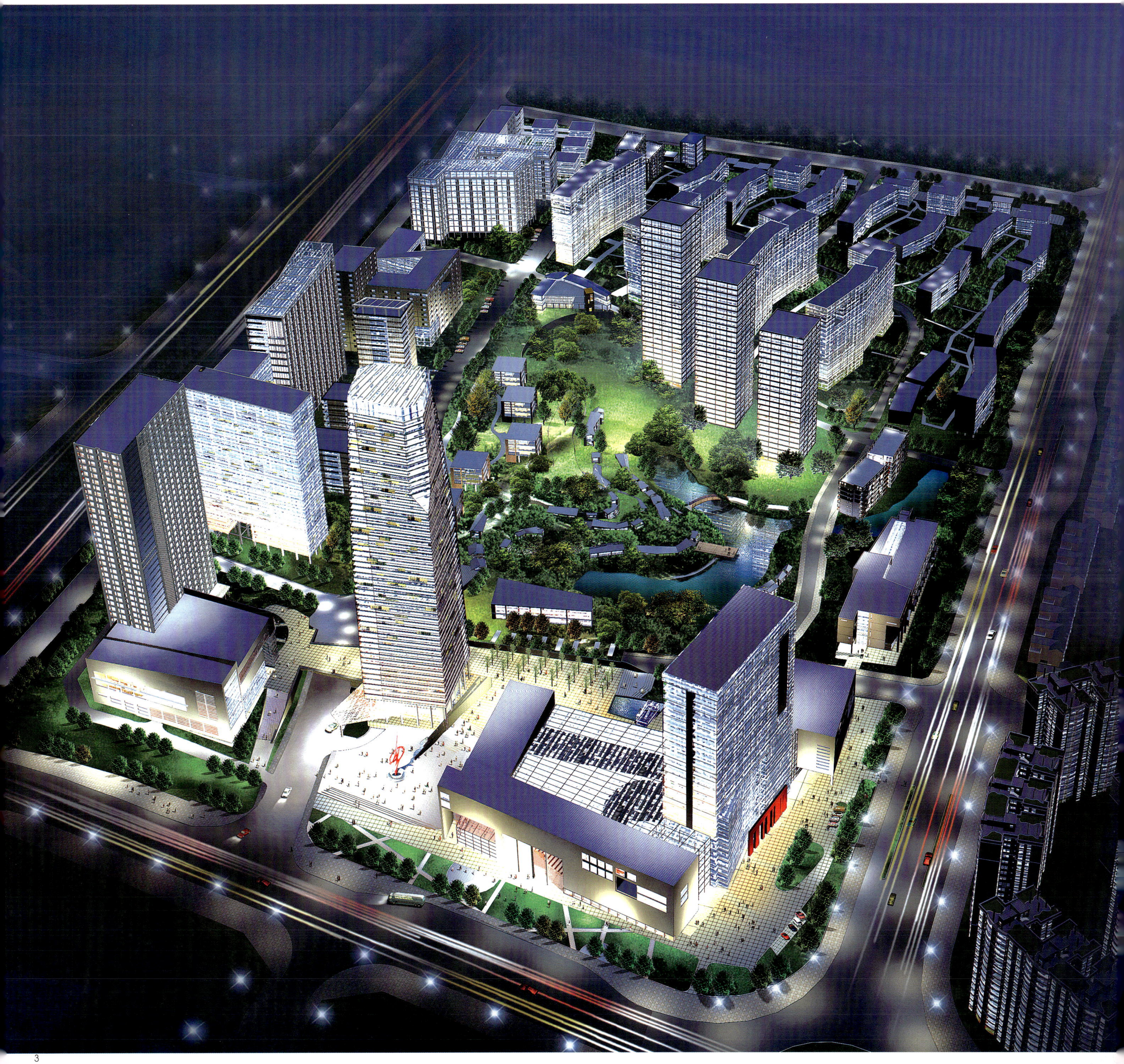
3

N

① 茶座外休闲廊
② 土丘
③ 小水景
④ 林下休闲空间
⑤ 聚乙烯植草格
⑥ 整形灌木
⑦ 垂吊植物
⑧ 草坪
⑨ 树阵
⑩ 景观水池
⑪ 消防道
⑫ 草坪
⑬ 灌木丛
⑭ 土丘
⑮ 可上人草坪

4. 塔楼总体平面图
5. 超五星级酒店效果图
6. 酒店群效果图

了一周售罄的销售神话，开创了写字楼销售新纪元。目前方正集团、中铁集团等国内知名企业已强势入驻，全面升级北部新区的商务氛围，奠定重庆财富中心写字楼里程碑。

国际星级酒店群

强大的群聚效应，打造尊贵商务之旅，近8万平方米国际酒店群由五星级酒店、三星级酒店组成，对于商务社交空间尚属匮乏的北部新区而言，强大的群聚效应将带动整个区域的商务平台，周边日益增加的行政办公机构与企业提供大量商旅消费群体，国际星级酒店群的一站式全天候商务配套成就尊崇商务之旅。为了满足重庆特别是北部新区迅速增长的商务活动需求，五星级酒店将建成国际五星级标准，配置有客房、餐饮、商务中心、会议中心等相关功能，酒店共有约298套标准间。总建筑面积达2万平方米的三星级酒店按照国际品牌酒店标准设计，为旅客提供高效、快捷和舒适的服务，满足不同商务居住条件。三星级酒店共有标准间客房289套。目前工程结构已封顶，预计于2010年初开业。

科技创业园

共分为A、B、C区，建设用地面积5.5万平方米，总建筑面积14万平方米，总投资高达2.2亿元。科技创业园是整个财富中心项目最重要部分之一，致力于吸引海内外高科技专业人才，提供高新技术创业的孵化平台。科技创业园主要是以高新技术公司服务为对象的办公楼及与之配套各项服务设施。

财富汇

重庆财富中心商务组团的核心物业之一，紧邻五星级、三星级国际酒店群和北部新区地标性写字楼。总建筑面积约为23500平方米，仅有336套单位，户型面积为40-80平方米商务公寓，自由空间随意组合，具有办公、居家、休闲、会议等多种功能。

财富国际大厦

总规模约6万平方米国际甲级写字楼以建筑高度与高品质形象，成就北部新区的商务新地标。同时，在硬件设施、车位配比等方面也将扮演顶级商务配套的角色。目前，正处于设计阶段，已于2008年7月中旬开工。

财富购物中心

打造精英圈层的消费场所，14万平方米体验式消费中心，集购物、餐饮、休闲、娱乐、旅游等多种现代都市商业功能于一体，打造一个辐射整个北部新区甚至重庆的大型购物中心，提升北部新区商务配套环境。财富购物中心于2007年8月开工。目前，主体已完工，现正进行内部装修阶段。商业招商已全面展开，已与众多国际知名品牌商家取得意向。

财富大厦B座

3万平方米5A级生态写字楼于2007年1月开工，2007年11月底正式销售，创造

5

从建筑的布局和设计上，充分营造舒适的绿色办公环境、低楼层的开敞式办公设计，人于自然的和谐共处规划，加上与财富中心整体建筑风格相协调，创造出以人为本、尊重环境、舒适优美的办公空间。

科技创业园于2006年3月开工，已于2007年4月及11月分期两次如期交付使用，对日软件外包基地、西部航空等知名企业已签约入驻。

国际居住区

集电梯洋房、板楼、高层塔楼一体的28万平方米住宅组团，遵从人本主义原则，关注品质细节。理性主义下的灰白两色建筑、尺度合理的空间布置，为追求品质生活与精神回归的新贵阶层度身订做。

财富中心国际公寓于2004年5月开工，共发了1784套国际公寓及洋房，并于2007年6月全面交付使用。如今，高层塔楼、板楼、多层花园洋房矗立在项目东侧，住区规划合理、园林布局和谐、建筑错落有致。

花园洋房——重庆财富中心·邻馆

总户数:396户。其中，D区共10栋4-5层的低层住宅，含顶部跃层花园户型;E区为9栋6-9层的多层电梯住宅。所有住宅均为一梯两户的条式设计，利用折线和住宅间的方位变化，形成大小的宅间院落。

户型特色：呈现城市花园特点（两层通高的花园平台式露台、两层通高的观景客厅）。

社区配套：会所、中央湿地公园、社区小商业、图书馆等。

写字楼

国际甲级写字楼、总部基地、博士创业园彰显企业品牌实力，凌驾于西部至高商务平台，抢占涉外先机。

酒店

由世界著名设计公司澳大利亚WOODHEAD担纲设计，特邀全球知名酒店管理机构入驻。以数十年的酒店管理经验，面向全球高端客户提供优越、专业、细致的服务，业务遍布世界各地，享有卓越声誉。

国际公寓\高层塔楼观景住宅

俯瞰财富中心全景的三栋点式高层塔楼，环绕中央湿地公园，沿坡地起伏轮廓，由北至南错落矗立。

利用地势的高低落差，形成强烈的视觉节奏感，成为中央景观商务区的视线焦点；现代简洁的造型语言和色彩语言，体现活力时尚的涉外居住区风尚。

塔楼产品全面秉承承袭与创新的设计思想，在集成板楼户型优点的基础上，进一步优化细节，根据客户需求灵活调整，使新户型的居住功能更加完善，布局紧凑明快，保证每户厨卫采光通透，主居室均朝向最佳景观面。

园林景观

大型的绿色草坪、景观大道、喷泉水景、湖滨公园、大型室外游泳池、郁郁葱葱的高大植被和花园等。

7/8．写字楼效果图
9．商业中心效果图
10．会所效果图

7

8

Project address: 16 Xingguang Boulevard, New Memorial archway, New North Zone, Yubei District, Chongqing.

Developer: Chongqing HKI Gaoke Estate Development Co., Ltd.

Architectural design: German AXIS Design Group

　　　　　　China Southwest Architectural Design & Research Institute

　　　　　　Chongqing Architectural Design Institute

　　　　　　China CTDI Engineering Corporation

Conceptual design: Britain BENOY

　　　　　　Australia WOODHEAD

　　　　　　Hong Kong Husband Retail Consulting

Landscape design: La Cime INTERNATIONALE PTE.LTD.

Floor space: 340,000 m^2

Gross floor area: 750,000 m^2

Plot Ratio: 2.3

Total units: 396

Parking space: 200

9

Project Chongqing Fortune Plaza is located in the core section of New North Zone with a total scale of 800,000 square meters. It is a super large scale comprehensive construction combining landmark international Grade-A office buildings, 5 A ecological office buildings, star grade hotel group, international shopping center, high-tech business incubator and international apartments as a whole. And it has been honored as "Chongqing's most investment-worthy real estate 2007" and "The most satisfactory real estate for purchasing in 2007".

Chongqing Fortune Plaza is by HKI Properties Ltd together with top professional team German AXIS Group, Hong Kong HRC, British BENOY and Australia WOODHEAD with multiple functions inside the comprehensive architecture, such as modern business, fashionable shopping, catering and entertainment, and homebodies, etc. 24-hour glamorous city elite life is perfectly brought in line with the city's whole day operation.

Guarding the golden gateway in the new town area, Fortune Plaza takes up the most mature core section with more than 100 municipal authorities such as Bureau of Finance, Bureau of Local Taxation, Development and Reform Commission and so on, and with building of high and new industry covering an area of 2,000,000 square meters, Chongqing's noble residential quarter covering near 10,000,000 square meters, 5 foreign-related hotels and city central parks. Fortune Plaza fully deserves a business core of the New North Zone with such advantageous location and wonderful external surroundings.

Business group consisting of five-star hotel, three-star hotel, international Grade-A office buildings, 5 A ecological office buildings, large shopping center, service apartments, multi-functional catering and entertainment center as well as high-tech business incubator can meet the all kinds of demand of business elites for office, living, shopping, leisure, entertainment and social activities, share first class business support service with them so that they can shorten the time switched between their roles in life and in their workplace while possessing the possibility to work effectively and lead a high quality life.

Fortune Way

One of the main properties of Chongqing Fortune Plaza's business group, right next to the five- and three-star hotel groups and landmark office buildings in the New North Zone. It totally covers an area of 23,500 square meters with only 336 units including business apartments of 40 to 80 square meters, inside which free space is allowed to composite at random for various purposes such as business, living, leisure and meeting.

Fortune International Building

It is a new business landmark of the New North Zone with international Grade-A office building of a total scale of about 60,000 square

10

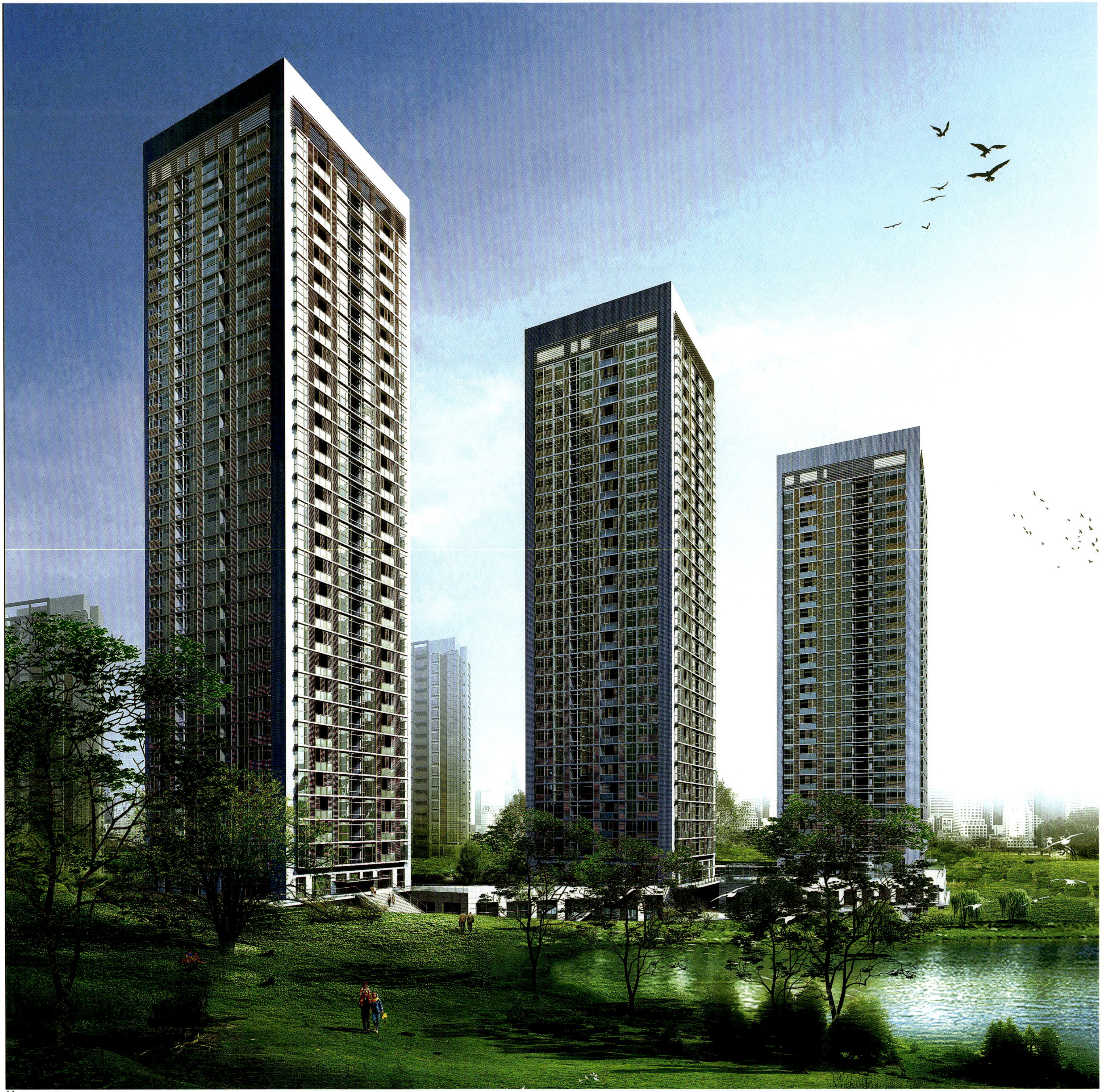

11

meters which impresses you with its construction height and high quality image. Meanwhile, it will support the business service in facilities and space matching. It is under design at present and has been started in mid-July in 2008.

Fortune shopping center

It is aim to serve consumers in the elite circle with an experiential center of 140,000 square meter for consuming lots of kinds of modern city business functions combining shopping, dining, leisure, entertainment and travelling as a whole therefore to make a large radiant shopping center in the whole New North Zone even in Chongqing and to improve New North Zone's business environment. The shopping center has been started in August, 2007. So far its main body has been finished and inside is being completed. It is also inviting business investment in all-round way and has gained support from lots of international well-known business partners.

Fortune Building Block B

5A ecological office buildings of 30,000 square meters have been started in January, 2007 and have officially been on sell in late November, 2007, creating a week's good selling result and opening a new era for selling of office buildings. So far domestically well-known enterprises such as Founder Group and China Railway Construction Group have entered, completely promoting the business atmosphere of the New North Zone and settling milestone of the office buildings for Chongqing Fortune Plaza.

International star grade hotel groups

Strong bunching effect brings you an honorable business trip with international hotel groups of 80,000 square meters made up of five- and three-star hotels. To the New North Zone which lacks business and social space, such strong bunching effect will motivate the business platform of the whole area and more and more administrative authorities and enterprises will off large number of consumers both for business and travelling. Then to meet the demand for increasing business activities of Chongqing especially the New North Zone, the five-star hotel will be equipped function of guest rooms, catering, business centers and meeting centers in line with international five-star standard. There are about 298 standard suites in the hotel. And according to the standard design of international brad hotels, the three-star of 20,000 square meters provides for tourists with effective, quick and comfortable service, meets different business living conditions. There are 289 standard guest rooms in this hotel. The structure or the project has been top ping-out at present, and is to be in practice in 2010.

High–tech business incubator

It is divided into block A, B, and C with land-using area of 55,000 square meters, total floor area of 140,000 square meters and total investment costs up to 220,000,000 Yuan. High-tech business incubator is one the most important parts of the whole Fortune Plaza, devoting to attracting high-tech

12

11．高层公寓效果图
12．多层公寓实景图

13

and put into use in June, 2007. Nowadays, apartments of tower building, slab-type apartment buildings and multi-layer western houses stand in the east of the project, in reasonable layout, in harmony with the gardens layout and in picturesque disorder.

Garden houses – Chongqing Fortune Plaza· Ling Hall

Total units: 396. There are ten 4- to 5-storyed low-rise residential buildings in Block D including top duplex apartment housing type; nine 6- to 9-storyed multi-stories elevator dwelling buildings in Block E. Design of two residents share one floor is taken, then polyline and direction variation between residences is used to form courtyards in different sizes.

Housing type feature: turn on like a city garden with platform patios of 2-story-high and feature living room of 2-story-high.

Community support services such as chambers, central wet land park, community business and library and so on.

personnel from home and abroad and providing incubation platform for Hi-tech Entrepreneurship.

In terms of the layout and design of the architecture, we fully create comfortable and green office environment, open design of low-rise buildings. Harmony between man and nature, balanced with Fortune plaza's general architectural style demonstrate gorgeous humanized office space in respect of the environment.

High-tech business incubator has been started in March, 2006 and has been delivered and put into use respectively in April and November, 2007. Famous enterprise like NTT DATA and China West Air have signed contracts and entered.

International residential area

Residential group of 280,000 square meters is composed of western houses, slab-type apartment building, and Apartments of tower building. Subject to the principle of humanization, quality and details are paid attention to. Rationalism architecture of grey and white and reasonable space arrangement is specially customized for Lower Uppers who seek quality life and return of soul.

International Apartment of Fortune Plaza was started in May, 2004 including 1784 International apartments and western hoses, delivered

14

Office building

International Grade-A office building, headquarters base, Doctors' pioneer park show tremendous enterprises' brands, superior to western business platform, and provide quick foreign-related opportunities.

Hotel

It is designed under the instruction of world famous design company Australia WOODHEAD, specially inviting global well-known hotel management organization to come aboard. Therefore over-decades' hotel management experience is made good use of to provide for global top-end clients with excellent, professional and detailed service. We run business all over the world and win excellent reputation.

International apartment\ apartments of tower building with landscape

Here are three dot mode apartments of tower building for you to overlook the whole Fortune Plaza, surrounding the central wet-land park and rolling along the slope land from the south to the north.

Strong vision rhythm formed by the undulating terrain becomes a vision focus.

15

16

Modern brief language of shapes and colors shows foreign-related residential area's vitality and sense of fashion.

Design rationale of succession and innovation is succeeded in the tower building; further optimizes details based on the advantage of slab-type buildings and improves the living function of the new housing type flexibly according to the demand of the clients. The layout is therefore close and bright, spacious lighting of kitchen in each residents and best aspect toward landscape in each main bedroom is guaranteed.

Garden landscape

There are large grass, feature avenues, spring features, lake parks, large outdoor swimming pools, green vegetarian and gardens and so on.

13/14．曲桥景观实景图
15/16．园林景观

1

总规划面积：198 000 m²
总建筑面积：360 000 m²
建筑密度：20%
容 积 率：1.5
绿 化 率：35%
住宅总户数：2762
停 车 位：1377

1. 河边高层
2. 夜景
3. 社区全景

2

宁波东部新城安置住宅（一期）

开 发 商：东部新城开发建设指挥部
项 目 地 址：宁波东部新城
规 划 设 计：DC国际（上海·新加坡）

项目概况

东部新城拆迁安置房位于东部新城核心区东北角，占地面积19.77公顷，地上住宅及配套设施(包括12班幼儿园,商业会所,超市,净菜市场等)建筑面积30万平方米，地下和半地下停车库6万平方米，计划安置拆迁居民11000人。68幢住宅中规划了18层高层住宅12幢、11层高层住宅14幢，其余为6层住宅，大大丰富了小区空间层次。根据对区域内拆迁房屋现状和未来居住需求的综合分析，项目套型面积主要控制在60—130平方米之间，其中60—80平方米套型为多层住宅，90—100平方米套型占总户数的40%。

设计概念

居住岛——社区的细胞。
细胞与细胞的间隙——分裂空间与联系空间——缝隙与桥。
设计概念基于"城市特征"的目的之上，将空间上和交通上

的私人空间与公共空间分离，适当抬高居住岛（细胞）的高度，1.2—1.5米不等，除了供消防车和自行车以及任务交通（搬家及救护）进出的坡道外，这些庭院具有花园式庭院的氛围。

组团(Block)是居住区的细胞。理想的组团必须同时具有鲜明的共同特征和格局特色的形态。共同的特征易于形成整个小区的建筑文化，使居民能够产生对整个居住区的归属感，而不同的形态则增加小区内各个组团之间的可识别性。设计中各个组团采用共同的、具有围合感的布局形态，设计不同色彩与材料的建筑单体。

公共设施的集约性与共享性

设计者把社区服务中心的概念扩大化，基于城市中街道和广场的原型，作为给社区居民提供另一种生活的媒介。结合现有河道的整治，形成开放的贡献给城市的空间，同时在社区的整体形

态上，有了活跃的中心。商业集中布置，减少了对社区内居民的干扰，中心的半围合广场，提供大面积、多种类活动的可能。同时带来集中的经济效益与设施使用的集约最大化。

建筑本身

在单体设计当中，可以提取的设计元素是：开洞，抽象画，空中花园，色彩，廊架。建筑大面积的露台提供了垂直绿化的可能，使小区整体的景观体系由二维走向三维。露台在小高层的建筑体量上以构图元素的形式出现，错落有致，形成不同一般的视觉体验。

建筑的外墙与内墙，通过覆以颜色丰富的涂料，使其更具有地区特征，这不仅仅是一种艺术形式，也不是城市空间的简单装饰，而更多的是一种进入。

Project address: Ningbo Eastern New City

Conceptual design: DC Alliance (Shanghai·Singapore)

Planning area: 198 000 m²

Gross building coverage: 360 000 m²

Building density: 20%

Plot ratio: 1.5

Afforesting rate: 35%

Total residential units: 2 762

Parking space: 1 377

Project profile:

Eastern New City Residential Area is located in the northeastern corner of the core area of the Eastern New City, covering an area of 19.77 hectare, and there are residences with its necessary facilities (including 12 classes of kindergartens, business chambers, supermarkets and markets, etc.) of 300,000 square meters, underground and semi-underground parking garage of 60,000 square meters to settle 11,000 removal residents according to the plan. The project is started in January, 2005 and finished in June, 2007. High-rise settlement concept is introduced for the first time in this project, among the 68 residences there are 12 18-storyed residences, 14 11-storyed residences and the rest are all 6-storyed, which make space layers of the residential quarter abundant. According to the in-depth investigations of existing removal housing situation and the population structure within the area, and to the analyses of previous data of removal and settlement as well as demand for future living, areas of the types of flat are controlled between 60 to 130 square meters, among which the 60 to 80 square meters ones are multi-layer residences, taking up 50% of total units and the 90 to 100 square meters ones take up 40% of total.

Designing concept

Living Island – cell of the community

Intervals between cells-splitting and connecting space-gap and bridge.

The design concept is based on purpose of City Feature, separating private and public space in terms of space and transportation and

3

4

5

6

7

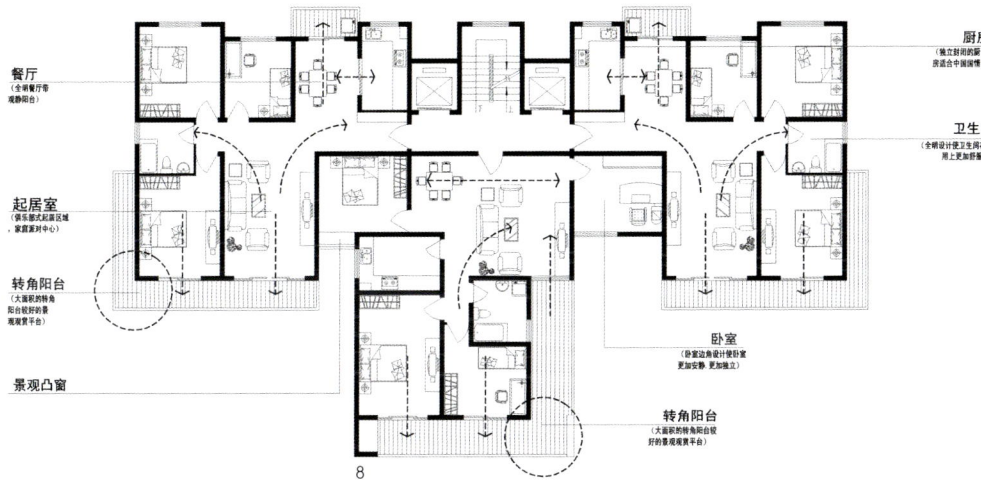

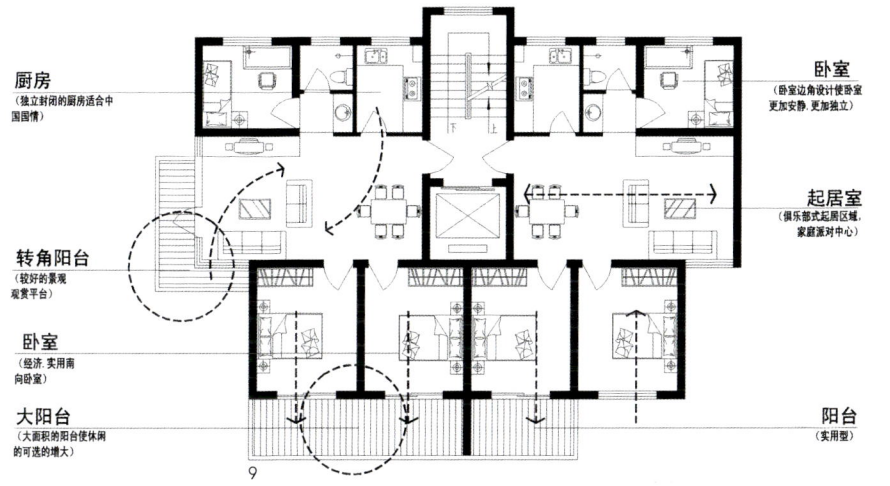

厨房
（独立封闭的厨房适合中国国情）

卫生间
（全明设计使卫浴间使用上更加舒畅）

卧室
（卧室边角设计使卧室更加安静、更加独立）

转角阳台
（大面积的特角阳台有较好的景观观赏平台）

餐厅
（全明餐厅带观景阳台）

起居室
（俱乐部式起居区域，家庭派对中心）

转角阳台
（大面积的转角阳台有较好的景观观赏平台）

景观凸窗

8

厨房
（独立封闭的厨房适合中国国情）

转角阳台
（较好的景观观赏平台）

卧室
（经济实用南向卧室）

大阳台
（大面积的阳台使休闲的可选的增大）

卧室
（卧室边角设计使卧室更加安静、更加独立）

起居室
（俱乐部式起居区域，家庭派对中心）

阳台
（实用型）

9

10

GREEN SLOPE   GARAGE   GLASS   GARAGE   GREEN SLOPE WALKWAY   GARAGE   GLASS   GARAGE   GREEN SLOPE

11

12

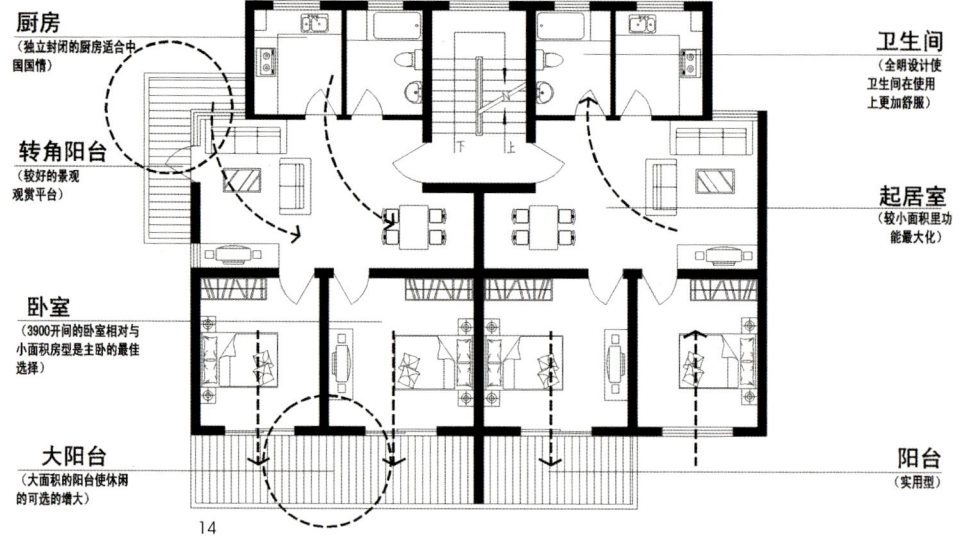

厨房
(独立封闭的厨房适合中国国情)

转角阳台
(较好的景观观赏平台)

卧室
(3900开间的卧室相对与小面积房型是主卧的最佳选择)

大阳台
(大面积的阳台使休闲的可选的增大)

卫生间
(全明设计使卫生间在使用上更加舒服)

起居室
(较小面积里功能最大化)

阳台
(实用型)

13

14

15

raising the height of the living islands (the cells) properly like 1.2 meters to 1.5 meters. All these courtyards are in the atmosphere of garden courtyards except the slope road for fire-fighting trucks, bicycles and missionary transportation (such as housing moving and rescue).

Block is the cells of the residential area, and ideal blocks should meanwhile have distinctive common features and shapes in special layout. The former one can be easy to form architectural culture for the whole residential quarter, providing residents with sense of belonging of the living area, and the latter one makes each block more recognizable. Finally in the design, common layout of enclosure is adopted in each block and architectural individual are design in different colors and with different materials.

Intensive and shared utilities

Based on the original shapes of the streets and plazas in the city, the designers expand the concept of community service center to offer another kind of life conveyor to the residents. So open space is given by combining the existing rivers and there is an active center due to the general shape of the community. Residents can suffer less disturbance thanks to the concentrative arrangement of the business and possibilities of large areas and many kinds of activities is offered by the semi-enclosure central plaza which also bring concentrative economic effect and maximum use of the facilities.

The construction itself

Extractable designing elements in the individual design are: punch, abstract paintings, gardens on the air, colors, and porches and so on. Besides, patios of large areas make vertical greening possible, so the general landscape system of the residential quarter is changed from two-dimension into three-dimension. Patios appear in the small high-rise construction mass in the form of sketching elements and in picturesque disorder, giving you particular vision feeling.

Inside and outside walls of the constructions are featured more locally by painted with different colors. Such is not only a kind of art, not only a simple decoration of the city space, but more of is a deeper reach.

16

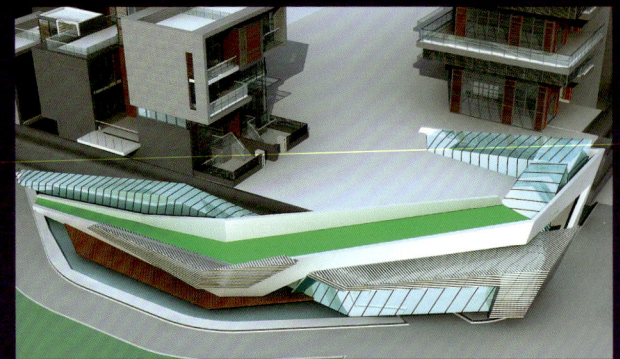

住宅标准层高：

Townhouse：3.3 m
叠 院 别 墅：3-3.3 m
叠 加 别 墅：3.2 m
高 层 住 宅：3-3.15 m
架空层车库层高：4.5 m
架空层大堂层高：7.2 m
地下停车场：南区：560个 北区：520个
总 户 数：556套

1

卓越·维港名苑

发 展 商 ：深圳卓越房地产开发有限公司
项 目 地 址 ：深圳市南山区科苑大道西，招商路与工业八路之间
建 筑 设 计 ：机械工业部深圳设计院
景 观 设 计 ：ACLA LIMITED
总占地面积：64 668 m²
总建筑面积：190 308 m²
总住宅面积：103 686 m²
容 积 率 ：2
绿 化 率 ：46%

2

1. 建筑效果图
2. 售楼处模型图
3. 售楼处实景图

深圳卓越·维港名苑位于蛇口东填海区，这里紧邻深圳的15公里滨海长廊，被六大公园环绕，是深圳别具一格的景观区域。周围有强大的立体化交通、规划的金融中心、世界级酒店的配置、顶级运动场、高端艺术配套、规划的城市中心以及罕见的城市自然资源等7大城市罕见资源。

深圳卓越·维港名苑偏执于产品的极致打造，在卓越，确实有一批产品主义的偏执狂，他们称维港为"作品"而非工业化的产品。即使是高层建筑，深圳卓越·维港也运用了金属色氟碳三涂表面处理、大面积双片钢化中空LOW-E玻璃幕墙、津巴布韦黑

越人说，深圳卓越·维港不怕挑剔的眼光。

规划概念

1. 高层组团的锯齿状规划使得小区园林移步换景，并保证了每栋高层具有独立的三角形活动空间。

2. 多层次架空，形成组团园林相结合的台地园林。

3. 景观概念源于风格派代表、新造型主义大师蒙特里安的艺

在建筑师们的细心考虑下，经过百余次的实地测绘和风洞模型实验，结合深圳历年来不同的季风和温差数据，卓越·维港名苑将建筑平面朝向整体东偏南55度，突破传统的建筑正南朝向，只为拥有更多的日光、更好的通风和更开阔的取景，兑现完美主义的追求理想。

Townhouse南区因地制宜，错落布置形成三个组团园林，平均20米以上的超宽楼距和前后大花园使客户享受到更多的深圳湾阳光，利用人造高差形成的"中间高、边缘低"的组团布局更好地保护了客户的隐私，避免视线干扰。

卓越集团
中国领先的综合地产运营商

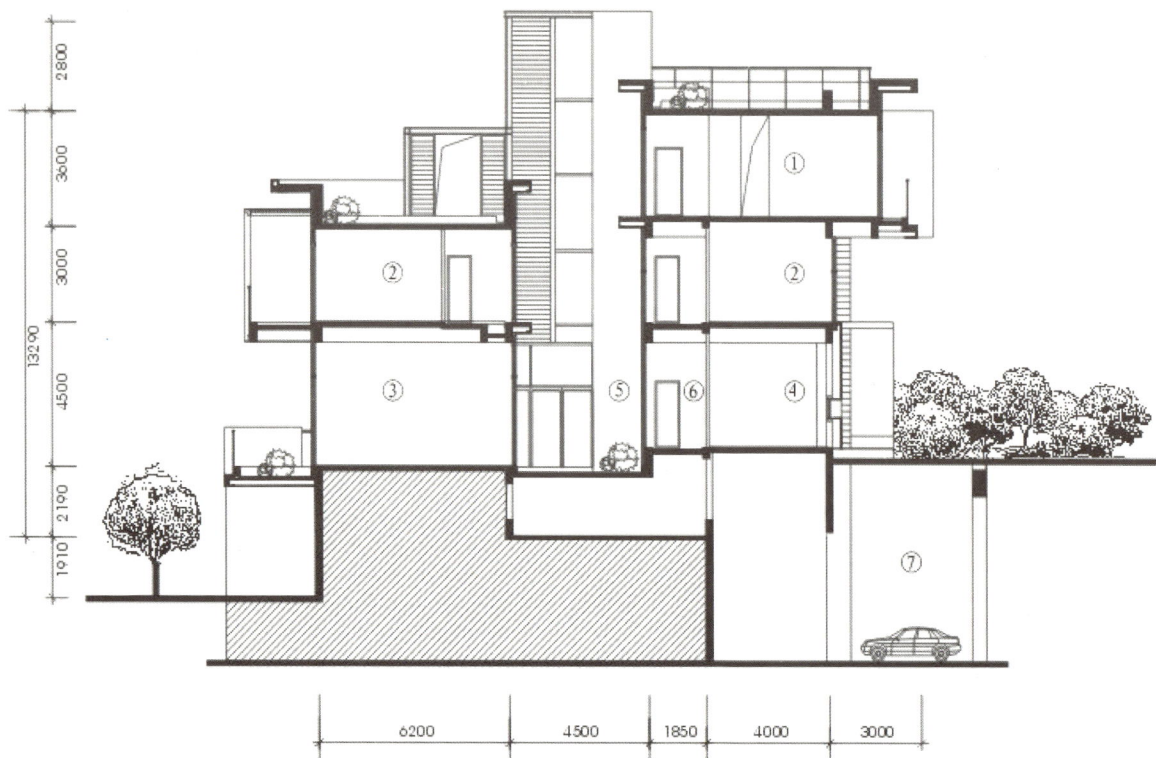

景观设计

卓越·维港的景观结合相关的景观元素、出色的细部设计、精心挑选的小品设施、繁盛有序的植栽展示一个现代、高雅，沉稳而内敛的别墅社区。

入口广场

作为小区出入口的公共广场空间设置为人车分流，利用双层地下车库形成的场地高差，把入口设置成错落有致的标志性社区景观。挑空6米高差的观景廊台，可将整个入口广场及市政公园尽收眼底，长达60米的叠级水景，从小区一直延伸至广场。伴着若有若无的流水声，穿行在"S"型的高档热带植物围合成的阶梯中，花草亮丽而香气悠远，原创的雕塑艺术，讲述着土地沧海桑田变化的故事……

组团园林

卓越·维港在设计的布局上，以抽象派的几何构图形式来呼应建筑设计的立面风格，以极简主义的手法，捕捉建筑最本质的形象和特质。建筑群落分隔出来的众多现代空间，满足了居住者独处及聚会的双重需要。安静的水景、细微的铺装变化、原创艺术雕塑、营造出一个沉思中的景观。

这个项目简单的景观元素精确的比例和适宜的尺度，创造了一个以人为本的实用景观。设计借助建筑所营造出来的两个连续的"Z"型景观面，把三个不同标高的场所连接起来。景观高差变化带来强烈的视觉感：湛蓝天空倒映在涌泉里，高档石材纹理的变化丰富景观的构成，花草植栽展示自然季节变化，同时，结合艺术、空间，缓解了建筑间狭窄场所给人带来的压迫感，传递了建筑与景观，自然与工艺的巧妙融合。

在园林文化内涵和空间灵魂的塑造上，卓越·维港本着卓越集团"生活美学"的建筑理念，以意象性的表达方式，在雕塑小品上以抽象造型，诠释了生活的美学品位和维港的艺术气质。

① 主卧室　② 卧室　③ 客厅　④ 餐厅　⑤ 庭院　⑥ 电梯厅　⑦ 车库

5

6

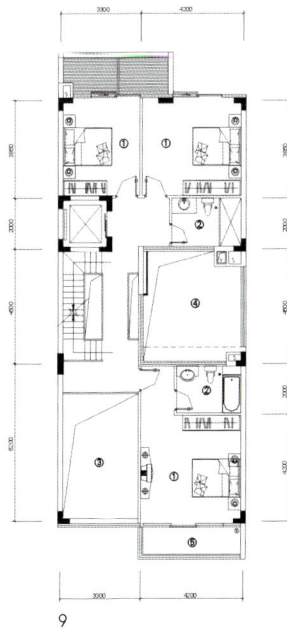

7

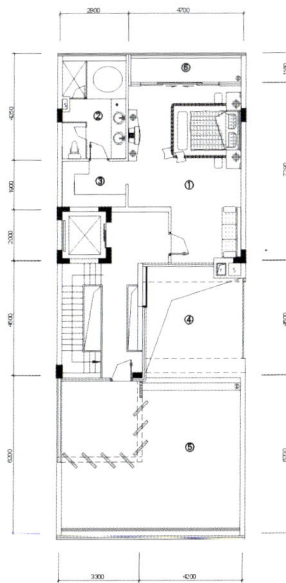

4. 鸟瞰图

5. 立面图

6/7. 效果图

8. 一层平面图

9. 二层平面图

10. 三层平面图

8

9

10

Floor area: 64 668 m^2
Building area: 190 308 m^2
Greening rate: 46%
Volume rate: 2
The number of dwellings: 556

Being a special feature area of Shenzhen, Excellence•Victorious Harbour is located in Shekou Dongtiansea District, right next to the coastal promenade which is 15 kilometers away from the city, and surrounded by 6 large parks. Nearby are the seven rare resources including three-dimensional traffic, planned finacial centers, world class hotels, top stadiums, high-end arts, planned city ceter and rare natural resources.

Greatest efforts has been made to mould the production, and there are indeed some paranoids of productionism who insist on claiming that Victorious Harbour is a work of art rather than prodution of non-industrialization. The use of surface processing with Metal colour fluorocarbon, double low-e glass curtain walls of large areas, Zimbabwe Black stones and Finland anti-corrosion wood,etc. are realized even in the high-rise buildingss. According to Excellence's staff, Victorious Harbour stands firmly to accept challenge, for their product turns out to be so unique and excellent.

Planning concept:

1.High-rise buildings groups planned in the shape of sawtooth present residents various views as they walk around the residential quater, they also ensure independent triangle space in each high-rise buildings block.

2.Mesa gardens combined by groups and gardens thanks to multi-level overhead.

3.The feature concept comes from Mondrian, representative of the de Stijl group and neoplasticism with his aritistic expression, segregating different garden groups geometrically.

With architects consideration, constructions of Victorious Harbour are set generally 55 degrees southeast according to hundreds of times of on-the – spot surveying and mapping as well as wind tunnel modelling experiment, combined with Shenzhen's

12

different monsoons and statistic of temperature range over the years. Such design breaking through south orientation of traditional constrution, for more sunlight, better ventilation and broader viewing, therefore realzing the pursuit of perfectionism.

Southern area of Townhouse is made good use of to form 3 group gardens separately. Extremely wide distance of over 20 meters on average, large front and back gardens allow residents more sunshine while group layout due to the man-made difference in height, low in the edge and convex in the middle, protecting clients' privacy and avoiding vision interference.

11. 售楼处夜景图
12. 售楼处全景
13. 售楼处室内

13

Design of the features

Relevant feature elements, excellenct detailed design, ornaments that are aboratively chosen and lush plants demonstrate a modern, elegant villa community.

Entry plaza

As public space of exit and ectrance of the residential quarter, the entry plaza is set to segregate pedestrians and vehicles. And difference in the field generated by the 2-storey underground garage makes the entry a symbolic community feature. Also, the whole entry plaza and municipal park can be seen from the feature gallery of 6 meters' high and there is a decked water feature of 60 meters' long extends along the residential quater to the plaza.

Please enjoy the wonderful feeling of the land while walking along the stair in the shape of "S", decorated by high range tropical plants with company of the soft murmur of the river, the fragrant flowers and the original sculptures.

Group gardens

Geometrical construction of abstractionism is used to echo elevation stlye in terms of the layout of Victorious Harbour, capturing the most natural image and characteristic of the construction. Life away from the noisy city proves the healing function of the static features. There are also lots of modern space separated from the construction groups to satisfy the need of the residents of being alone or having a party. Peaceful water feature, subtle pavement changes and original aritistic sculpture create a feature for your rumination.

Brief feature elements, accurate scale make up this practical project. Two feature plains in the shape of "Z" connect three areas of different highlevels. Such elevation difference brings strong sense of vision: reflection of clear blue sky in the water springs, veins of high quality sones enriching the features, plants demonstrating natural season changes. Meanwhile they are combined with arts and space to release feelings of oppresion resulting from the narrowness between the constructions, and to pass on the ingeniously mergence of constrction and feture, nature and crafts.

Finally, in terms of the mould of the garden culture, Excellence has illustrated its aesthetical tatse and artistical quality of Victorious Harbour in a imagist way in the spirit of the construction concept of "life aesthetics".

14. 楼盘实景图
15. 景观手绘图
16/17. 社区景观效果图
18. 社区景观实景图

16

17

18

1

中茵·皇冠国际

发 展 商：苏州中茵置业有限公司

项目地址：中茵皇冠国际位于园区星港街178号

景观设计：美国MBC园林景观设计公司
　　　　　广州太合景观设计有限公司

容 积 率：2.5

绿 化 率：49%

2

"中茵·皇冠国际"坐落在苏州工业园区CBD核心圈内，东南面临金鸡湖，南接香樟公园，四周环水，自然环境优越，由高尚居住区即中茵国际花园、高级涉外酒店式服务公寓即中茵皇冠国际公寓、园区唯一的五星级酒店即新苏国际大酒店共同组成，开发定位是苏州及世界级高贵精品社区。

"中茵·皇冠国际"由荣获中国房地产协会"中国房地产十佳顶级金字品牌"的中茵集团倾情打造，绝版水域天堂之高尚社区，获得中国住宅产业博览会"中国住宅十大名盘"、中国房地产协会"中国房地产最佳豪宅规划楼盘"等一系列殊荣。

"中茵·皇冠国际"是苏州第一个豪宅社区，为财富金字塔的顶层成功人士量身定做，倡导新居所空间，创造奢华生活。诚邀全球知名物业管理公司世邦·魏理仕为物业管理顾问，容五星级酒店为社区会馆，独设私家豪华游艇俱乐部，室内外多功能会所等设施，住宅实行全封闭智能化管理，一卡通系统，为业主提供更人性化、智能化的生活居住环境。

设计理念

要打造苏州及全世界级的高贵社会居住环境，遵循以下原则：以人为本的人性化共外空间，处处考虑人的适用性、使用方便，愉悦、舒适、恬静、清雅。

景观艺术化

形状、内容艺术化处理，结合一切艺术手段，运用现代材料、造型，整体是一个艺术品，局部也是一个艺术品，是一种高雅的文化艺术氛围。

人文景观自然化

中国造园手法是来源于自然而高于自然，在尺寸天地中运用现代材料、现代构图、空间要求、突出来于自然、生活于自然的一贯原则，在人密集的居住区造现代的自然化的室外空间，使人在景中，人景交融。

设计主题

因地制宜，利用基地内水系发达，规划以水为骨架，主题是"水域天堂"。以河、湖、溪流、水地叠水、涌、喷泉，立体的、多层次的、多样性的水景造法，使与人最具亲和力的水体贯穿于每个角落，如一串串的明珠，在丽日和风下波光粼粼，玉光闪烁，犹如"水玉天堂"。

设计手法——装饰化的景观

以中国造园中理水手法，结合西方及现代水景设施造法，用装饰化处理景观。采用室内装饰的手法，亭、台、楼、廊、桥、林各种造园元素呼应组合。装饰的艺术造型精细入扣。

设计成高贵社区，给人感觉高雅、精致、自然、时代，突出时代性，代表富有、高品位、有文化修养，是休闲天地。整小区按现代高档和高品位打造景观，强调工艺性、细部装饰、铺装、构筑物、景石、植物，一草一物，无不是遵循自然法则，人工精雕，体现人工的手艺及造物的精致。

规划布局

规划需要根据功能把小区分为居住区，酒店区。

1. 居住区

居住区又分为外围与内部二部分：

(1)外围区是小区的外部视觉点，目的是造生态的自然式，吸引人的视点，把人的注意力吸引到小区内，使人共鸣、驻足。用浓密深绿的香樟树(与香樟园对应)与韵律节点的水景(喷叠水、雕整墙)变化相统一，精细、坚硬的花岗岩铺装与水的柔、植物的绿对比，外看是风景区，高耸的建筑是点在绿海中，通过水景节点，通透外围与小区内，围墙景观化处理，通透处为景墙其它为浓密的乔木花草造档。

(2)区内

a. 中轴区

从主入口一直到会所入口，是一规则轴线景观，强调中心线景观，从入口到植、水池、涌泉，到中部水晶景观泳池，是双层泳池设计。二层用玻璃构筑，晶莹通透，人置身其中，更清凉，视野开阔。泳池是喷泉，叠水瀑布。其功能是泳池，实是水

4

景。在会所，利用地下车库，造下陷式水景，人行其间上，有如
凌波踏步。

b．四个组团区

　　四个组团以景观式立体阳光车库为主景，小区整个地下全
是车库，采光、通风欠缺。为了解决这问题又能造景，设阳光车
库，用叠布、水景观景廊造竖向景观，人进入车库同样看到景
致，也是地面上的景观。

c．架空层

　　为了小区景观整体性，楼一层架立处理。架空层景观用室内
装饰的手法，求精致、整洁，是室内装饰的延伸.

2．酒店区

　　与居住区仅一河之隔，东南面为金鸡湖，四周环水。在一岛
上，利用与居住区相隔之河，既是分隔又是联系的细带，既独立
又统一。利用其地势比小区高，造瀑布，叠水。斜坡用梯级状流
线种植地，既可护坡又具韵律节奏。从小区方向看到是一个山水
的竖向景观丰富的景观，整个酒店区是一个天然的风景区式，以
植物造景为主，是浓绿密林、鲜花盛开、水中漂浮、波光水影、
水天一色的风景区。

5

Volume Rate: 2.5

Green: 49%

1. Introduction

Join—in the Grand is located in Singapore Industrial Park. The Jinji Lake is on its southeast. Xiangzhang park is on the south. Water is running around the apartment. It has the beautifull location. It will become the top living area in Suzhou and world. This apartment will combine With a five star hote1. The hotel is having the wonderful garden view.

2. Design Idea

Follow the below rules to make a top—leve I apartment in Suzhou and world.

a. Humanity Area

Consider on convenience, health, happy, comfortable and quite for people.

b. Art View

Make the shape and content as art design. Use the modern material and sculpt to make an art product.

6

7

c. Natural

The style of chinese garden structure is made from nature but much better than nature, using modern material to make a modern structure in a small space, this small space requires naturally. We have special features on our nature design, following the rules of living in the nature, to make a modern with nature outspace, it's the combine of human and view.

3. Subject of Design

Different Places are having different Plans. The subject is going to do a "Haven of Water" use lakes, rivers, and fountains to make varieties of water view. In the apartment area can see water in every corner. The water like the pearls, looks beautiful. We can call it "Heaven of Jade" also.

4. Design Method Decorated View

Mix the method with China and Western to use water as the decoration. Can Use gloriette, stage, building, corridor, bridge and tree to decorate our area.

They will make the area looks more top level.

8. 泳池一角
9. 泳池手绘
10. 泳池鸟瞰
11. 水晶泳池区景观剖面图
12. 泳池上的亭子
13. 泳池过道

5. Design Style

This is a top—level apartment. It Will give the beautiful and modern feeling to everyone. It looks high stranded and has taste. It is a leisure world. We will make the good View through build-i ngs, stones and plants.

6. Structure

We will divide the area as living atea and hotel area.

Living area

There will be inside and out side in this area.

a. outside

This is the out View of the building. It Will build the nature style to arttact people's focus. Use the grasses, trees and fountain to show the beauty of this atea. Use the green plants to make people to feel they are in the green sea.

b. Inside
a) Middle

From the main entrance to apartment entrance, there is a regular view.

The guests can see plants, pool, fountain and swimming pool. The swimming pool is two level designs. The swimming pool is making from glass. People will feel cool when they swim inside.

b) 4 Group Area

The main view for this atea is sunshine garage. Underground of apartment is garage. Because there is mo view underground, so we use the special method to make our customers can have sane view as on the ground.

c) Empty Level

Because the consistency of vi ew, we make level one set up.

Hotel Atea

Only one lake distance between apartment and hotel. The Jingji Lake is on i ts Southeast. Water is around the hote1. The hotel will have a wonderful lake view. Hotel is a staircase shape. Hotel is a natural view of the area. Around hotel there will have lot of paints and flowers. A hotel on the warer it is wonderful view in Suzhou.

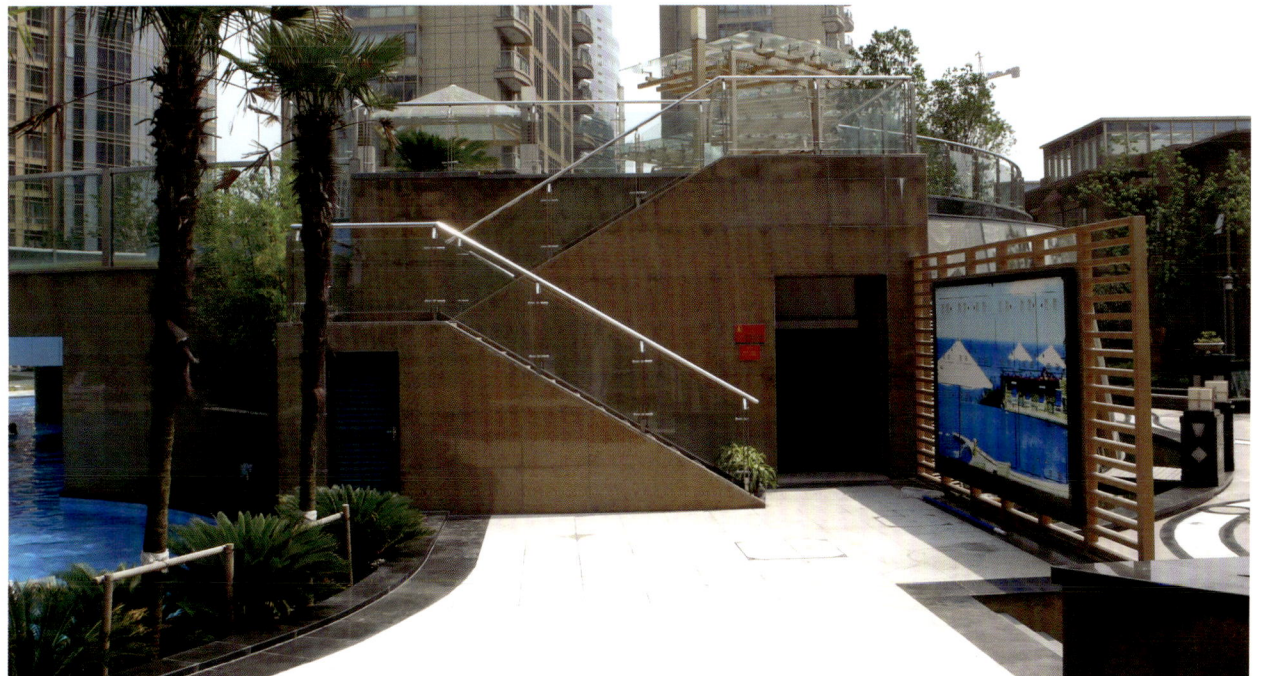

14

15

129

16

17

18

19

20

21

22

23

21/22/23．瀑布景观
24．瀑布．亭景剖面图
25．瀑布．亭景手绘图

26

27

28

29

30

31

26. 阳光车库水景区手绘图
27. 时尚的喷水柱
28. 玻璃结构的入口
29/30/31. 水石结合的景观

1

东莞宏远江南第一城

发 展 商 ：广东宏远集团房地产开发公司
项目地址 ：东莞南城区金丰路10号（江南雅筑西侧）
建 筑 设 计 ：加拿大AEL建筑景观设计有限公司
占地面积 ：200 000 m²
总建筑面积：400 000 m²
容 积 率 ：1.8
绿 化 率 ：56%

2

宏远地产倾力打造的东莞江南第一城2008年6月15日开盘，项目位于南城区金丰路金丰体育公园，此次共推出单位200套，其中包括250m²联排别墅、200m²叠加别墅、110-150m²情景美墅、100m²宽景洋房等。项目主力消费者为社会精英、工商个体户、高级白领。

此项目最大的特色就是采用中式江南风情园林,市区首席中式大宅，用料名贵，精心设计。作为江南雅筑的升级版，其中式风格在整个东莞楼盘的风格中可谓颇具特色。小区内部曲径通幽、梨花院落、柳絮池塘、桃花映水、竹林幽静、鹅浮绿水、小桥流水、湖色春光等让你目不暇接。

江南第一城作为运河沿岸唯一拥有别墅产品的社区，距离东莞CBD段不过十分钟的路程，其项目地理位置得天独厚，周边配套设施完善，使用率超高。

1/2/3．建筑立面局部
4．总体鸟瞰图
5．总平面图
6/7/8．叠加别墅建筑立面
9．叠加别墅实景

3

4

Project location: 10 Jingfeng Road, Nancheng District, Dongguan (West of Jiangnan Yazhu)

Developer: Guangdong Winnerway Real Estate Development Co., Ltd.

Architectural design: Canada AEL Architectural Landscape Design Co., Ltd.

Floor area: 200 000 m²

Gross building area: 400 000 m²

Plot ratio: 1.8

Afforesting: 56%

Dongguan Jiangnan First Town, specially promoted by Winnerway Real Estate, is opened on June 15th, 2008. The project is located in Jinfeng Sports Park, Jinfeng Road, in Nancheng District. There are totally 200 units in this project, including townhouses of 250m², superimpose houses of 200m², villas with scene of 110 to 150m² and villas with wide scene of 100m², etc. Leading consumer groups of the project are social elites, Individual enterprises and superior white collars.

The best feature of the project is its Southern Chinese riverside scenery style garden. Chief Chinese downtown houses are ingeniously designed with expensive materials, and as a updated version of Jiangnan Yazhu, its Chinese style has been a special one in all of Dongguan's estate. Inside the residential quarter there are winding paths will lead you to serene sports, courtyard with fragrant flowers, ponds with catkins, waters reflecting pitch flowers, quiet bamboo thickets, small bridges and murmuring streams, and lake with spring scenery and so on.

As the only community possessing villa production along the bank of the canal, there is only a distance within 10 minutes between Jiangnan First Town and Dongguan CBD road section. With such unique geologic location and well-improved necessary facilities around, it is of high use.

6

7

8

9

10

11

12

13

14

15

13/14．情景美墅立面

15．会所

16．会所入口

16

1

占地面积：83 365 m²
总建筑面积：74 748 m²
建筑密度：23.3%
容 积 率：0.78
绿 化 率：40%
总 户 数：205
停 车 位：240

1. 效果图
2. 鸟瞰图

惠州润园一期

发 展 商：惠州市润宇置业投资有限公司
项 目 地 址：惠州市著名的红花湖风景区
景 观 设 计：泛亚（香港）景观设计有限公司
建 筑 设 计：深圳市水木清建筑设计事务所
深圳清华苑建筑设计有限公司

　　惠州润园一期位于惠州市新区西南方的丘陵地区；其北部有现状道路与城市衔接，其西面为红花湖公园及自然生态保护区；四周均为尚未开发的山地。

规划理念：健康住区

　　为了营造健康的居住环境，强调有充足的阳光、自然风，尽量保护自然地貌、植被与水源。合理利用自然条件，扩大人与自然的联系，引水开湖，优化居住环境。提供健康的环境保障，设有公共健身设施、家政服务系统、室内外公共活动场所等以保证健康硬件建设，对全体住户的健康宣传教育行动是和谐社区的不可分离的软件。室外环境创造良好的社交空间以营造尊老爱幼气围，室内环境注重私密性，尊重居者个性化心理要求，建设和谐社区。追求健康的社区经济，尽可能减少土方挖填，保留山体及植被，节省投资，同时引水开湖，优化居住环境为地产增值创造条件。

布局特点

　　1. 充分利用山景之自然景观资源：所有住宅兼顾自然山景均为南向、东南向布局。

　　2. 开发水岸空间以营造优质生活品质：以平湖为主体、结合会所、汇聚广场、依山自然岸、湖心岛等主题，形成主题鲜明的中心景观；营造区内湖、溪、潭、瀑景观，以动静水景穿插，辅以完善的团组绿化。

3. 细分地形条件，高密度区分布在临北面市政衔接口，节省综合管线投资；将局部基地平整便可开发的别墅类型结合山体布置，注重保护自然生态，减少自然植被的破坏。综合考虑叠加式住宅、连排别墅的均好性。

平面与竖向构成

整体布局围绕中心湖，充分利用水岸空间与自然山景，依山傍水，自然和谐。北部利用自然平地，布置密度较大的高层住宅，依山低密度住宅均布置在南部，建筑形态在顺应自然地势的前提下呈北高南低之势，有利与日照、自然通风与城市噪音控制。

住宅设计

1. 叠加住宅的4、6、8层采用了一梯两户2层一户的叠加式设计，使得每一户均有良好的通风，采光条件。

2. 所有公共楼梯、电梯间直接对外通风、采光。

3. 起居室、主卧室及厨房均与此类型建筑面积配合，设计得比较豪华。

4. 每户的起居室及主卧室均为南向朝向，可以充分享受阳光。

5. 每户均设有入户花园，且朝向主景观面，与小区公共环境对话，使住户在空中也能够享受庭院生活。

6. 每户的餐厅与起居室均南北通透，通风、采光及视野均很好。

7. 多层住宅和2—3层的别墅采用异形柱框架结构；平面均采用多用普通钢筋混凝土梁板结构，根据建筑需要在起居室和餐厅采用大板结构，避免露梁影响观感，为住户提供室内空间变化的最大灵活性。

住宅立面设计

住宅立面以简洁明快为基本原则，在强调平面功能的同时，尽量避免多余的装饰，使建筑能够很好地融入周围的自然环境之中。一层的外装饰采用暖灰色的石材贴面或仿石材的喷涂，其余墙面采用浅米黄及白色喷涂效果，顶部局部设计的构架（屋顶遮阳花架）采用乳白色喷涂，营造出温暖柔和的居住空间。立面局部的装饰线角、空透的阳台栏杆、凸窗、遮阳板及空调机位置等的精心设计，使整体效果简洁而富有细节。多种单元的自由组合、局部尽端单元的变化形成起伏的总体轮廓，即避免了行列式布置形成的单调的外观，同时也与背景山体形成呼应。

3

交通组织

1. 小区路网在北端与城市交通衔接。

2. 小区内道路分成三级；小区级道路：路面10米宽，其中机动车路面7米宽，建筑控制红线大于12米；组团级道路为7米，其中机动车路面5米宽，建筑控制红线大于8米；宅间小路路面2.5米宽。组团内人车分流。

3. 合理与安全布置停车设施。别墅区随户停车，其它类型住宅利用地下室与半地下室就近停车。

4. 消防设计：按规范设有宽度均不小于4米、坡度均不大于10%的消防车道，转弯内径为12米。消防车道荷载为30吨。

排洪设计

道路一侧设排水沟，收集区域雨水；标高变化之挡土墙上部设截洪沟，收集山体雨水；严格控制与设定道路纵坡，通过道路、水沟引至水景湖与潭，再越过南侧与东南侧排至自然生成的山间排洪沟。

别墅设计

1. 豪华湖心岛别墅具有较好的日照、景观及视野，总建筑面积从450—550 m^2。

2. 连排别墅根据面积大小共有四种套型设计，较大的套型设置内庭院，在直接通风采光基础上更为住户提供良好的私密花园。

3. 将小区水景引入各户观景平台内，并与户外的内庭院相联系，达到内外景致浑然一体的效果。

4. 起居室及餐厅方正宽敞，主卧室及次卧室（父母房）均设计成套房，具有独立的卫生间。众多阳台及露台的设计，既丰富了别墅外观造型，也使使用功能更加舒适、自然。

5. 前院、侧庭、景观内院、后院等众多庭园的设计，使得别墅的每一使用空间均具有很好的外部景观、通风及采光，最大限度地提高了舒适性。

景观绿化

根据规划布局特点，采用集中与分散相结合，点、线、面相结合的绿地系统。围绕中心湖、别墅小湖景区设置小区中心绿地，设置有健身设施与游泳池的会所以及老年人、儿童活动场地；各类型建筑群组内设置团组绿地，设置儿童游戏设施和适于成人游憩活动。

4

Plot area: 83 365 m²

Gross building area: 74 748 m²

Building densityL: 23.3%

Plot ratio: 0.78

Afforesting rate: 40%

Total unit: 205

Parking space: 240

3/5．水边的别墅

4．内庭景观

6．对称规整的造型

5

6

Huizhou Runyuan Phase 1 is located in the southwest country of the Huizhou New Area, in its north there are roads connected to the city, in its west are the Honghuahu Park and the natural and ecological preservation area, and around it are the virgin mountains.

Planning concept: Healthy residential area

In order to build a healthy living environment, sufficient sunlight, natural wind is emphasized and natural feature, vegetarian and water resources are protected. They are reasonably made use of to expand to connection between man and nature. The water is introduced in to become a lake, optimizing living environment. Healthy environment is guaranteed, so there are hardware constructions such as public fitting facilities, housekeeping service system, indoor and outdoor public activity field. Outdoor environment has created good social space and atmosphere of politeness for elderly and care for children while indoor environment pay attention to respect for owners' privacy and psychology, but both of them are aimed to construct a harmonious community. To pursue a healthy community economy, earth excavation is reduced as much as possible, saving the hills and the vegetarian and investment.

Layout feature

1. Make full use of the mountain landscape: all the residences with mountain landscapes have southern aspect and southeastern layout.

2. Develop space along the water bank to create good life quality:

7

Design of the residence

1. There are two residences sharing one stair and two floors for one residence in the 4th, 6th and 8th floors of the Superimpose villas, so every house gets good ventilation and lighting condition.

2. All the public stairs, lifts get direct ventilation and lighting.

3. Living room, master bedrooms and kitchens are luxuriously designed in harmony with the areas of this type of construction.

4. Living rooms and master bedrooms of every residence have southern aspect which allows owners to enjoy sunshine.

5. There are entry garden in every house, facing the main landscape and the public environment, so the residents can enjoy courtyard life even living in high place.

6. Dining room and living room of every residence has good view, ventilation, and lighting.

7. Multi-stories dwelling buildings and 2- to 3-storyed villas are in column structure. In terms of plane, reinforced concrete beam slab construction is mostly adopted. According to the requirement for construction, large panel

take the slack water as main body, combine the chambers, merge the plazas and take mountains and island in the center as themes to form a distinctive central landscape; create lakes, streams, ponds and waterfalls inside the residential quarter dynamically and statically, then decorate them with perfect greening block groups.

3. Classify the terrain condition in detail, so the high-density area is set at the junction of the municipal government in the north side, saving investment for comprehensive pipeline; level part of the base so as to develop villa that are combined with the mountain body, pay attention to the protection of the ecology and fewer destruction of natural vegetarian. Besides, take homogeneity of Superimpose villas and townhouses into general consideration.

Plane and vertical formation

General layout revolves around the central lake, making full use of the water bank space, lying against the mountains and along the water, naturally and harmoniously; natural terrace in the north is used to arrange high-rise residences of high density while low-density residences are arranged in the south against the mountains. The shape of the construction turns on a trend of high in the north and low in the south on the basis of the natural terrain, which is good for sunshine, ventilation and the control for city noise.

8

9

10

structure is used in the living room to avoid disturbance for sense of viewing.

Elevation design of the residences

Elevation of the residences is in the principle of briefness and brightness. While function of the planes is emphasized, excessive decorations are avoided as many as possible so that the construction can well merge into its natural surroundings. Warm and soft living space is built because on the first floor, outer decorations are stone slices in dove color or ashlars' spray and the rest are sprayed in golden fleece and white colors, then part of the tops (sun shading flower shelf in the roofs) is milk-white sprayed. Decorating moldings, spacious handrails, projected windows, sun shading boards and air conditioners of the elevation are elaborately designed, making the general effect brief and abundant in details. Changes of multi-unit free matches and part of the dead-end units make the general outline, avoiding monotonous appearance by the functional determinant arrangement while echoing with the mountains in background.

Design of the villas

1. Luxurious villas in the center of the lake get better sunshine, landscape and views, with gross building areas of 450 to 550 m².

2. According to areas there are 4 types of townhouse and larger types are set with inner courtyards, and residents are provided with private garden on the basis of direct ventilation and lighting.

3. Water feature of the residential quarter is introduced into each landscape platform of every house while being connected with the outdoor inner courtyards. Thus, outer and inner landscapes are integrated.

4. Living rooms and dining rooms are square and spacious. Master bedrooms and sub bedroom (for aging parents) are suites with independent washrooms and design of many balconies and patios both enriches the outward appearance of the villas and makes their functions more comfortable and natural.

5. Design of the front yards, side yards, landscape inner yards and backyards provides every space of the villas with good outer landscape, ventilation and lighting and maximally improves comfortable sense.

Landscape greening

According to the feature of the layout, greening system of combination of concentration and dispersal, combination of point, line and plane is adopted. Revolving the central lake and the lake feature of the villas there sets a central grass of the residential quarter and fitting facilities, swimming club and field for the elderly and children in every type of construction group.

Organized transportation

1. The road network of the residential quarter is connected to the city transportation in the north.

2. The road inside the quarter is divided into 3 grads: roads for the quarter with road surface of 10 meters' wide including the 7 meters' wide motor road and controlling red line of the construction more than 12 meters; roads for the blocks of 7 meters, including 5 meters' wide motor road surface and controlling red lines of the construction more than 8 meters; roads for between the residence of 2.5 meters' wide. Pedestrians and vehicles are parted inside the blocks.

Design for draining

One side of the road is set with cut drains to collect rainwater; changes in the ground elevation are that the retaining walls are set with gaps to collect rainwater from the mountain body. Longitudinal slopes of the road are strictly controlled and set, and are introduced by the roads, the gutters to the lakes and the ponds, over the south and southeast side to the natural water gutter between the mountains.

12

7．河边一景
8．漂亮的外景
9．湖心岛别墅全景图
10．內庭景观
11．楼距合理阳光充足
12/13．联排别墅外景

13

1. 建筑立面效果图
2. 局部立面实景
3. 组院式双拼别墅效果

1

南沙境界

投 资 商：番禺富门花园房地产有限公司
项目地址：广州南沙区黄阁镇市南公路290号
设计机构：BDCL(博德西奥)国际建筑设计有限公司
占地面积：213 334 m²
建筑面积：约200 000 m²
容 积 率：0.93
绿 化 率：43.1%
总 户 数：1300

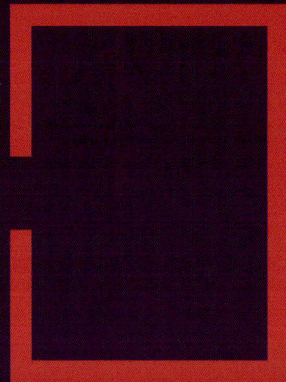

2

设计构思

　　充分体现温暖的人文主义的风格是这一居住建筑的设计追求，是现代主义意识形态的一种回归，是对人和自然的关怀。建筑一方面要满足居者对住宅的企望，对地域文化和空间更深层次的发掘，另一方面也是对空间的引导与提升，是对新的建筑风格、材料的尝试，是对生态手段的引入与贯彻。

设计特点

　　项目坐落于广州市自然山体公园大山·公园生态自然保护区

内，用地最大落差达50余米。设计依托起伏地形的山体优势，将保留的近约5万平方米的原生态山地作为"私家园林"景观，而丰富的观景公寓、花园洋房、双拼、叠拼、连排别墅等多种产品构成了分享大自然味道的场所。

　　连排别墅是很有特色的建筑，它的平面呈"吕"字形，首层餐厅和客厅由走廊相连，中间带有内花园，门口还有外花园，而屋顶是不规则的形状结构。在平坦的低密度区中心区的台地上，布置了该项目中的最高端产品：组院式双拼别墅。而远处半山坡

度加大的地方，设计师利用地形高差设计了叠拼产品。坐落于二者之间的是充满了"温暖的功能主义"色彩的连排。那些塔式与板式的高密度住宅多处于地势平坦地区，之间散落着一些花园洋房，这样利用不同地势的优点而设计不同类型特点的住宅，不仅是对不同人群的关注，更是对环境的尊重。

市场影响

　　南沙境界因其和谐的生态环境设计与独特的地理环境应用，多次获得中国原生山地别墅大奖和中国最佳山景别墅金奖。

Investor: NORINCO WK Properties Co., Ltd.

Designing Institute: BDCL Design International Ltd

Project location: 290, South City Road, Huangge Town, Nansha District, Guangzhou

Building coverage: about 200 000 m²

Floor area: 213 334 m²

Plot ratio: 0.93

Afforesting: 43.1%

Total units: 1 300

4．总平面图（图片由贝尔高林提供）

5．花园洋房立面

6．模型图

7/8．组团鸟瞰图

9．花园洋房实景

10/11．商业实景

12．北入口组团平面

13．双拼组团平面

14．花园洋房效果图

5

6

7

8

9

10

11

154

12

13

14

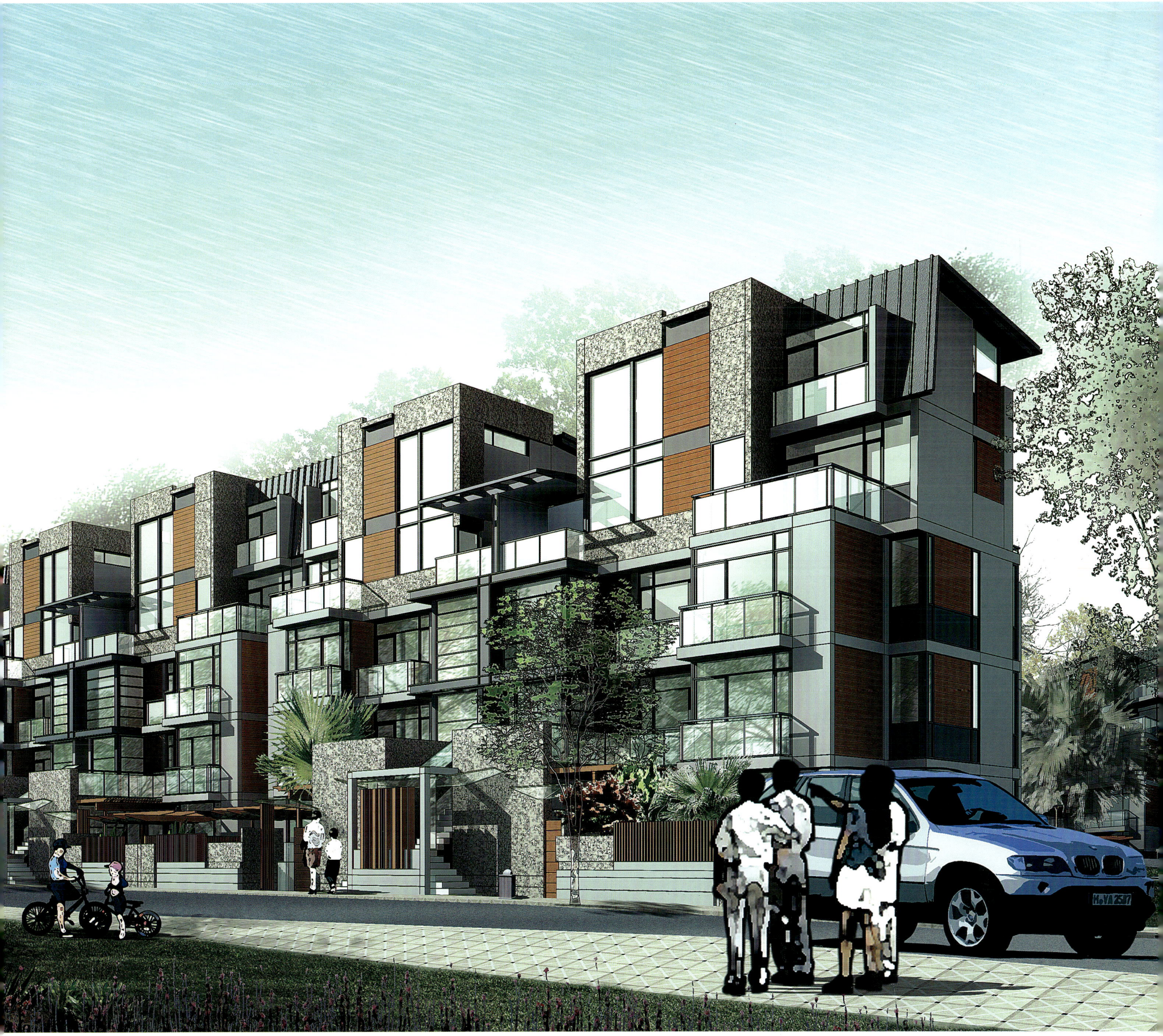

15

Design concept

This residential architecture is designed to fully demonstrate the style of warm humanism, which is a return from modernistic ideological form and care for man and nature. The architecture on the one hand meets the expectation of the residents to the houses and further searches regional culture and space. On the other had, it instructs and promotes the space, which is also a try for new architectural style and material as well as a introduction with ecological method.

Designing features

The project is located inside the Ecology Nature Reservation of Dashanna Park, a natural mountain park in Guangzhou. Against the mountain terrain, near 80 acreage of original ecological mountain is reserved to make private garden landscape, so various kinds of houses like landscape apartments, western garden houses, semi-detached houses, cityhouses and townhouses have constitute a place for people to enjoy nature. Townhouses, special architectures, turn on a plane in the shape of Chinese character "lv". Dining rooms and the living rooms on the first floor are connected with the corridors which there are inner gardens in the middle and outer garden at the doorstep, and their roofs are in irregular structure. On the flat terrace of centre of the low-density area there is the most high-end product of the project: semi-detached houses in groups. Far where slope of the half-mountain gets steeper there is cityhouse product designed according to such terrain. In between these two products are warm functional townhouses. Most of the tower type and slab type of high-rise dwelling residences are in flat area, among which some western garden houses are

16

dispersed. Such different types of residences designed according to advantages of different terrain are not only concern for different groups but also respect for the environment.

Market influence

The view point is an application of both its harmonious ecological and environmental design and its unique geography, which has help to win China Original Mountain Villa Reward and China's Best Mountain Landscape Villa Reward----The Golden Prized for many times.

15．会所
16．会所入口
17．商业街鸟瞰

17

1

1. 三期高层效果图
2. 园林水景
3. 三期鸟瞰图

江苏昆山和兴东城花苑

开　发　商：昆山和兴房地产开发有限公司
项　目　地　址：江苏昆山市前进东路333号
建筑规划设计：东南大学建筑设计院深圳分院
总 用 地 面 积：137 200 m²
总 建 筑 面 积：206 100 m²
容　积　率：1.5
绿　化　率：43%
总　户　数：1400
停　车　位：600 – 700

2

　　江苏昆山市和兴东城总建筑面积超过20万平方米，是南通和兴地产在昆山开发的第一个项目，也是整个昆山东部地区首个大型商业住宅项目。该项目地处东部副中心的中心位置，"双中心"的地理优势更使其具有得天独厚的生活便利。小区四面环水，以绿色生态、景观优美、户外休闲三大体系构筑洞庭别业成为江南水乡风采的理想都市人家。设计"户户有景"、"家家尝水"的人与自然交融的优质居住社区。小区规划将绿化系统作为小区的呼吸系统，充分利用自然要素，构筑中心绿化、环河绿化、庭院绿化，并结合三面环水的独特环境设计了滨水生态绿化带，与中心绿化、住宅庭院相互渗透，共同组成多层次的园林绿

化系统。同时，尊崇朴素自然观与艺术导向性的景观形象系统，强调安全保障与休闲舒适的户外活动系统相互结合，共同打造美景住宅典范。

　　和兴东城处在未来昆山东部核心区域，在这片热土上，开发区管委会办公楼、时代大厦、商会大楼等一批标志性建筑和功能性项目将在此崛起。和兴东城周边，20万平方米水上公园、大型公建绿地、时代大厦、商会大楼、开发区国际会展中心、开发区配套大酒店以及晨曦园小学、开发区高级中心等等，已将和兴东城层层环绕。其中国际会展中心项目建筑面积达32万平方米，

拥有4个大型展览馆，会展中心旁边还有148米高的星级帝豪大酒店和华敏翰尊国际大厦两幢高层建筑，全部建成后，将形成集会展、国际商务、高档酒店、行政办公、文化娱乐为一体的多功能综合服务区。其中开发区会所已经封顶；行政副中心、水上公园开工在即；晨曦园小学、开发区高级中学已经开学。如此全面的大配置，为每个昆山人展开了一幅清晰的未来生活蓝图。该项目分三期开发，一期15幢多层已基本售罄，二期为小高层和别墅，2006年6月11日开盘，三期是高层住宅及部分商业用房。

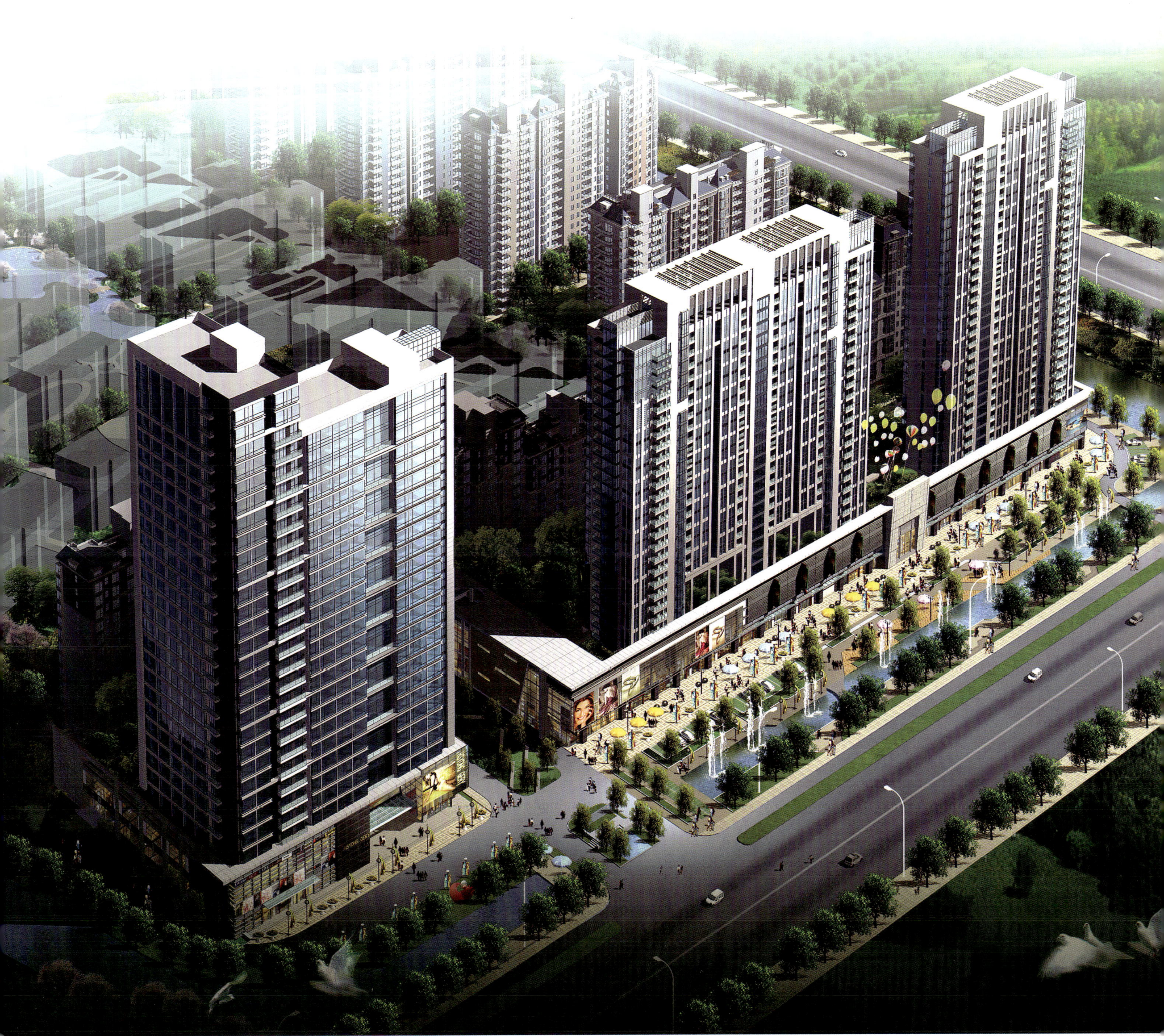

3

Developer: 333 East Qianjin Road, Kushan, Jiangsu

Architectural planning & design: Architectural Design &Research Institute Of Southeast University (Shenzhen Branch)

Gross plot area: 137 200 m²

Gross building area: 206 100 m²

Plot ratio: 1.5

Afforesting rate: 43%

Total units: 1 400

Parking space: 600 to 700

Jiangsu Kushan Hexing East Town covers an area of over 200,000 square meters. It is Nantong Hexing Real Estate's first project in Kushan as well as a first large commercial residential project in the whole eastern area in Kushan. The project is in the centre of the sub centre in the eastern area, and location of double-center has made it win more unique convenience for life. The residential quarter faces waters in four directions, and it has become an ideal residence of Jiangnan Watertown style by the construction of Dongting Special Property consisting of greening ecology, beautiful landscape and outdoor leisure. It is design to be a good quality residential community where man and nature live in harmony, and it will be like there are landscapes in every house and every resident enjoy the water. The plan takes the greening system as a respiratory, makes full use of the natural elements to construct the central greening, river greening and courtyard greening and combine the special environment of facing waters in three sides to design the waterfront ecologically greening belt, interpenetrating with the center greening and the residential courtyards, and all of these make up a multi-layer garden greening system. Meanwhile, landscape image system of plain and natural view and guidance quality is revered, combination of safety security and comfortable outdoor activity system is emphasized to together build residential example with beautiful scenery.

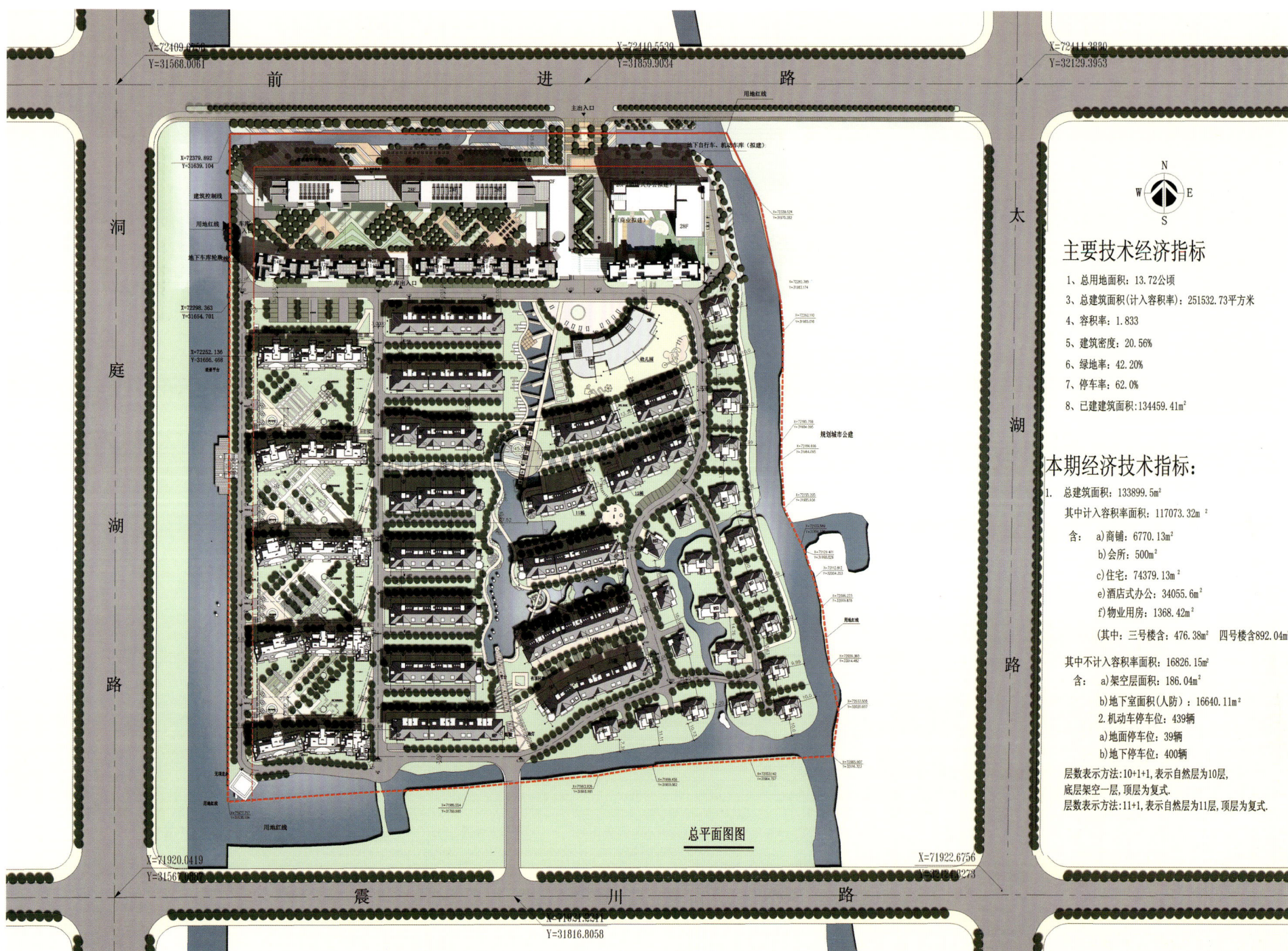

总平面图图

主要技术经济指标

1、总用地面积：13.72公顷

3、总建筑面积(计入容积率)：251532.73平方米

4、容积率：1.833

5、建筑密度：20.56%

6、绿地率：42.20%

7、停车率：62.0%

8、已建建筑面积：134459.41m²

本期经济技术指标：

1. 总建筑面积：133899.5m²

其中计入容积率面积：117073.32m²

含： a)商铺：6770.13m²

b)会所：500m²

c)住宅：74379.13m²

e)酒店式办公：34055.6m²

f)物业用房：1368.42m²

(其中：三号楼含：476.38m² 四号楼含892.04m²)

其中不计入容积率面积：16826.15m²

含： a)架空层面积：186.04m²

b)地下室面积(人防)：16640.11m²

2.机动车停车位：439辆

a)地面停车位：39辆

b)地下停车位：400辆

层数表示方法：10+1，表示自然层为10层，底层架空一层，顶层为复式.

层数表示方法：11+1，表示自然层为11层，顶层为复式.

Hexing East Town is located in the future core area of Kushan eastern area. Group of symbolic constructions and practical projects will be on this hot land, such as office building of management committee of the development zone, Front-Station and Commercial Building and so on. Around Hexing East Town there are aquatic park of 300 acreage, large public grass, Front-Station, commercial building, Development Zone International Conference & Exhibition Center, hotels of the development zone, Chenxiyuan primary school, Advanced Center of the development zone, etc, which have revolved around it. Project International Conference & Exhibition Center covers an area of 320,000 square meters, possessing 4 large exhibition complex, and beside it there are two tall building, one is the star-class Dihao Hotel of 148 meter's high, the other is Huaminhanzun International Building. All the construction completed, the project will become a multi-functional and comprehensive service area, integrating exhibition, international business, high-class hotels, execution and office as well as cultural and entertainment as a whole. Chamber of the development zone has been completed; executive sub center and Aquatic Park are to start soon; Chenxiyuan primary school and senior school of the development zone have opened. Such comprehensive distribution has show a clear future life blueprint for every Kushan residents. The project is developed in 3 phases, the first one is 15 multi-layer residences which have been sold out, the second one is small high-rise residential buildings and villas which has been opened on 11th, July, 2006, and the last one is high-rise dwelling buildings and partly for business.

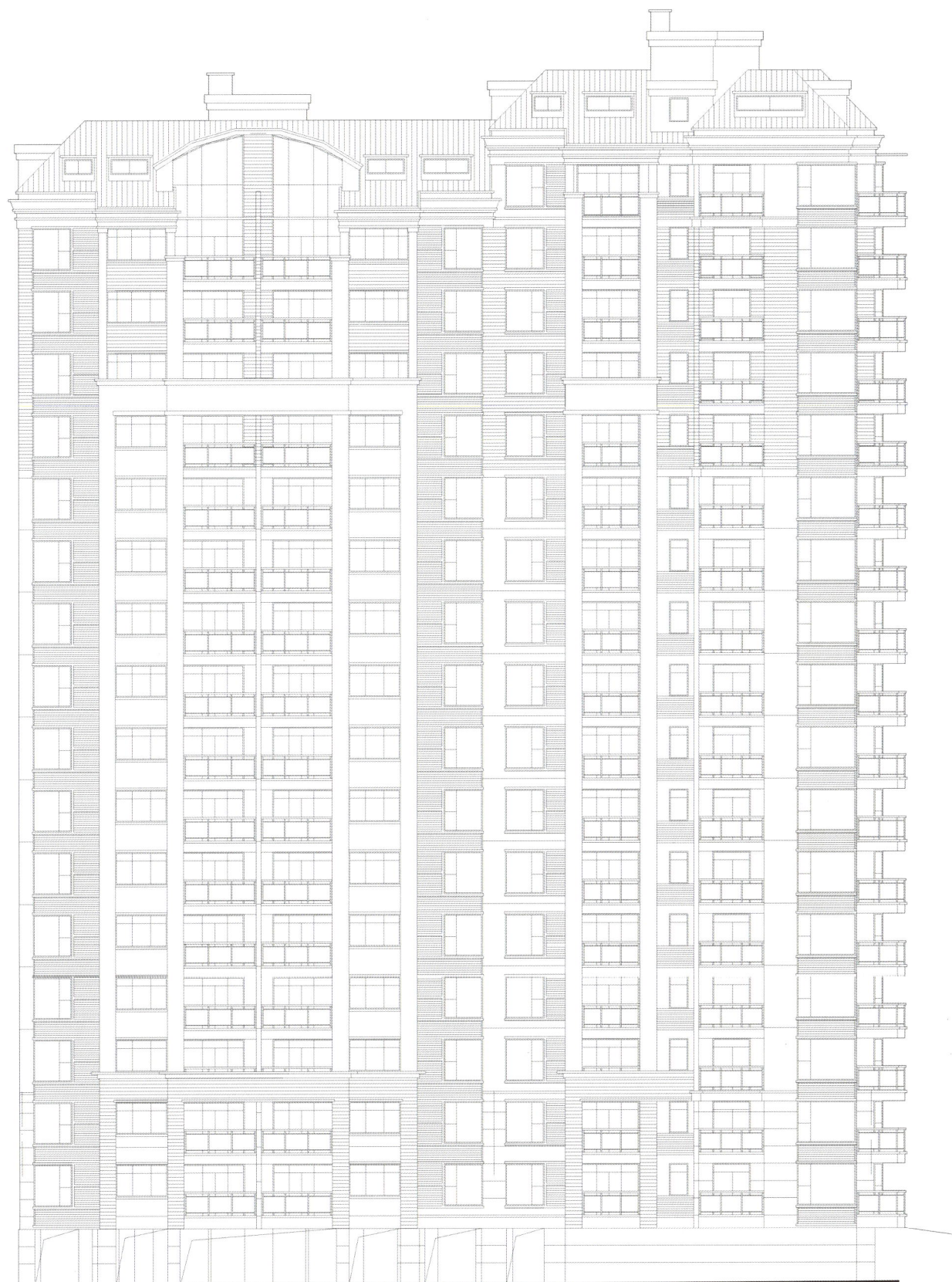

6

5

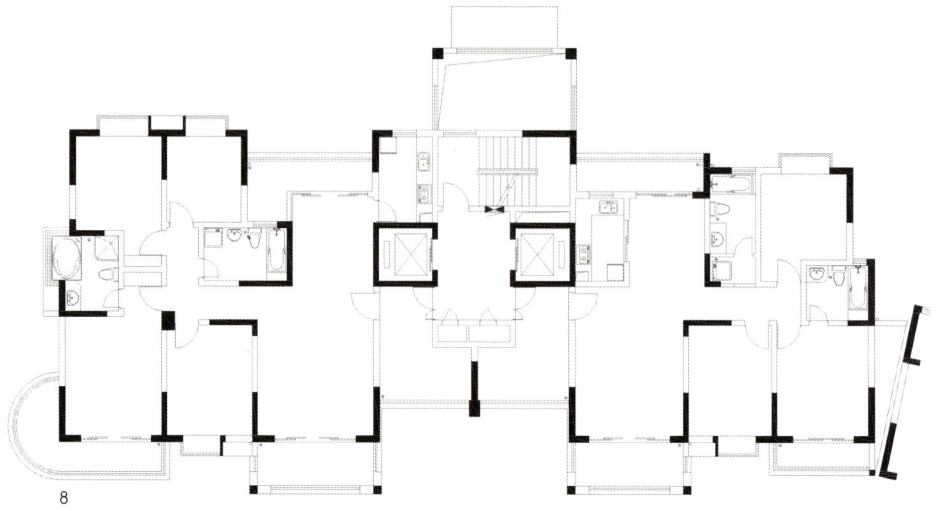

7

8

9

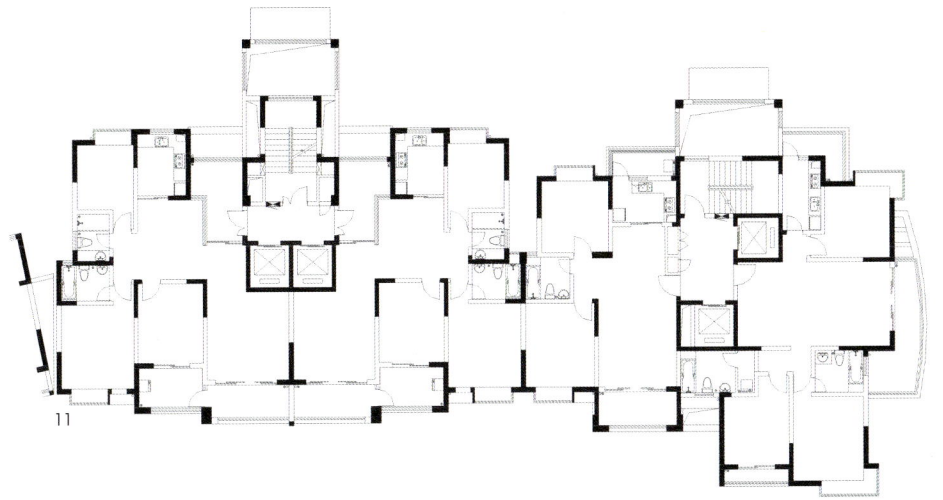

10

11

12

13．多层立面图

14/15/16．多层水景

17/18/19/20．多层平面图

21．中庭实景

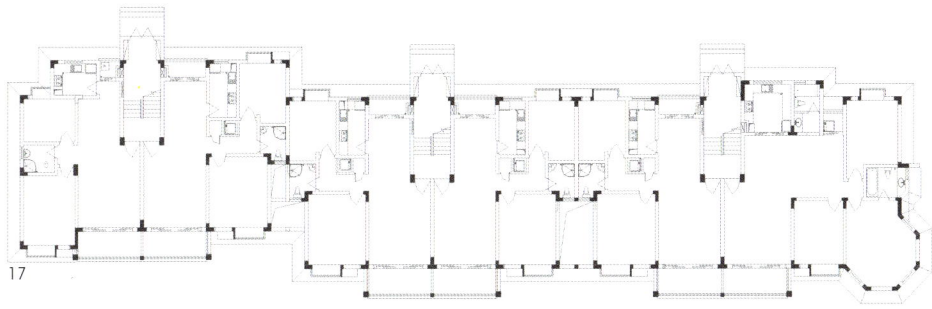
17

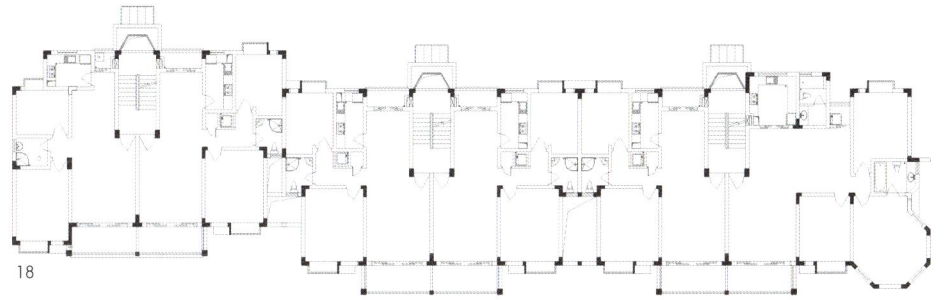
18

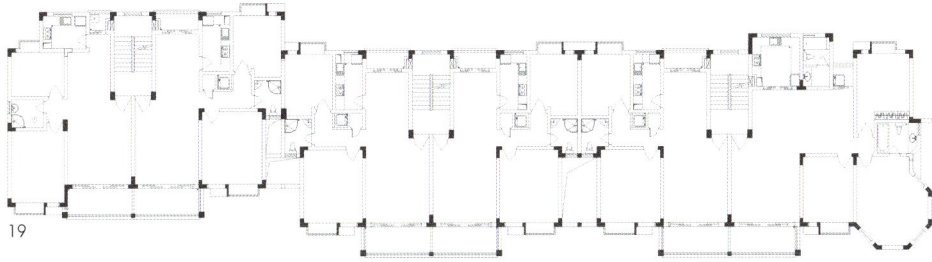
19

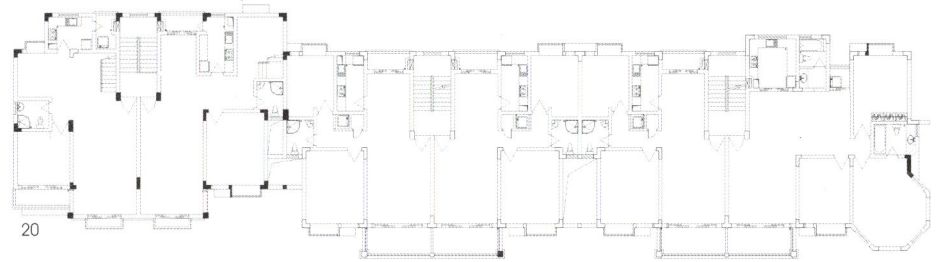
20

21

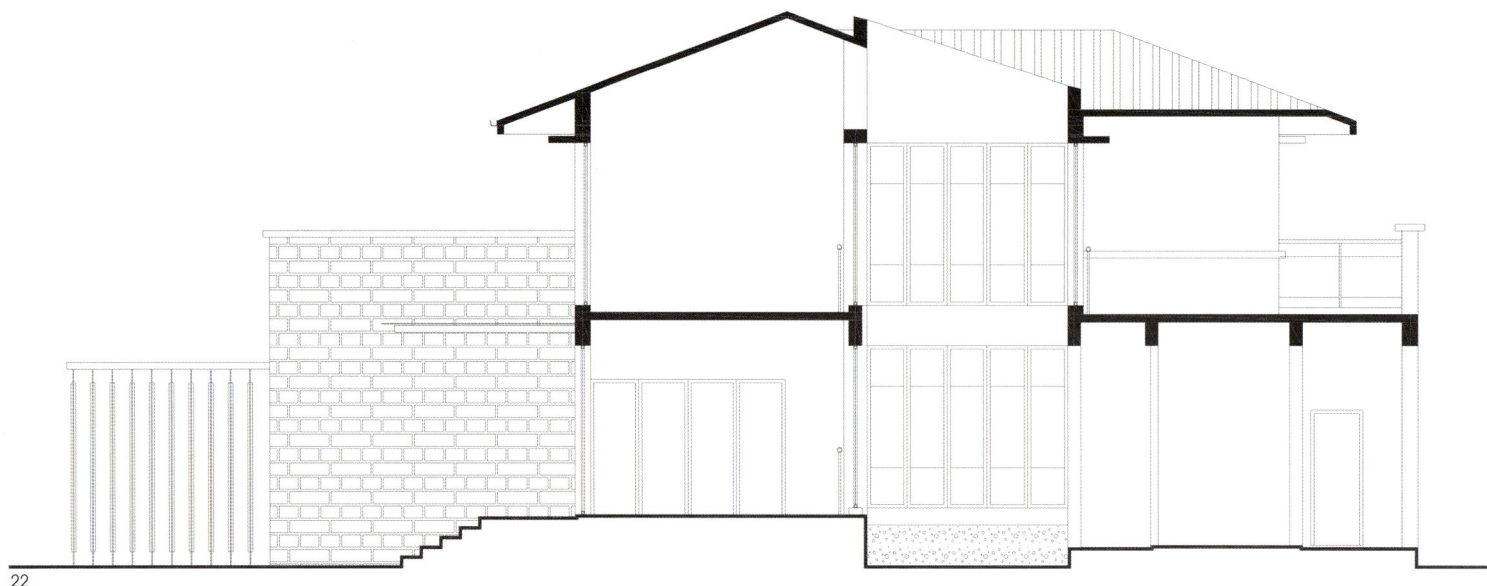

22

23

24

25

26

汕头阳光海岸

开　发　商：广东龙光（集团）有限公司
项 目 地 址：汕头市东区四十街区
建筑规划设计：东南大学建筑设计院深圳分院
总占地面积：约400 000 m²
总建筑面积：约700 000 m²
容　积　率：1.9
建 筑 密 度：30%
绿 地 率：35%

"阳光海岸"位于汕头市东区四十街区，长平路东段，东临泰山路，南接韩江路，西临黄山路，项目总建设面积约70万平方米(包括地下及半地下室)，总投资额近20亿元。

整个小区以人工湖为中心景区，布置联排别墅及独立别墅，四周布置高层、小高层和多层住宅。小区配套有大型地下停车场，小学及幼儿园各一所，市场一个，主会所及泛会所，室内、外泳池，架空层及社区活动场所多个等，配套齐全。

设计理念

是按国际最新健康设计理念设计的大型居住区，注重居住环境的健康性、自然环境的亲和性及住区的环境保护及健康环境的保障。

布局设计

在空间设计上北高南低，充分利用南向及海景资源；东西向中高两边低，加大中心区价值与区域住宅面积；利用人造中心湖景一、二期以中心主入口绿轴对称观，围而不合，空间灵动。

建筑单体设计

均为一梯二户，正南北向，有很好的自然通风采光，前入户花园的设计为住户提供更多的活动空间，户型配比上有小高层、多层、别墅、"2+2"复式别墅，提供多样化的选择。

造型设计

建筑采用三种色调的双坡型屋面，层层叠叠，天际线自然错落有秩，墙身运用三种颜色的穿插变化，丰富里面效果；色彩上下部运用高档石材，提升建筑品质。

设计特点

1. 在总体规划上，以国际最新的健康设计理念进行规划设计，即在符合住宅基本要求的基础上，突出健康要素，以人类居住健康的可持续发展为理念，满足居住者生理、心理和社会多层次的需求，为居住者营造健康、安全、舒适和环保的高品质住宅和社区。整个规划围绕"健康"二字展开，是具体化和实用化的体现，包括人居环境的健康性、自然环境的亲和性、住区的环境保护、健康的软硬件设施保障等。总体布局上，紧贴潮汕居住文化，南低北高，南向开敞，有利于整个小区的通风采光，同时最大限度地利用了南向与海景。在东西向上，中高边低，加大中心区价值，提升了空间的利用价值，尽量消除沿路环境噪音及景观视线的影响，充分利用中心湖区景观，围而不合，通而不透，空间自由活泼。

2. 投入巨资从韩江支流引入新津河水营造整个小区的水环境，在小区中心形成一个人工湖，成为小区的一大亮点。引水工程总长约4公里，取水源头位于金叶岛北侧，通过原水净化中心一系列工艺流程处理，水源由取水加压泵房提升后，通过输水管道沿新津河堤和长平路到达小区西北角引入小区，水质经环保部门检测，其ph值、浑浊度、总碱度、溶解氧(DO)和水中可降解有机物含量(BOD5)等各项指标均保证水质可满足小区用水及湖体水质要求，河水受海水潮汐的影响不大，基本为淡水，由西北向东南顺着地形高差流入中央人工湖，最后对小区东南侧的黄厝围污水沟进行冲淤后排入大海，通过每周定期换水(2万立方米/次)及日常流动循环(循环量约6000立方米/天)，营造了水的景观，保证了小区水质的清洁卫生。同时充分利用引水进行清洁、冲污、绿化，节约了自来水资源，这是小区基于节能、环保、生态方面的设计。

3. 交通组织方面。合理组织路网，人车分流。除中心湖区外，机动车在进入小区前都进入地下，最大限度地减少对小区的影响，车位配备充足，户均达到1.2个停车位左右。地下及半地下车库通过微地形处理、顶部开天井处理、边界交接处边坡处理，使得车库通风采光良好，进出顺畅，室内外空间延伸，车库与自然景观交融，使其真正成为"阳光车库"；将室外绿化及水景引进地下，或者将地下的绿色延伸到地面，并可通过架空层贯通上下，给人以方便舒适之感。

4. 在中心湖区主入口轴线上设置四幢圆形高层建筑，使小区空间布置趋于平衡，成为主入口的对景，单体平面外圆内方，180度开阔视野，有优美的沿湖岸景及无敌海景，是整个小区的标志性建筑。沿湖2+2四层小洋楼住宅，以及湖岛中的连排别墅和独立别墅，临近湖系水面，中心湖占地面积达5万平方米，结合30多米宽近1000米长的环湖生态走廊及环湖健康跑道，成为整个小区的中央景观区，蜿蜒水道，前庭后院，水榭平台，露台花园，雕空中庭，使人与自然环境的交流达到最大的亲和性。

5. 单体套型平面方正、宽敞、实用，南北通透，户户向南，基本做到明厨明厕，通风采光良好，延引潮汕地区"前庭后院"的居住文脉，入户大花园，空中大露台，公共私密分明，动态静态分区；室内跃式设计及复式设计，丰富了室内空间，少一些平淡，多一些变化，通过一些个性化设计，设置生活小阳台、储藏间、阳光房、景观餐厅、厨房餐厅与阳台的微妙结合、观光电梯等，为使用者的考虑细致入微。在立面造型设计上，通过线条色彩的变化，构架造型、屋顶造型、复式天窗的变化，将新古典主义与现代简约风格融会贯通，与园区的景观绿化和谐统一，令人赏心悦目，历久弥新。利用外墙遮阳设施、空调架设计、屋顶遮阳隔热层设计、沿路中空玻璃设置等办法，既不影响建筑物外观，又能有效控制对热量和声音的吸收，达到隔热隔声的目的，有效改善光环境、声环境及热环境。在建筑立体上采用镂空形式，设置空中花园，让住户更加接近绿色及与室外的交流，改善住宅区小环境的通风采光条件；架空层的设计形成开敞空间，与建筑群体结合，创造适宜人群活动的场所，营造潮汕文化风情，重塑邻里交往的社区生活空间；尽量利用水系及绿地，减少硬地布置，以绿色和水面作为小区小气候的调节介质，调整夏天炎热状况，保持室内热稳定性；通过可调节植被体系，南侧种植落叶乔木，夏天枝繁叶茂起遮阳作用，冬季落叶稀疏将阳光引入室内，北侧种植常青树，冬季可挡风及引导风向。在部分建筑如四层小洋楼、连排别墅、独立别墅中设立太阳能热水系统，将光能转为热能，环保节能，通过太阳能光电系统，可将太阳能转化为室内外公共用电及庭院灯用电等。

6. 在景观设计中，小区以法式园林为基调，揉合现代休闲景观设计元素，整个景观体系由"一环二轴三线"景观结构构成。

"一环"即中心湖环湖的生态环廊，长1公里，宽30米，环绕近5万平方的绿色湖岛，远可眺海，近可观湖，以密植绿化、环湖慢跑径、沙地、健康步道、亲水平台、休闲观景台等组成，是健康运动的天然场所，也是自然生态的真实写照。"二轴"包括入口的中心景观轴，横向文化景观轴及斜向奥林匹克公园景观轴等。中心景观轴是整个小区主入口通向中央景观区的主轴线，有入口广场、迎宾广场、鸽子广场、下沉式休闲泛会所、望湖台、树阵、几何水景、涌泉叠水，灯光地面、彩幻光雕、长廊画亭等。横向文化景观轴是连接一二期组团及中心景观轴的一条文化风情长廊，通过廊下架空、水边小筑等将潮汕文化特色与异域人文风情融入社区的文化生活当中。斜向奥林匹克公园以健康运动为主题，将奥林匹克文化及精神在这里重现，"生命不止，运动不息"，网球场、羽毛球场、体育雕塑广场、滑板旱冰场、武术天地、太极天地、棋牌活动、器械健身等等，无不处处透出浓浓的运动气息。"三线"是贯穿于小区中组团与组团之间、景观节点与节点间的景观带，通过点线面的结合，达到景观资源的贯穿与共享。

通过小区VIP星级会所的配套，包括室内恒温泳池、室外游泳池、水吧、酒吧、网吧、茶座、桑拿健身中心、美容美发美体中心、棋牌书画音乐活动室、图书阅览视听室、咖啡厅、商务洽谈会议室、多功能厅、社区服务诊所、银行、商业名店等等，引进中海物业管理公司合作管理，为住户提供了周到、细致、高尚的管家式星级服务，并通过对整个楼盘的形象包装策划整合，提升了整个社区的生活品质、健康内涵和文化品位，首创汕头领航湖海生活的崭新模式。

4

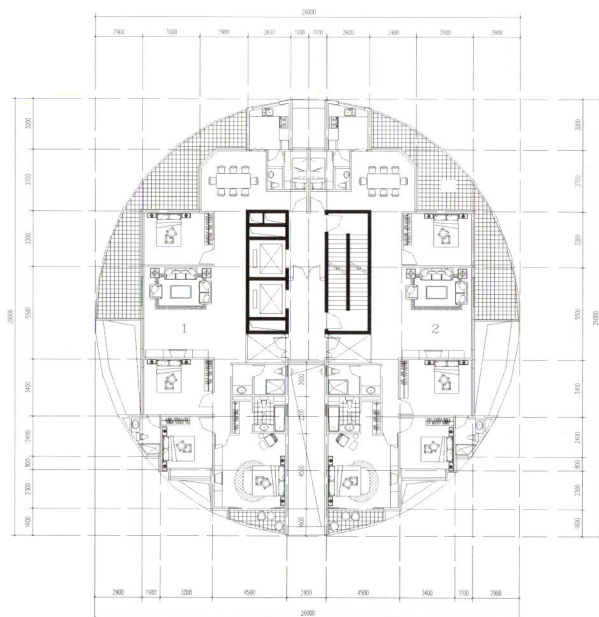

3

5

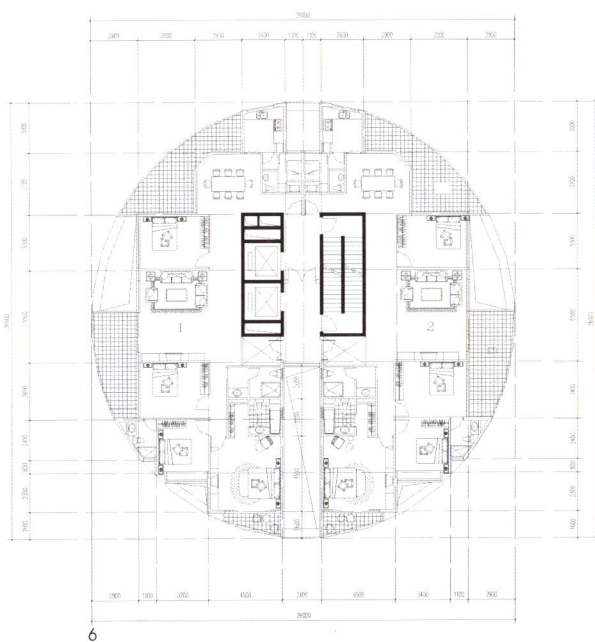

6

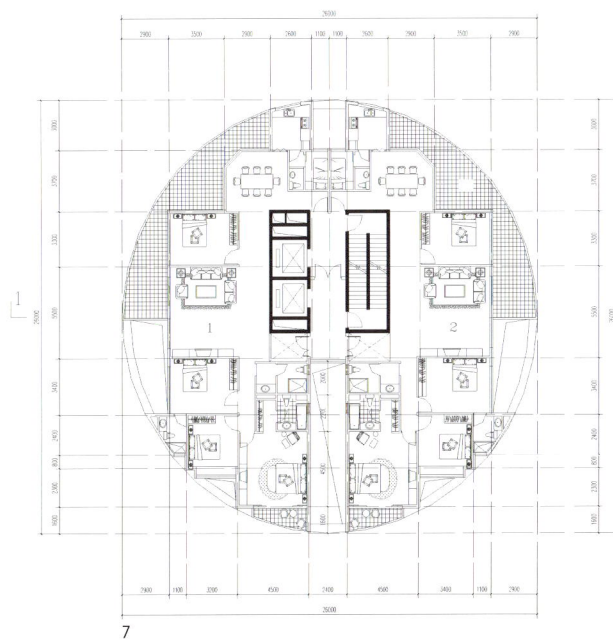

7

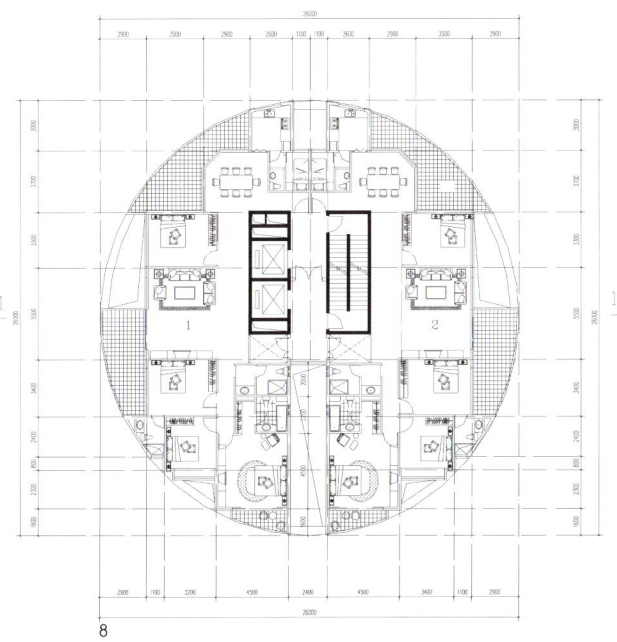

8

Designing concept

It's a large-scale residential area based on the latest healthy international design concept which focuses on the healthy living conditions, natural environment and environment protection and guarantee for the surroundings.

Layout

The designers make full use of sea resources and the southern aspect to make the residences higher in the north and lower in the south. In the west-eastern direction the residences are higher in the middle, which increases the value of the central area and enlarges the area of them. Making use of the manmade central lake features and taking main green central entry axis, Phase 1 and 2 are symmetrical. It is surrounded but not closed and very spacious.

Design of single room type

Two residents share one stair, and they have square north-southern aspect and natural ventilation and lighting. The garden before the entry of the house provides more activity space for residents. In terms of the proportion of the house types, there are small high-rise, multi-story, villas, and 2+2 duplex villas for you to choose.

9

Project location: Block Forty, Eastern Area, Shantou

Developer: Guangdong Logan (Group) Co., Ltd.

Architectural design: Shenzhen Thinker Design & Consult Pty Ltd.

Gross floor area: about 400 000 m^2

Gross building area: about 70 000 m^2

Plot ratio: 1.9

Construction density: 30%

Afforesting rate: 35%

Sunshine Coast is located in the western area block forty, Shantou City ,which stands on the western of Changping Road, east to Taishan Road, South to Hanjiang Road, west to Huangshan Road. The total area of the project is about forty hectare and the area of construction is about 700,000 m, including underground and semi-underground rooms. Its plot ratio is 1.9, construction density is about 30% and afforesting rate is up to 35%, and it takes a total investment of 2 billion Yuan.

The man-made lake is the center landscape of the whole community in which townhouses and separate villas are built, and around it there are high-rise, small high-rise and multi-storied dwelling residences. There are complete amenities such as a large parking space, a primary school, a kindergarten, a market, chambers, indoor and outdoor swimming pools, open floors and other activity rooms, etc. in the community.

10

Modeling design

1.Double slope house elevation of three color tones are adopted in the architectures in various layers. You can see the natural skyline in picturesque disorder while three colors changes naturally on the wall. All the walls are built with high class stones which promotes the quality of the architectures.

On the overall planning, it is designed by the new concept which is not only new but healthy through the world, that is on the basis of suiting requirement, concentrating on the element of health, based on the sustainable development conception about people's healthy life, further more, contenting residential physiological and psychological life and the need of social life, which creating a healthy, secure, comfortable and not environmental but high quality community. All the planning surround health, reflecting the concrete and the practical side, which including people's health when living, natural Affinity, hoe to protect the environment, and the facilities for health and so on. Overall the layout, it closes to the Chaoshan culture, south higher but north lower, the south side faces sunshine, which is good to the light and ventilation, meanwhile, to a large extent, taking advantage of facing to south and sea view. From the east and west side, it is higher in the middle and lower in the edge, which increasing value of center, promoting the value of space, tried to get rid of the noise along the road and the effect of a line of sight, which makes full use of the central lake scenery, leading space freedom and lively.

2. We spend large sum of money to introduce water from the branch river of the Hanjiang River, the Xinjin River to create a water surrounding for the community, forming a manmade lake in the center which become a bright spot of the community. The diversion project is about 4 kilometers long. From its source in the north side of the Jinye Island, the water has been processed through a series of flows, brought up by the district pond and pressure pump house, through the water pipe and along the bank of Xinjin River and Changpng Road; it is introduced into the community, reaching the northwest corner. Quality of the water has been tested by the environment protection department, so with its PH value, turbidity□total alkalinity□ DO□and BOD5 it is qualified enough to meet the requirement for water of the community and the lake. The river water is fresh water, hardly affected by the tides and it goes into the central manmade lake in the direction from north-west to south-east due to the distance produced by such terrain. At the end it rushes through the Huangcuowei bilge drainage in the south-east side of the community and flows into the sea. Every week the water is refreshed regularly (20,000 cubic meters per time) and it circulates about 6,000 cubic meters per day, so water features are created. Meanwhile, in terms of design of energy-saving, environment protection and ecology, to save the top-water resources, the water introduced is fully used to do some cleaning, greening, and wash the dirt.

3. The transportation organization. The community road network is reasonably organized and pedestrians and vehicles are separated. Before entering the community, all the motor vehicles should go through the underground, maximally reducing their bad effects on the community. There are enough parking spaces, about 1.2 per residents on average. The underground garage has good lighting and ventilation thanks to the processing of the slight terrain, courtyard opened on the top, slopes on the intersection of the boarder, so you can get easy access to and out of the underground garage. Also, you can enjoy such sunshine garage

11

12

174

13

14

15

because it has been integrated with the natural landscapes. The outdoor greenness and water features are introduced into the underground, or the greenness underground is extended through the open floor so that people will feel comfortable.

4. Four round high-rise architectures are built on the main entry axis of the central lake area. They balance the space layout of the community, and become symmetrical landscape of the main entry. Each single construction looks round from outside but square from inside. They have a broad view spreads about 180 degrees with beautiful bank landscape along the bank and unique sea view, which make them remarkable construction of the whole community. The 2+2 four-storied western houses and the townhouses and detached houses on the lake are near the surface of the lake system. The central lake takes up an area of up to 75 hectares, and along with the round-lake ecological corridor of about 30 meters wide and 1,000 meters long, and the round-lake healthy track, becomes a central landscape area of the whole community. People are allowed in large scale to be in nature with these winding water paths, front and back yards, platforms, patio gardens and hollowly carved middle courts.

5. Single Room Type gets a square, spacious and practical elevation. Every residence has a southern aspect, good ventilation and lighting, bright kitchen and washrooms. It is Chaoshan residential context that most houses has a front and back yard, so here you can see a big garden as you enter the house, then a large patio in the air. Thus, public and private space is clearly divided and so do the dynamic and static areas. Indoor space is enriched by the jumping and duplex design, the more changes there are, the fewer monotones there will be. Then, there are some individual designs showing our care for users, such as setting a small balcony, a storage room, sunshine room, landscape dining room, area that combines the kitchen and balcony, and sight-seeing elevator and so on. New classicism and modern brief style are merged together, in harmonious with the landscape greening of the garden, making you feel satisfied. Design of sunshading on the outer walls, air conditioner, layers of sunshading and insulation on the roof, and hollow glasses along the road have help to effectively absorb the heat and the noise while not affecting the outward appearance of the construction, so the environment of the light, sound and heat is well improved. Garden in the air in the construction allows residents to be more closed to the nature, and small environment of the residential area such as lighting condition is improved. Spacious space due to the open floor is combined with the architectural groups to create place for activities and Chaoshan cultural style, and recreate community living space for neighborhoods. The water system and the green lands are used as much as possible and hard grounds are reduced. On the one hand, the water is regarded as the adjusting conveyor of the climate of the community

to adjust to the hot summer and keep stable indoor temperature, on the other hand the vegetarian is also adjustable, deciduous trees are planted in the south side so that it can be a sun shade in the summer while in the winter gap between the loose leaves can let out space for the sunshine, evergreen trees planted in the north sides can prevent and lead the wind in winter. Parts of the construction such as four-floor western houses, townhouses, detached houses are equipped with solar heating system which transfers the light energy into heat energy and environment protecting energy. The solar energy can also be transferred into indoor and outdoor public electricity and one for courtyard by the solar lighting and electrical system.

6. In the landscape design, we take French garden as basic style, combing it with modern leisure landscape designing elements. The whole landscape system is made up of a landscape structure of one ring, two axes and three lines. One ring refers to an ecological corridor of the central ring lake. It is a kilometers long and 20 meters wide, surrounding the green lake island of near 50 thousands large. You can watch the sea here in distance and lake near you. There are dense plants, jogging track surrounding the lake, sand fields, healthy walking path, water front platform and leisure landscape platform on the lake, so it is a natural and healthy area for sports and also a reflection of the nature. Two axes are the central landscape axes at the entry, including the transverse cultural landscape axis and syncline Olympic Park landscape axis. The central landscape axis is a main one of the community entry to the central landscape area, and there are entry plaza, welcoming plaza, pigeon square, sunken

16

leisure clubs, lake-watching platform, tree squads, geometrical water features, springs, lighting ground, magic color light scribe and art gallery and so on. The transverse cultural landscape axis is a cultural style gallery connecting Phase 1and 2 and the central landscape axis. It merges special Chaoshan culture and exotic culture into daily cultural life of the community by the hollow corridors, small architectures along the waters. The syncline Olympic Park landscape axis reproduces the culture of Olympic and spirit of sports with its main theme of healthy sports, where there is life, there is sport. Tennis courts, badminton courts, sports and sculpture square, skating field, area for Wushu, Taiji, and for chess and cards, gyms, etc, shows atmosphere of sports everywhere. Three lines is the landscape belt throughout the groups, the nodes of the landscapes. By the combination of the node, line and plane, the landscape resources can be therefore shared.

China Overseas Property Management Company are invited in collaboration on managing the supporting services of the VIP Chambers of the community, which includes thermostatic indoor swimming pools, outdoor ones, water bars, pubs, net bars, tea houses, sauna gyms, salons, , rooms for playing checks, doing writing, drawing and music, reading and Audio-visual Rooms, cafeterias, business negotiation meeting rooms, multi-functional halls, community clinics, banks and brand business shops and so on. This whole management has provided for the residents with considerate and noble attentive housekeeping services. By packaging the image of the whole property, we have promoted the living quality, healthy insight and cultural taste of the community, creating a new pattern of life on by the water in Shantou.

17

13/15．內庭通路
14．园林水景
16．別墅区全景
17．別墅水景

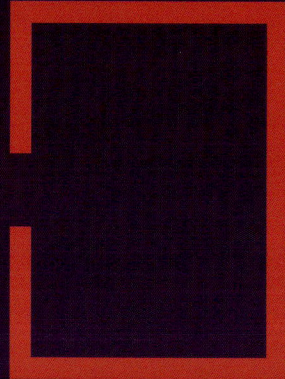

1

1．叠加效果图
2．叠加平一层平面图
3．总平面图

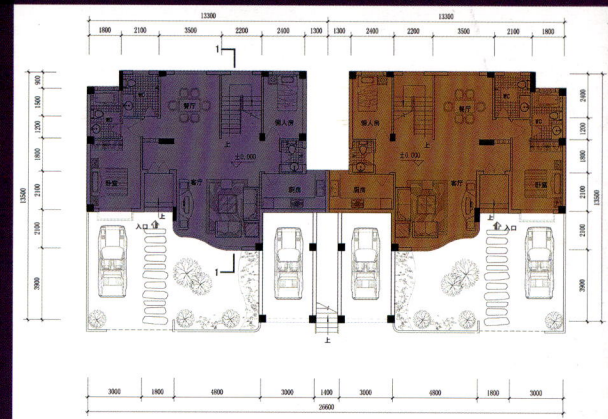

2

高鑫园·黄河原生墅规划

规 划 设 计： 优山美地·布里斯亚太联合规划设计公司
福州市规划院
用 地 面 积： 758 000 m²
建 筑 面 积： 600 000 m²

黄河原生墅规划项目位于河南省郑州市连霍高速路与新107辅道交叉口以北约2.5公里处，西临新107辅道，东临京黄公路，北距黄河3公里。该区块属黄河湿地地带，地下水位高（约1.5米左右），该地块原为养殖用地，地块内鱼塘较多，植被较丰富。

规划以"人本、自然、超前、融合、科技、实际、安全"为中心原则，以整体社会效益、经济效益与环境效益三者统一为基准点，着意刻画优质生态环境，为居民塑造都市中自然优美、舒适、便捷、卫生安全的怡然栖息之地。设计中充分利用临近黄河的生态环境形成的环境优势，实现人与自然环境、建筑群体的有机融合。

整体定位：城市近郊的低层高档社区。

社区远景：环境优雅，提供健康的生活方式，远离嘈杂拥挤的街道。自然环境受到保护，人们与鸟类及自然景观和谐相处，并从中受益。人们可以到此周末休闲或远足郊游，远离城市尘嚣。

社区由交通主路加三条生态景观走廊和水系组成空间骨架，组团相对独立，以中心绿地为核心；沿湖堤景观有中心公园、带状绿地、融合为一体，不同的草地，不同的水体，繁多的植物，将天然景观和人工景观有机融合。采用核心放射的形式，把小区分为若干组团，核心区是围绕中央景观区的小区服务配套设施和商业设施，居住组团环绕在外围。居住组团由放射状的绿色通廊和主干道以及水系自然分割。西侧沿107国道辅路设立100米绿化隔离带，西侧主路口南端为湿地景观区，北端为幼儿园、小学和小区主停车场形成的功能序列。这一布局最突出的特点是：所有

区域能与社区中心有便捷的联系，几条生态轴线自然将社区划分为均等的几个组团。

居住组团由景观轴水系道路系统自然划分相互独立，互不干扰。社区以西入口广场为中心，安排了会所、超级市场；商业遵循便捷的原则，布置在西侧核心区；三条生态绿带由中国古典园林和西方园林相结合，形成丰富的空间层次；休闲娱乐设施布置在组团的中心；组团的公共空间由开放的公共空间（小区西侧绿化核心区）进入半公共空间（住宅半围合空间），再进入半私密空间（住宅单元），再进入私密空间（户内），使组团中心与组团中心之间，互相串通、延伸，成为完美的网络；公共停车场的设置利用地形的高差营造半地下停车场，使得空间领域的变化更加明显。

柳林镇贺庄村

祭城镇北录庄村

N

京
黄
公
路

一
零
七
辅
道

20   60
0   40   80   100

图例
1. 西入口
2. 东南入口
3. 东入口
4. 会所
5. 主入口广场
6. 五星级酒店
7. 购物中心
8. 邮局、银行、卫生所
9. 商业步行街
10. 露天市场、地下停车场
11. 小学
12. 幼儿园
13. 儿童游乐场
14. 篮球、网球场
15. 门卫管理
16. 水处理厂
17. 垃圾中转站
18. 湿地公园
19. 水系
20. 地下停车场
21. 停车场
22. 放射性绿化带
23. 防护林带
24. 滨水平台
25. 室外游泳池

3

4

4．鸟瞰图
5．景观分析图
6．景观节点分析图
7．大双拼效果图
8．大双拼一层平面图

5

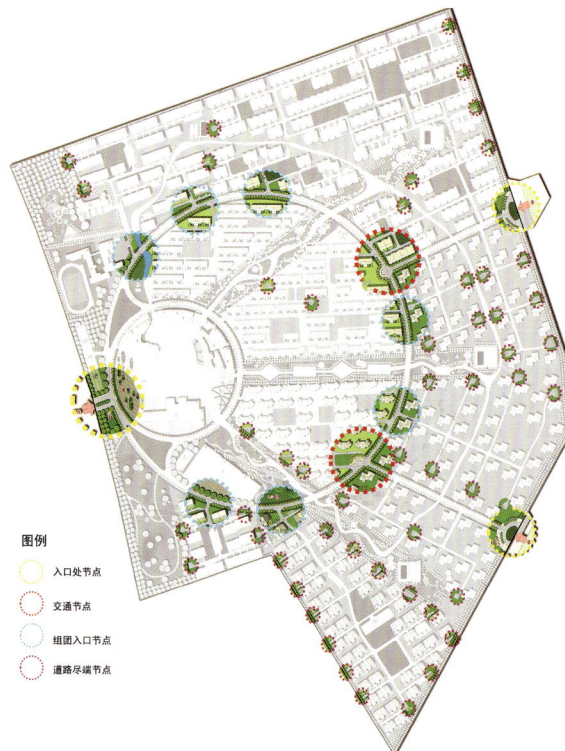

6

7

Plan: USMD+Polis Pacific, inc

Fuzhou Planning Design & Research Institute

Floor area: 758 000 m²

Building area: 600 000 m²

The project of Yellow River Original Villa is located 2.5km from the North of the junction of Lianhuo Expressway and the New 107 Des Voeus in Zhengzhou,Henan. Right in its West is the New 107 Des Voeus, Jinghuang Road is in the East and the Yellow River is 3km away from it. The block belongs to the Yellow River wetlands where the underground water level is about 15m. It was originately for raising fowl with a lot fish ponds and abundant plants.

With "Huam,nature advance, integration, technology, practice, safty" as the central principle of the plan and reunification of overall social benefis, economic benefit and environmental benefit as the standard, we aim to mold a comfortable and high quality living environment for residents.

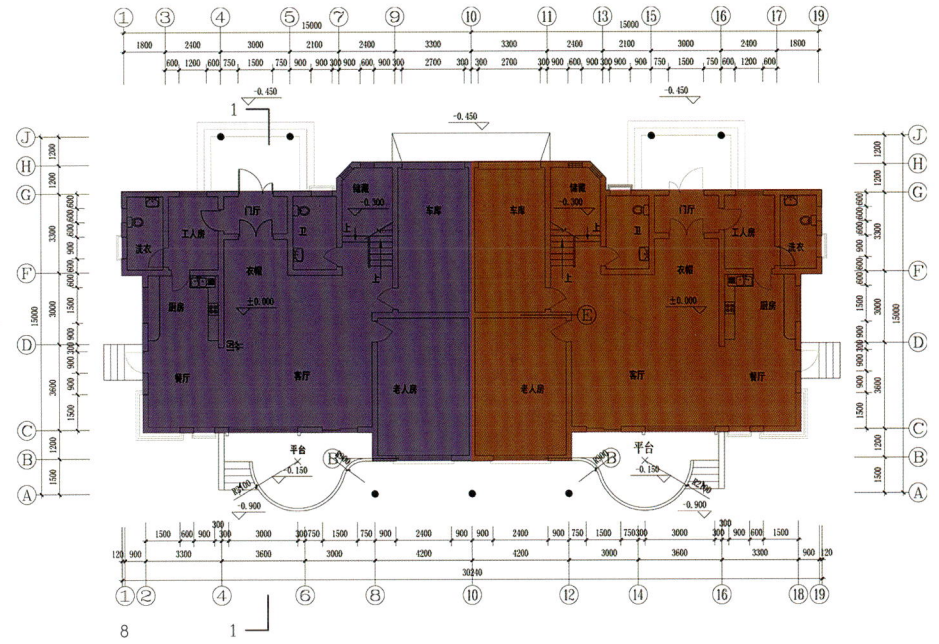

9

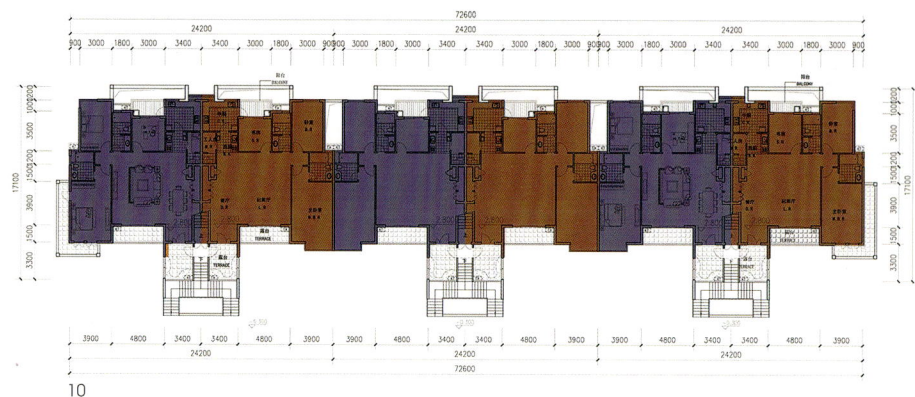

10

9. 花园洋房一层平面图
10. 花园洋房二层平面图
11. 花园洋房效果图
12. 联排效果图
13. 联排一层平面图

11

12

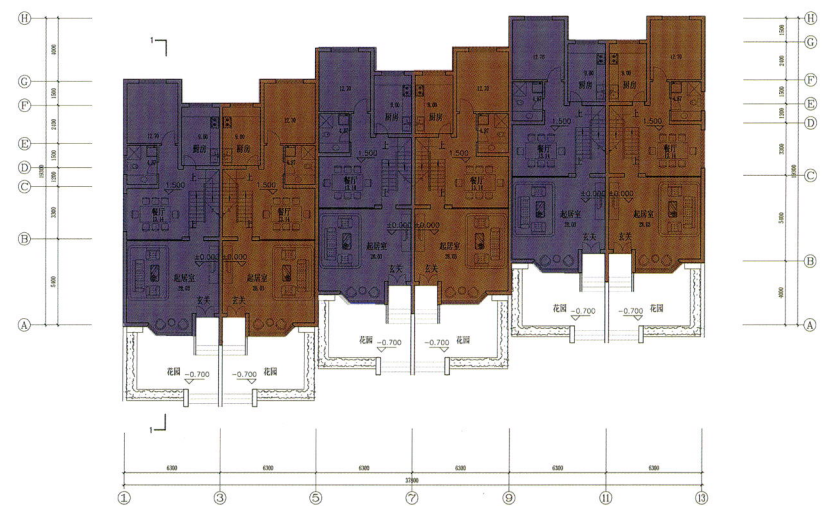

13

The advantage of being near to the ecological environment of Yellow River is made fully use of to have we human beings, nature and the construction groups organically fit into a harmonious picture.

Overall orientation: suburban low-storey community of high quality.

Future lood for the community: elegant environment far away from the noisy and crowded streets, healthy living-style are provided. Harmonious society of people, birds and natural landscapes being protected can be beneficial. People can com here for the weekengds or go hiking to escape from the cities.

Space framework of the community consists of the main traffic road, 3 ecological feature corridors and the water features, they are relatively independent from each other, revolving around the central lawn. Along the lake feature there are central park, green belts integrated as one. Different lawns, water and numorous plants organically form either natrural or artificial features. In the form of radiation from one core, the block service and commercial facilities revolving around the central feature park is set as a core, which is circulated by the block that made up of several groups naturally separated from green way, main roads and the water. Also, in the West along the 107 Des Voeus a 100-mertre-long greening segregating belt is set up, and South of the West Main Road is the wetland feature, the North is a function order consists of kindergardens, primary

14

15

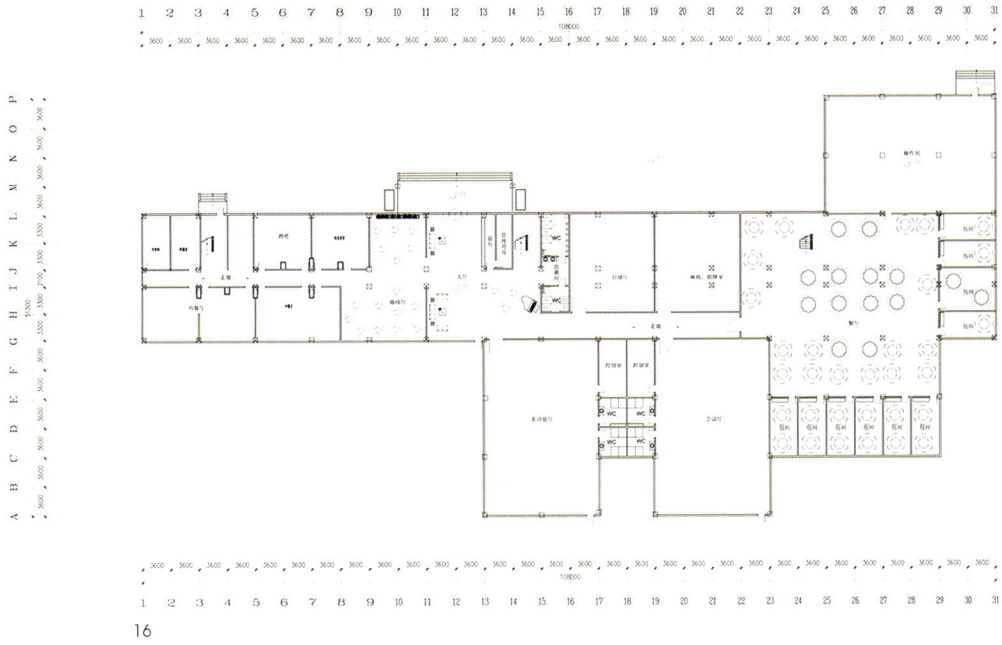

16

17

14. 退台花园洋房效果图
15. 小双拼效果图
16. 酒店一层平面图
17. 酒店效果图
18. 公建及商业效果图

![landscape perspective]

18

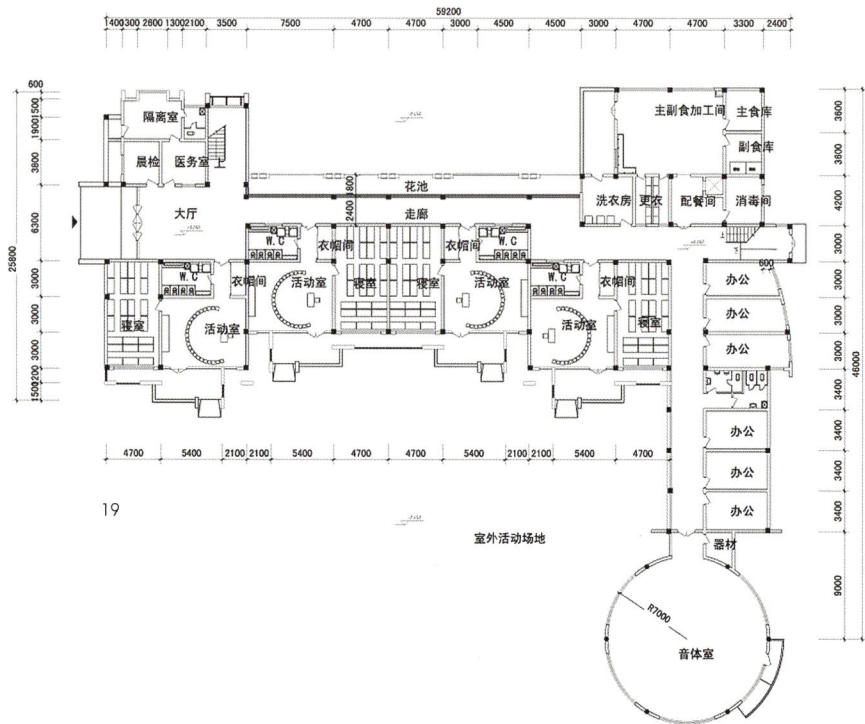

19

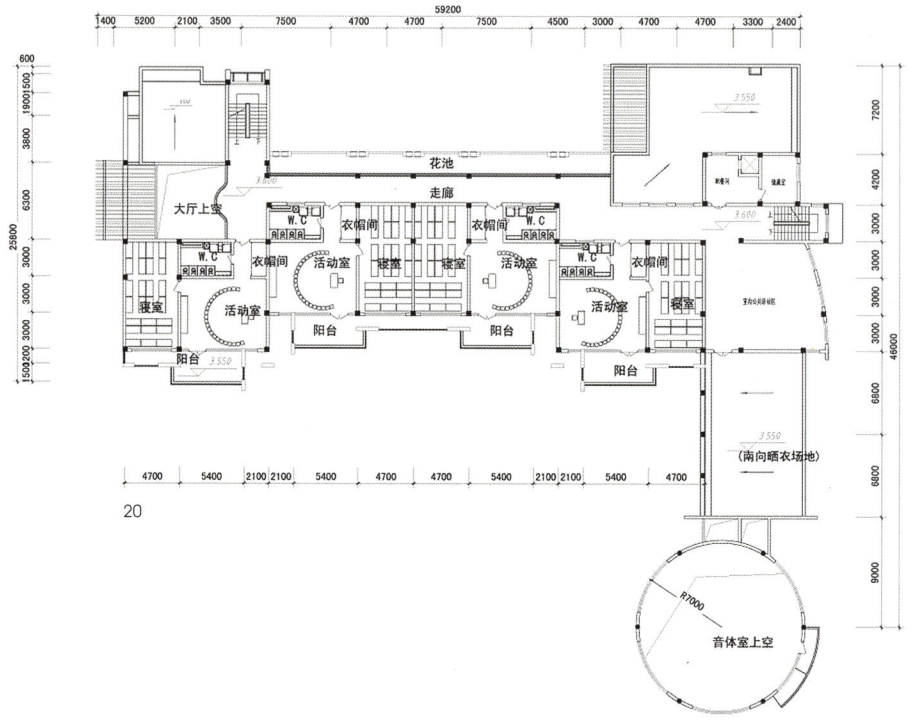

20

21

schools and the block's main parking lot. The most outstanding feature of such arrangement is that all areas can get quik touch with the community centre and several ecological asixes that naturally devides the community into several equal groups.

This housing groups is relatively independent from each other due to the feature axises water road system. The West entry plaza is the community's centre, with chambers and supermarkets; then the commercial area is set in the West core area according to the principle of convenience; 3 ecological green belts combining the form of Chinese classical garden and of the Western garden, add to the abundance of the layers of space; facilities for leisure and entertainment are in the centre of the groups. From open public space (the West core green area of the block) to the semi-overt space (semi-enlosure space of the houses), less private space (the residentail units), finally to the private space (indoors), such arrangement becomes a perfect network.. Variation of space is more clear with the spare space set as a public parking lot between up and under ground uniquely.

22

23

1. 体育馆白天透视
2. 体育场夜景鸟瞰
3. 体育场白天鸟瞰

昆明星耀体育运动城

开 发 商：昆明星耀房地产开发有限公司
建筑设计：澳大利亚U&A设计国际集团
　　　　　北京澳亚中元建筑设计咨询有限公司
主设计师：潘泰、潘洋、邓黔、熊艺

　　昆明星耀体育运动城是为2007年昆明举办残运会设计的，项目位于昆明市东南部，官渡镇、矣六乡用地范围内，总项目用地约173万平方米。规划内容主要分为2大部分：80万平方米的体育运动城建设用地及93万平方米的房地产开发项目用地。

　　运动城用地功能包括国际标准比赛场馆及其有关配套设施，其主要需解决的问题是赛时及赛后的可延续性。房地产开发项目主要为居住房地产开发。为昆明提供体育设施完善的健康休闲居

住区。本规划方案就是围绕上述两大功能要求展开，将体育比赛与赛后的运营利用有机地结合起来，以求成为昆明市体育、文化、休闲、娱乐及居住的永久性亮点。用地沿昆洛路及广福路两大城市主路，成不规则多边锯齿形。用地中央沿东西主向贯穿一条40米规划路，将用地分成两大主块。

　　规划布局：由于本开发项目的综合性功能要求，规划设计不同于一般居住区或体育城规划。本设计方案力图充分借鉴以往中

外成功规划理论及成果，平衡各种综合因素，以求达到技术、艺术及实际可行的国际水平。规划的一条南北向的主轴景观带将不规则的用地一分为二，左侧为房地产项目，右侧为体育城用地，其中房地产规划部分又被东西向的规划用地划分为南北两块，以开发的周期性分为一期至六期，南侧的一期用地在西侧端点处设为主入口。结合入口广场有一条500米长的景观带，沿景观带设置了小区中心会所。

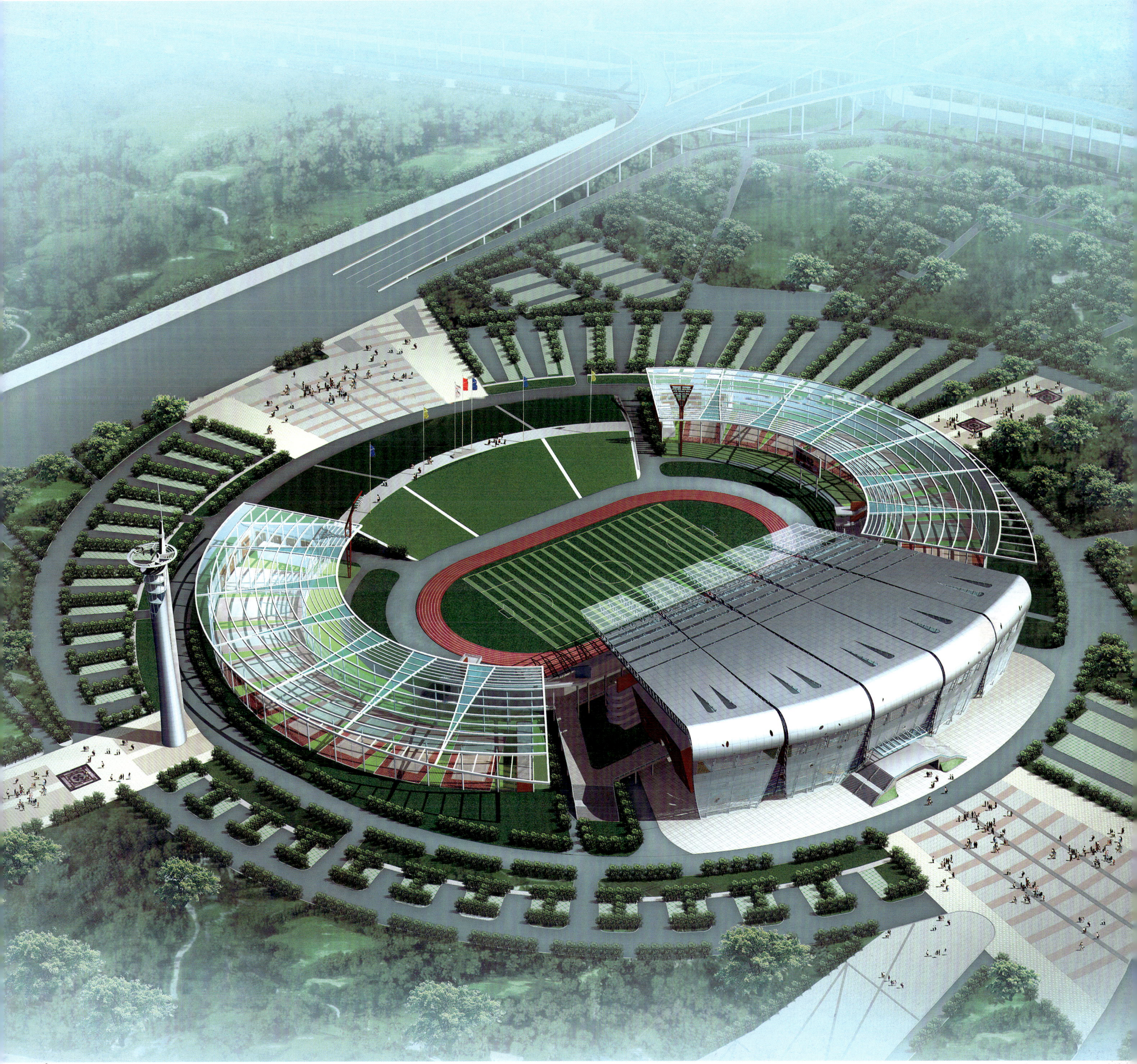

3

N

5

4. 总平面图
5. 鸟瞰图
6. 办公楼效果图一
7. 办公楼效果图二

4

6

7

8

9

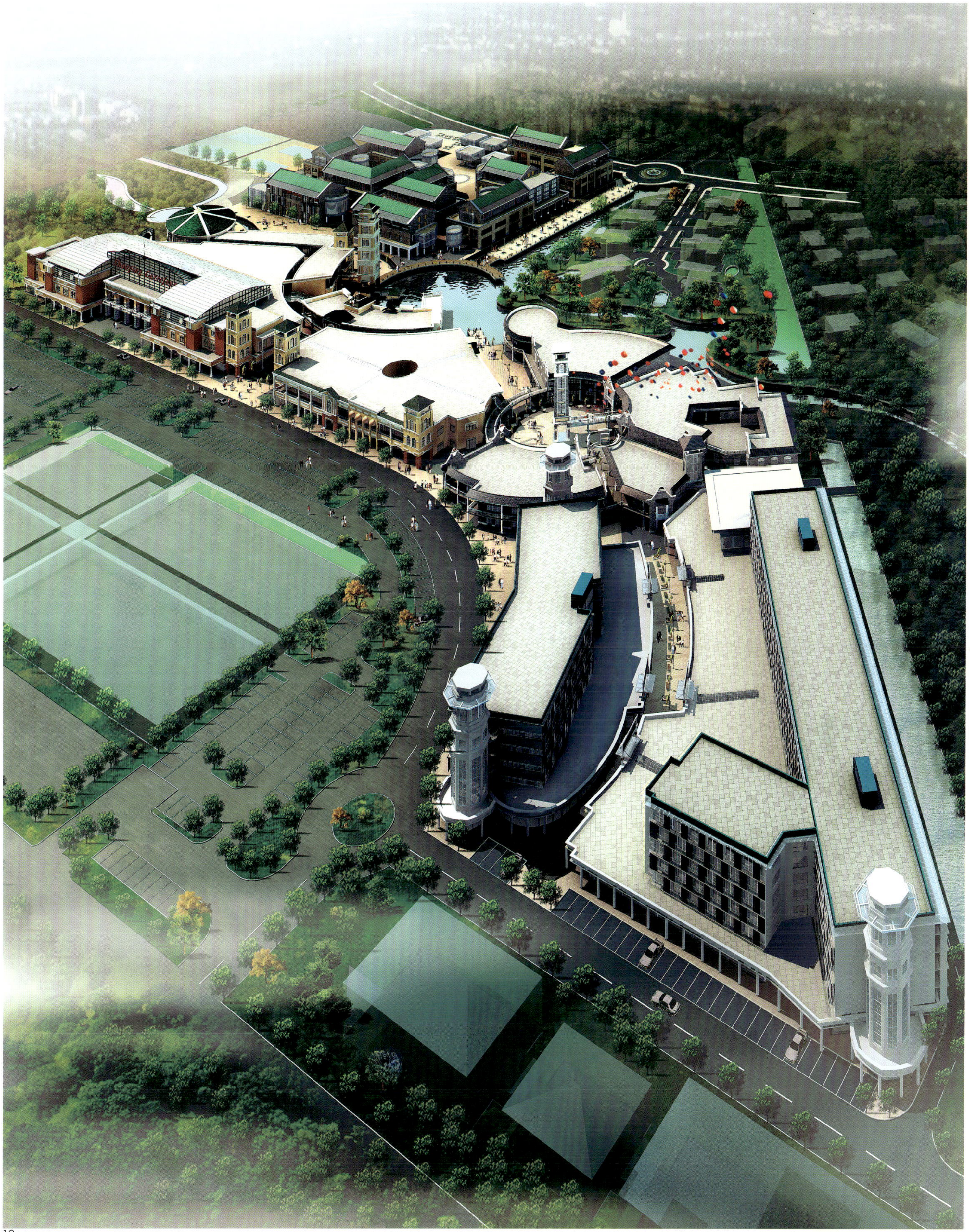

10

8. 商业街效果图一
9. 商业街效果图二
10. 商业街鸟瞰图

Developer: Kunming Shining Star Real Estate Development Co., Ltd.

Architecture Planning and Design: Universal Atelier International Group

Kunming Shining Star Sports Center is specificially designed for Kunming2007 Paralymic Games, which is located in Southeast of Kunming City .It's in the scope of Guandu Village and Ailiu Village with about 1.73 million square meters. This project is mainly divided into two part: 800,000 square meters for the construction sites and 930,000 square meters for real estae development.

The functions of the Sports Center include the international standard conpetition vanue and a complete set of facilities, which is aimed to deal with the continuity after the race. Real estate development projects is primarily for residents providing healthy and leisure residential area with sports facilities for Kunming. The plan focuses on the two main functions mentioned above. It organically integrateds competitions and the operation which focuses on physical training, culture, leisure, entertainment and residential area. The land of the Sport Center along the two main road, Kunluo Road and Guangfu Road is irregular multilateral serration. The center of the land along east and west runs through a fourty meters planned road which is divided the land into two parts. The other land is 400 acres which is located in south Guangfu Road and stand alone.

Programme layout: According to the requirement of synthetic functions, the design of this project is different from the ordinary residential area or Sports Center programme. This design plan aims to use the successful theories and achievements at home and abroad for reference, balance a combination of factors, to achieve technology, are and practical, international standards. Planning a landscape the main north-south band of irregular land will be divided into two,left for 1,400 acres of real estate projects, the right to 1,200 acres of land for the Sports Center, art of the real estate planning space is divided into two that are in north and south, of which are the   .A south side issue of the end point. Office is located in the west of the main entrance. Entrance plaza, a combination of a 500-meter0long with landscape of district center set up chambers.

11

12

11．图书馆与办公楼效果图

12．教学楼与学生宿舍效果图

13．幼儿园效果图

14

15

16

17

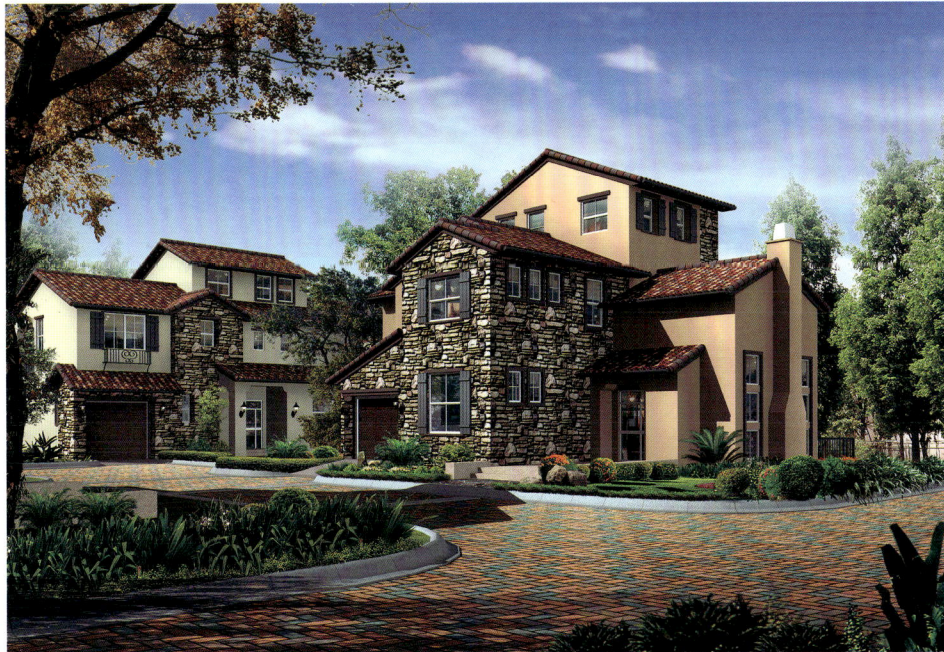

18

19

196

1

广东江海花园二期

开　发　商：江门市东华房地产开发有限公司
项 目 地 点：广东省江门市
建筑规划设计：澳洲高臣建筑事务所
总建筑面积：652 480 m^2
总占地面积：388 732 m^2
容　积　率：1.99
建 筑 密 度：27.3%
绿　化　率：39.8%
规划居住总户数：3030户

2

　　"江海花园"二期项目位于江门市中心城区，基地东临银泉路，南至五邑路，西起东海路，北侧为麻园路，整个地块较为方正。地块东北侧临已建成的江海花园一期。项目规划立足于高起点、重点体现江门市中心城区未来的居住方式。

　　本次规划将小区设置为"一轴一环"的结构布局：一轴为南北向景观主轴与斜向环境景观主轴组成的"S"形轴线；一环为连接各组团的中心环路，突出中心区域的地理优势，通过由点到线再到面，逐层带动各个组团之间的联系和谐发展，整体提高其产品的档次。规划采用突出院落的处理手法。以院落为基本单位，通过在组团内各种院落中灵活地布置绿化、硬地以及休闲设施，创造出丰富的宅前院落空间，并保证整个小区的协调一致性，体现"永续发展的生态社区"。

　　该项目分四期开发，一、二期以双拼住宅、联排住宅和情景洋房为主，三、四期以高层与小高层住宅为主。整个地块中心规划了一个中心水域，形成几个小型半岛，沿小岛周边布置双拼住宅，并将水面向一、二期其他组团延伸，不仅增加了小区内部的活泼感，同时也形成丰富的景观资源，达到资源共享。以南北向景观绿化主轴及斜向水域景观主轴为主线的"S"型轴线，以环状道路系统和生态绿化景观节点为纽带，串联起四期各个住宅组团。

　　一、二期以高尚住宅为主，设计中有私家花园，同时辅以水体景观，提高环境层次；三、四期以高层与小高层为主，规划中尽量扩大各组团的中心绿地，以提高大环境的质量，使该区居民真正生活在一个舒适的绿色环境之中。同时公共集中绿地与组团中心绿地相互呼应，扩大后期景观设计的可能性。景观主轴本身的设计在一、二期和三、四期也设计为完全不同的个性：一、二期高尚住宅区部分相对较窄，呈园林化趋势；三、四期高层区比较宽阔，呈都市化风格。

　　位于地块西南角的酒店与售楼部注重增加空间的参与性、情景的诱发性、聚合性，积极创造品位生活。商业布局分布在东南北三面，其间配以休闲广场和景观植物，形成商业步行街，突出显示了休闲性、享受性和文化性，符合现代居住和购物行为方式的特点。小学与幼儿园集中放在地块的东部，便于管理。

高尚住宅区建筑以简约的地中海建筑风格为主题，提取拱券、坡顶以及筒瓦等典型地中海的元素，立面造型丰富精致，采用质朴温暖的色彩，使建筑外立面色彩明快，既醒目又不过分张扬；所采用的柔和涂料，不产生反射光，不会晃眼，给人以踏实的感觉。各种类型的产品均形成独立的院落与平台，配以红瓦白墙，融入了阳光和活力，以及对于小拱券、文化石外墙、橙色坡屋顶、圆弧檐口等符号的抽象化利用，都表达出地中海风格的特征。建筑设计对称规整，高高低低，饶有趣味。同时注重基地所在城市的特色与文脉，充分利用地理优势，并使小区融入城市景观，规划和塑造未来江门的新城市形象。

1. 底商高层南立面图
2. 高层效果图
3. 总平面图
4. 售楼处效果图

3

4

6

Developer: Tung Wah Group Jiangmen City Real Estate Development Co., Ltd.

Project location: Jiangmen City, Guangdong Province

Architecture Planning and Design: Australian construction firm Hill High

A total construction area: 652 480 m^2

Total area: 388 732 m^2

Volume Rate: 1.99

Building density: 27.3%

Green: 39.8%

Planning, the total number of households living: 3030

"Jianghai Garden," two projects in the city centre Jiangmen City, the base east Yin-Quan Road, Wuyi Road, south, west and the East China Sea Road, Ma Yuen Road to the north, the whole block is Founder. A plot has been completed and the Northeast Side of the Garden, a Jianghai. Project planning based on a high starting point, the focus of Jiangmen City downtown future mode of living.

The planning area will be set to "axis of a part of" the structural layout: one for the north-south axis to the landscape and the main spindle oblique environmental landscape of the "S" shape axis; one of the groups connected to the central loop, Prominent regional centre of the geographical advantages, from the point through to the line and then to face, layers and various groups led the link between the harmonious development and overall improvement of the quality of their products. Planning, highlighted by the handling of the compound. To the compound as the basic unit, through the various groups within the compound of flexibility in the layout green, hard and leisure facilities, create a wealth of the former courtyard home space, and to ensure the coherence of the whole community, of "sustainable development Ecological Community.

The four sub-project development, one, two to Shuangpin homes, row houses and detached houses the main scenario, three, four to senior high-rise residential and small-based. The whole plot centres planning a centre waters, a few small peninsula and the surrounding islands along the layout Shuangpin residential, and 12 other water-oriented groups extension, not only increased the sense of community within the lively, but also a wealth of Landscape resources, to share resources. The main north-south landscape planting and landscape oblique waters of the main spindle "S"-axis to ring road system and eco-green landscape nodes as a link, all four series from the residential groups.

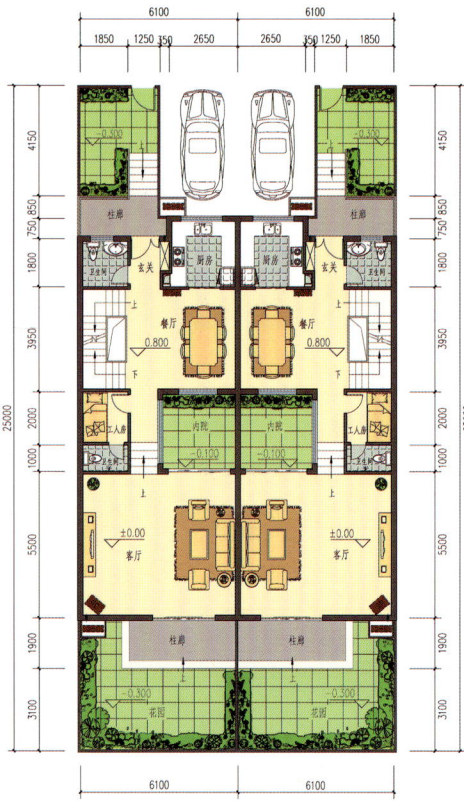

7

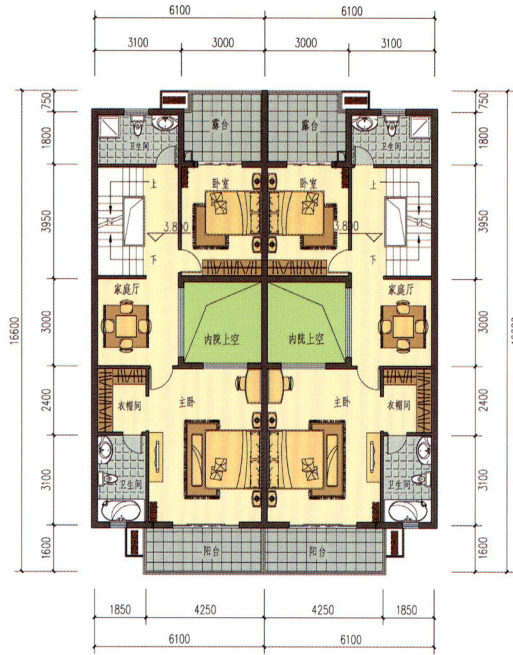

8

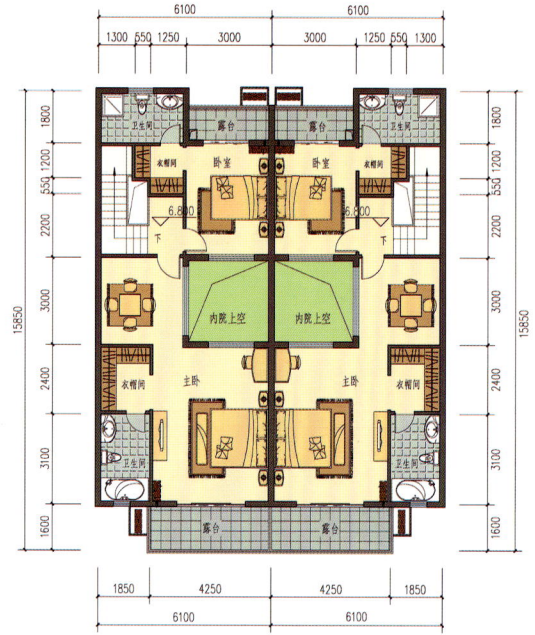

9

10

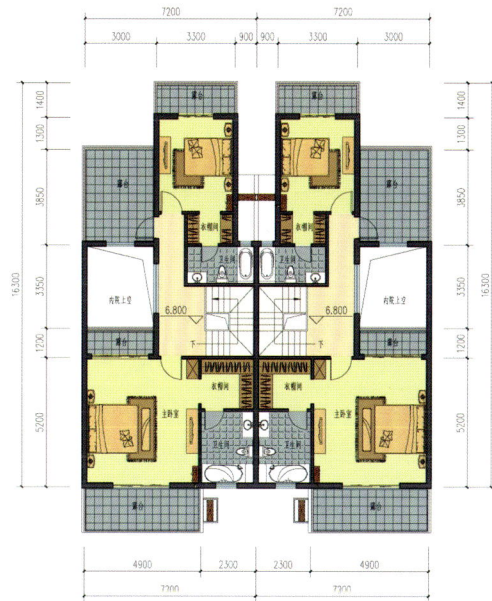

11

12

13

12 mainly residential with a noble, the design of a private garden, complemented by a water landscape, improve the environment level; 34 to mainly high-level and high-rise, planned to maximize the group's central green space, to improve the environment The quality, so that local residents really live in a comfortable green environment. At the same time focus on public green spaces and green groups echoed Center, expanding the possibility of the latter part of landscape design. Spindle landscape of the design in one or two and three, four also designed to be totally different personality: one, two residential areas of the relatively narrow noble, a garden of the trend; three, four senior relatively wide area, with Urbanization style.

In the block southwest of the hotel and sales increased emphasis on the participation of space, the scene of the induced polymerization, to create quality of life. Commercial distribution of the layout in three north-south east, which supported leisure Square and landscape plants, a commercial pedestrian street, highlight the recreational, and cultural enjoyment, with modern residential and shopping behavior characteristics. Primary schools and kindergartens focus on the block in the east, to focus on manageable source of trouble.

Noble simplicity of the residential construction to the Mediterranean architectural style as the theme, from Gongquan, Poding and Tongwa and other elements of a typical Mediterranean, elevation modeling exquisite rich, warm colors using simple, bright colors so that the Building facade, but also both the eye-catching The publicity; adopted by the softer coating, no reflected light, not dazzling, to give people the feeling at ease. All types of product are the courtyard and into an independent platform, supported by Hongwabaiqiang, into the sunshine and vigor, and for small Gongquan, cultural stone facades, sloping roof of the orange, arc Yankou symbols of the use of the

14

15

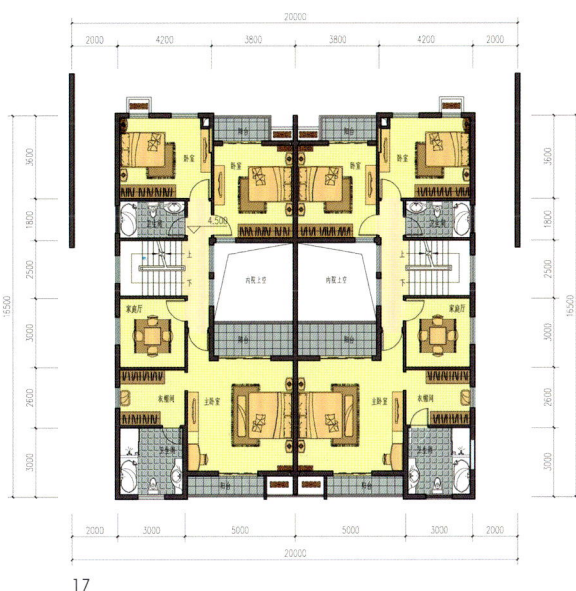

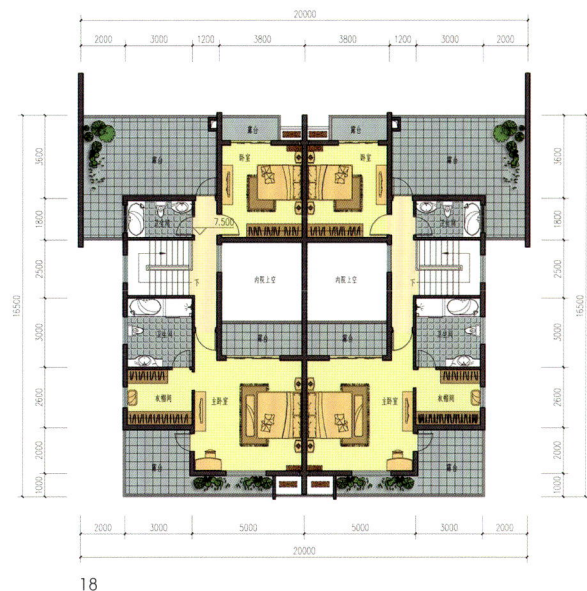

16

17

18

19

abstract , Have expressed a Mediterranean-style features. Structured symmetrical architectural design, Gaogaodidi, interesting. At the same time pay attention to base city with the characteristics of context and make full use of geographical advantages, and community integration into the urban landscape, planning and shape the future of the new Jiangmen city's image.

14．效果图
15．洋房效果图
16．A型双拼住宅一层平面图
17．A型双拼住宅二层平面图
18．A型双拼住宅三层平面图
19．A型双拼效果图

1

长沙中城丽景香山

开 发 商: 长沙中达房地产开发有限公司

项目地点: 长沙市雨花区体育新城

建筑设计: 深圳市立方建筑设计顾问有限公司

占地面积: 150 000 m²

总建筑面积: 400 000 m²

容 积 率: 2.3

绿 化 率: 40%

1. 透视图
2. 会所入口
3/4. 立面图
5. 鸟瞰图

2

3

4

5

6. 总平面图
7/8/9. 效果图

6

7

8

9

Developer: Changsha Zhongda Real Estate Development Co., Ltd.

Project location: Sports Center, Changsha

Architectural design: Shenzhen Cube-architects

Floor area: 150 000 m²

Building area: 400 000 m²

Greening rate: 40%

Volume rate: 2.3

With an intrinsically of 18.340 m² , land of Changsai Zhongcheng Lijing Xiangshan is located in Yuelu District, Changsha city, between Wangjiali Road and Gaoling Road. It's low in the west and gradually rises up towards northeast, maximum distance is near 6m. No construction and old trees is remained inside the field. The structure is planned to be a remained mountain with a north-south-towards grant axis connecting the west-east-towards courtyard. The main axis of the block consists of a sidewalk in the west side of the mountain, which is a main feature line that connects the north and the south, and the business street,main plaza for inner activities of the block, the mountain,etc that along with the sidewalk. That is a most wonderful point of the block.

10

11

12

209

The distance being not so obvious, the vertical design makes many slopes according to the layout, and sets ecological garage along with the feature, not only solvig problem of ventilation and lighting of the garage, but also saving energy. Traffic of the block is the form of pedestrian and vehicles being apart, motor vehicles enter the main undergroud express way directly through the block's round way, reaching all the feature. And the routes for emergency firefighting is set. The northsouth central axis in the block is the main feature area, shared by the whole block, and every group has its own group organizing center to meet residents' demand. Both sides of the entry's main aisle has all low commercial construction, showing a broad sight for you and the layout features is that rooom of small scaleinside the business block is major to create commercial atmoshpere, but large distance is set between residential houses. The block turns on the vision of splendid, and magnificent, New European classicalism is applied into all the construction , forming ternary form shape by the slope roofs, simplified eaves and material skintle,etc. And focus on the business street commercial atmosphere and design of the details in small scale, using Spanish Style to lay on the business street.

Body of the residential houses are all north-south-towards, with 2 residence on one floor. And inside the houses,lift hall are bright and cool enough to do housework. Distance between residential buildigs are broad enough to meet the demand of sunlight, and distance between houses. Entries of the block and the units have slope way and parking place for the disabled, leading systems also takes avyanadhana into account. Estate Management security also intelligentized.

10．交通分析图
11．空间分析图
12/13/14/15．建筑局部

13

14

15

16

17

16/17．立面设计
18/19/20/21．组团内景

18

19

20

21

1. 效果图
2. 立面图
3. 总平面图

昆明波西米亚凯旋花苑

项 目 地 址 ：昆明市西山福海乡陆家、杨家办事处
建 筑 设 计 ：澳大利亚U&A设计国际集团
　　　　　　　北京澳亚中元建筑设计咨询有限公司
总占地面积：69 286 m²
总建筑面积：57 334 m²

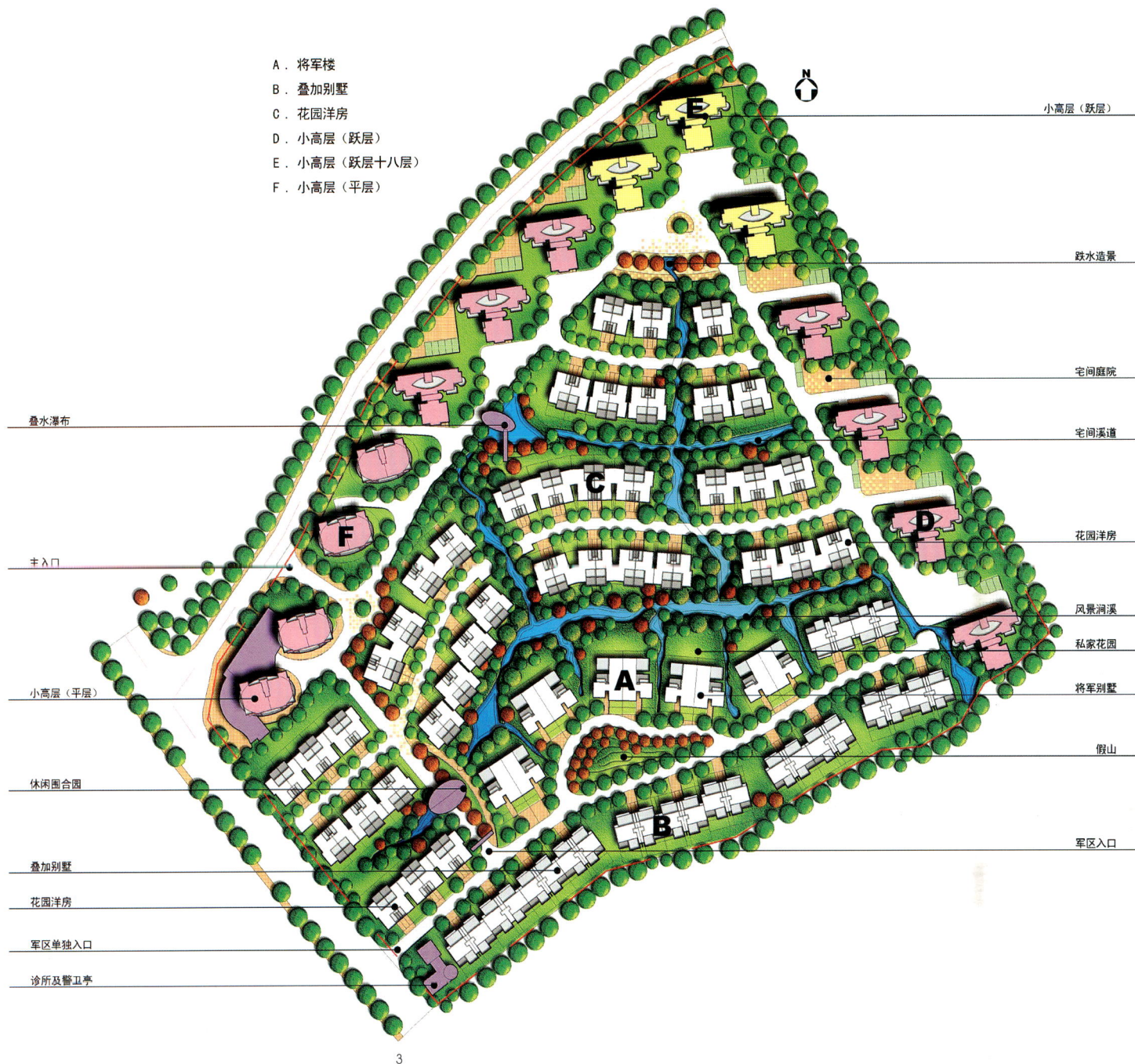

A．将军楼
B．叠加别墅
C．花园洋房
D．小高层（跃层）
E．小高层（跃层十八层）
F．小高层（平层）

小高层（跃层）
跌水造景
宅间庭院
宅间溪道
花园洋房
风景涧溪
私家花园
将军别墅
假山
军区入口

叠水瀑布
半入口
小高层（平层）
休闲围合园
叠加别墅
花园洋房
军区单独入口
诊所及警卫亭

3

Location: Fuhai Rural Land and Yang Office in Xishan District, Kunming City

Design: Australia U & A Design International Group, Australian Mesoproterozoic Beijing Architectural Design Consulting Ltd.

The total area of land: 69,286 m²

The total area of construction: 57,334 m²

The ratio of volume: 1

Triumph Garden is located in Fuhai Rural Land and Yang Office in Xishan District, which is by S.3rd Ring Road. It lies to the east of Fayuan Residential Quarter, the south of Southwest Hotel, the west of S.3rd Ring Road and the north of Western Tour Town, which is a comprehensive project that consists of residences, business and offices.

The design focuses on the peaceful feeling, and unifies social benefits, economic benefits and environmental benefits as a benchmark. It is intended to depict a peaceful holiday town landscapes and to create a natural, quiet and convenient place for residents to live in.

To avoid shaping into a strip space and increase more meridional dwellings, the land is divided into three groups according to the vertical flow direction.

214

4

5

4. 将军楼立面
5. 将军楼侧面
6. 将军楼透视
7. A户型一层平面图
8. B户型一层平面图
9. 将军楼透视二

6

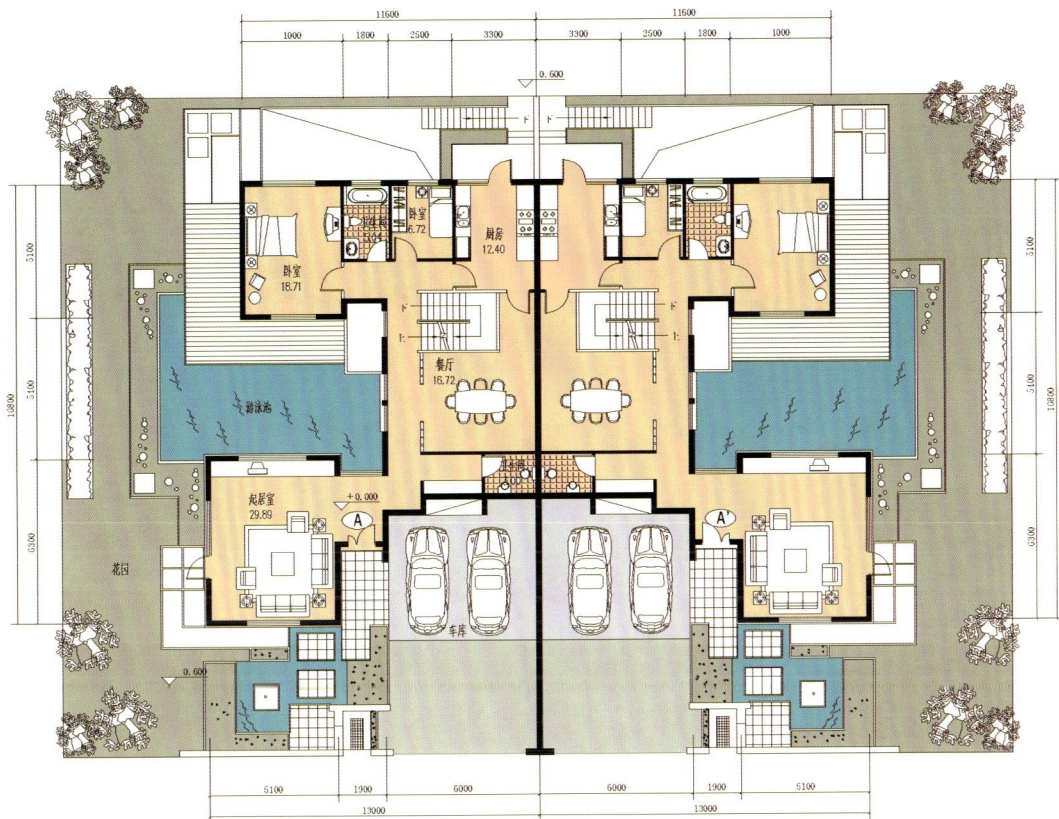

7

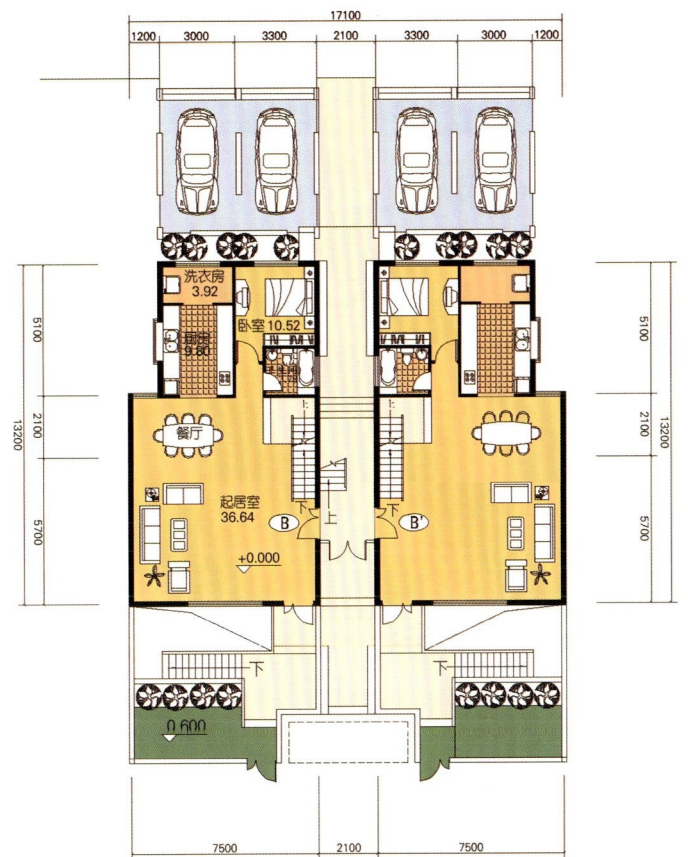

8

9

216

1

1. 夜景效果图
2. 中庭鸟瞰图
3. 总体鸟瞰图

田森·奥林春天

开 发 商：四川田森房地产开发有限公司
项 目 地 址：绵阳涪城区科创园区园艺路139号
规划建筑设计：北京越格建筑设计有限公司
设 计 师：李东
占 地 面 积：180 000 m²
总建筑面积：500 000 m²
建 筑 密 度：≤25%
容 积 率：2.5
绿 化 率：40%

2

"田森·奥林春天"项目位于绵阳市政府着力打造的未来第一高尚居住区，北临九州大道，南及西山风景名胜区，东依南山双语学校，西达西科大西山校区，占据园艺山制高点。毗邻原生态缓坡山林、雷锋湖、西山人文风景区、迎宾大道景观带等众多原生态绿化。周边汇聚倍特·林郡与中华坊等众多高端住宅项目，名校云集，人文居住气氛浓郁，生态环境清幽。"奥林春天"以其优越的社区环境，一流的建筑品质、超大规模的景观规划，金牌物管的贴心服务，缔造绵阳最高品质生活。规划设计打造百米超大尺度的主题庭院，营造舒适、健康和谐的山地生活空间，将自然的生活气息，朴实的建筑品格，作为整个小区的基调，利用多种户外健身运动场所，积极引领现代都市人热爱运动，追求健康的生活主张。

"以景为核"的分区建构

各区各组团之间用景观进行分隔，同时也作为共有资源而充分的将欣赏景观最大化，使每户每院都可以拥有属于自己的特色景观。而用于分割的景观带以"线"的形式汇集在中心的以"面"为特征的景观中心，加之小区内部重要的景观"节点"，于是形成了整个小区的景观网络收放有序的情趣和韵律。

以"院"演绎传统经典的空间格局

这其中有两个含义，一个是建筑格局本身的艺术魅力，另一个方面是文化习俗的浓缩提炼。项目设计有意将每个组团以"院"成组，并通过景观和建筑布局来形成有特色的新的空间围合，使空间既有归属感、私密性的同时，又增加了每个空间组团的可识别性和特色。

"以人为本"的人文精神规划模式

处处以区内的居民为出发点，设计在最大程度上方便住户，并使他们能充分接触自然环境。整个小区基本上形成了"人车分流"的规划结构，方便居民出行和保证交通安全的同时，也提供了更加专属的休闲娱乐的景观步行系统。

強調道路系統的"附加功能"
　　使道路除具有原始的出行功能外，復合一些其他的功能，在設計中著力整個道路及其交通的豐富蘊涵，體現三大功能：交通功能、交往功能、景觀功能。

綠化景觀網絡化、主題化
　　景觀規劃以庭院景觀為主題，致力於創造一個空間相互獨立、私密的豪華居所。項目以自然有機意象為主，以抽象幾何的營造系統為輔，滿足社區安全、便捷、優美、私密的要求。各種景觀元素的配置產生相互借景的效果，邊緣界面的營造及建築與

景觀的融合成為本案的重點，豐富的綠化使建築物能充分地與軟性景觀融合，通過地形、植被、水景的變化，營造出淳樸自然的都市田園風情。

塑造風景中的建築
　　風景是院落和道路的表層，它定義了空間的性格：公共、半私密、私密。風景定義了院落、道路和建築的界面，風景並不止於界面，它蔓延到公共庭院和建築的立面，它是建築內部空間的延續，並最終成為其中一部分。

公共配套設施的小品化、景觀化
　　將垃圾站，變電室，沿街小型公建等戶外的公建設施同綠化景觀結合起來，形體立面的變化作為建築群視線近景的豐富手段，使它們成為其中的一個部分。同時，作為一個有特色的景觀小品，它能豐富小區的戶外空間場所，使整個小區的景觀更加體現人性化、情趣化、精品化。

　　立面設計堅持建築立面的簡約原則。建築外觀採用簡約、整潔的處理手法，整體外觀以直線勾勒為主，細節之處輔助點、線、面交錯式設計，汲取簡約、時尚的現代建築風格，糅合簡單的幾何

图例：

1 24軌制國際雙語小學
2 16班制國際雙語幼兒園
3 VIP高級會所
4 老年活動站
5 24小時連鎖超市
6 便民超市
7 乾洗店
8 茶樓
9 醫療中心
10 中西時尚餐廳
11 游泳館
12 網球場
13 籃球場
14 戶外運動器材場
15 屋頂特色幕坪
16 過街天橋

17 紅酒坊
18 雪茄室
19 商務會議室
20 SPA水療中心
21 連鎖高級健身中心
22 高檔美容中心
23 名品書店
24 情景商業街
25 主題雕塑
26 音樂水泵
27 藝術走廊
28 自然氧吧
29 景觀小橋
30 堆石假山
31 藝術鋪裝
32 疊級水池

33 景觀小島
34 人造叢林
35 藝術小景
36 人工湖
37 古典連廊
38 休閒廣場
39 人工釣魚絲
40 游泳池
41 菱苣樹陣
42 依水起階
43 觀景平台
44 體育館

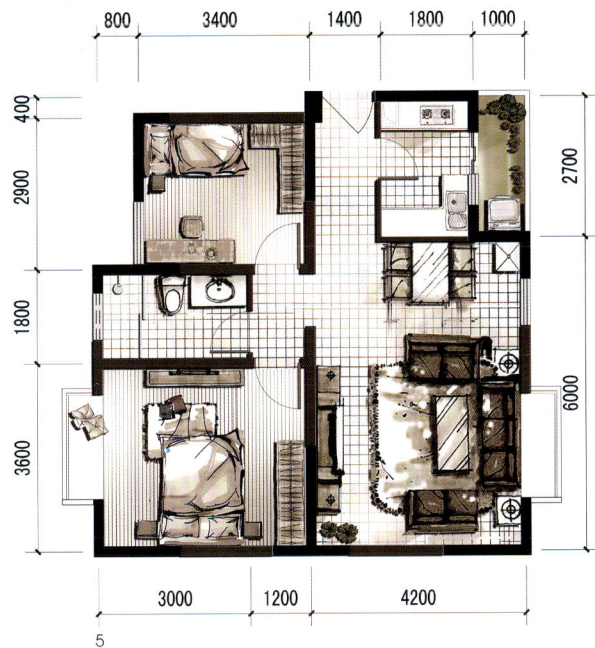

5

6

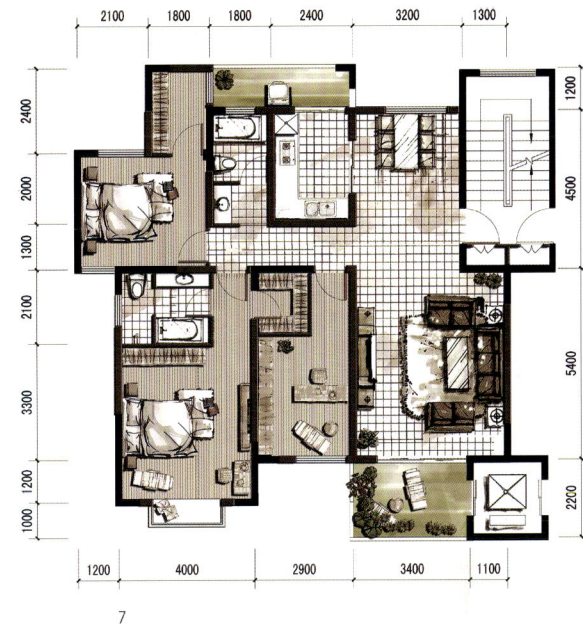

7

4．总平面图

5．B户型平面图

6．D户型平面图

7．E户型平面图

8．内庭效果图

与功能主义手法，在整洁的线条中又赋予整体多样的变化，再融合建筑的尊崇色调以及窗户玻璃金属般的质地，使建筑单体更加方正挺拔，耀眼而夺目。

8

9/11．会所入口

10．别墅入口

12．别墅局部

13．园林实景图

Floor area: 180 000 m²
Building area: 500 000 m²
Greening rate: 40%
Volume rate: 2.5

Project of "Tiansen•Aulin Sring" is located in the leading noble residential quarter that is forged with most efforts by the government of Mianyang City. It faces Jiuzhou Avenue in the north, West Mountain Landscape Area in the south, Nanshan Bilingual School in the east and Xike Campus of Daxishan in the west, taking up the commanding height of the Yuangyi Mountain. Also, it is next to quite a lot of original ecological place such as the original ecological gentle slope forest, Leifeng Lake, West Mountain Cultural Landscape Aea, Yingbing Avenue Landscape Area,etc. And with a forest of famous schools bounding around, it gets a strong cultural living atmosphere and quiet ecological environment. Aulin Spring aims to create life of highest quality with its advantages community environment, feature plan of top-class construction qualityand extraordinarily large scale and best logistic management service. The

plan is designed with theme courtyards of extraordinarily large scale which own comfortable, healthy, harmonious living space. And plain construction style with natural living atmosphere is taken as basic tone of the whole quarter, meanwhile many kinds of outdoor gyms are used to positively lead modern city inhabitants to the love for sports and pursuit for a healthy life.

Division structure of "feature as core". Each block and group is segregated by features which are also shared fully and maximumly as common resources, enabling every residents to have special views of their own. Meanwhile, these views area used for division is converged by a line form into the center with plane form feature center, making the whole quarter an ordered feature net.

Traditional and classical space framework is performed by courtyard with two definition, one of which is article glamour of the construction pattern itself, another is the extract of culture and costom. The project intendedly makis each group as yard and by such feature and layout to form new characteristic enclosure, not only adding to the sense of belonging and privacy, but the recogntion and that of the space group.

Planning pattern of human-oriented. Restidents is a stand-point of the design, maximum convenience and sufficient touch with natural environment is therefore considered. And basic structure of pedestrian and vehicle separation ensures convenience of residents and traffic safty. Also, exclusive walking system for leisure and entertainment is provided. Emphasis on attached function of the road system. Besides primary function of trip, roads are abundantly equipped with other functions like traffic, communication and features.

Networked and thematization of greening feature. Courtyard feature is set as a theme of the feature planning, devoting to creating a relatively independent, private and luxurious residential living place.

And with natural organic image as main part and abstract geometry system supporting it, the project is to meet the demand of community safty, comvenience, grace and privacy. Disposition of each feature element produces an effect of taking each other's advantages. Creation of edged interface, the convergence of construction and feature is key point of this project which is embodied like the following description: abundant greening integrateds the construction and soft landscape, meanwhile creates natural, simple garden city style with variation of terrain, vegetation, and water feature.

14

Construction with scenery. As a surface of the courtyards and roads, scenery of the quarter defines how space will be: public, seme-private and private. Not only does it define the courtyards, roads and surface of the construction, it also reaches for public courtyards and elevation of the construction. It extends the inner space of the construction and becomes part of it.

Public facilities.Out door public facilities like refuse storage station, transformer room are combined with greening views, making such variation of elevation of moled hull a way to make construction groups part of it. Meanwhile as a distinctive feture, it enriches the outdoor space of the block which is more humanized, more refined.

Design of the elevation holds to the principle of briefness. The outward appearance of the construction is brief and neat. And the wholness is outlined with straight line, and from point, line to plane to support in details. Concise, fashional modern construction style with simple geometry and functionism add to its multiple changes. In addtion, color tone of the contrucion and metal quality of the window galss makes construction unit more straight as well as shining.

14. 花园
15. 建筑立面
16. 水池景观
17/18. 水景小品

15

16

17

18

19

20

21

22

19/20．会所大堂
21/22．会所过道
23/24．样板房客厅
25/26．样板房卧室

23

24

25

26

1

厦门国际山庄

开 发 商：厦门国源房地产开发有限公司
项目地址：厦门市湖里区湖里虎头山公园南侧
占地面积：61 038 m²
建筑面积：18 906 m²
容 积 率：1.25
绿 化 率：44%
车 位：323个

2

　　国际山庄西临莱坑路，东面是已建好的住宅小区，南面为城市主要交通道路金山路，北面是虎头山，山庄依山而建，南北高差5米左右。山庄建筑为厦门高尚别墅区，设计巧妙利用山势，依山设景造势，有效利用空间，把山体融入山体，让山庄成为山体不可缺少的一部分。利用中西园林相结合的模式，以现代风格为主，中式园林为辅，借势强化山体的力量，配以柔性的水景，刚柔相济，声情并茂。以"山"为载体，用森林中的各种元素，加以提炼、美化，应用到设计中去。顺山势引水，势为水随，水到溪成，山水相依，人宅相扶，天地感通。人车分流，动静交替，刚柔相济，高低错落，层层美化，立体营造。大量应用抽象化手法，使景观文化内秀而深厚，意味悠长，耐人寻味。用"无声胜有声"的乐律表现模式来体现景观的灵动美，使整个环境温馨而浪漫，充分表达出人居环境的高雅和优美。小品风格上充分体现山林、都市的企业主题文化。

　　游泳池区利用现代小品景观的表现方法，如天鹅喷水、花盆喷泉等来表现游泳池的都市特色。同时巧妙使用曲线和浓密的植物使游泳池显得自然而生态以呼应整个山庄设计自然化、生态化的大趋势、大走向。泳池内的图案设计清新活泼，营造了轻松快乐的健身嬉戏氛围。木平台休息场所的设计则增添了泳池大环境温馨浪漫的情调。人工湖区在游泳池旁以连贯水系此起彼落、前后呼应。在人工湖里用一艘游船为模型组景，人们可以在这里闲聚观景，听流水潺潺，看湖光山色。湖边木平台为人水相亲搭建了自然舒适的平台，湖中心的两艘小船，既呼应了大船的设置又迎合了木平营造的温馨宁静，荡漾于波光艳潋中，又增添了生活的无限情趣。

　　活动广场区是生活区的功能配套区。合理利用空间，在架空楼层底，营建健身、休闲娱乐场所，以森林中叶子为设计元素

使铺砖的图案生动而又充满生机。在广场中设置了可供中老年打太极、挥洒曼妙舞姿的硬地广场，可供少壮者拼比技艺的乒乓球、羽毛球场，还有可供儿童嬉戏的游乐场所。闲余时候，无论男女老少均可在此找到自己活动的天地。茶吧休闲区让业主充分接触、沟通，营造一种良好邻居关系。

　　私家花园：在连排别墅之间不是有墙体简单分割而是用立体密植的植物，形成一道生态分割线，使花园隐逸于幽秘之中，延伸了居住者的私密空间。在花园的设计上主要以低矮植物为主（草坪）配以一棵大乔木，两三棵中小乔木，局部点缀香型草本植物以做到简洁明快，便于平常的管理。

3

1. 中庭实景图
2. 园林局部
3. 中庭鸟瞰图
4. 总平面图

4

6

7

5. 园林景观
6/7. 鸟瞰图

8.别墅实景
9.园林水景

Floor area: 61 038 m²
Floor area: 61 038 m²
Building area: 18 906 m²
Greening rate: 44%
Volume rate: 1.25

10. 小径通幽
11/12/13/14. 园林水景

10

233

International Manor faces Laikeng Road in the West, and there is a completed residential block in its East, the South is the main city traffic road Jinshan Road and the North is the Hutou Mountain. This manor is a noble villa block in Xiamen City, which is built against the mountain with a distance of about 5 metres between its North and South. The smart design makes use of space of the mountain to create features, effectively merge the manor into the mountain, making the manor a essential part for the mountain. With modern syle as major mpart and Chinese gardening second to it, softy of the water as well as strength of the mountains, the Manor becomes a combination of the Chinese and Western gardening patterns, powerfully and mildly, statically and dymanically. With "mountain" as conveyor, we extract and modify every kinds of elements in the forest to apply them into the design. Down the mountain comes the water, along the water goes the river,is is just with mountain and water being together as it is with harmony between people and the houses. Penestrian and vehicles are apart from each other, static and dynamic changes alternatively, vision of differen hights of houses, all are three dimensional. Also, elegance and beauty of the living environment is fully express with strong cultural feature, vivid silent picture, warmly and romantically. All the styles embody cultural themes about green mountains and city enterprises.

City specialty of swimming pools is embodied by modern feature like swan water, fountain pot,etc. Meanwhile, swimming pools with curves and lots of plants appeals to the whole design of the Manor, which tends to be natural ,ecological. Also, pictures inside the pools are fresh and active, creating relaxing atmosphere. Woodern rest platform add to the romantic and warm feelings of the whole environment. Man-made lake also fluently chimes in with the pools. There's a pleasure yecht acts as model feature groups, where people can watch the scenery for leisure, listen to running water singing. The natural and comfortable woodern platform and 2 boats in the middle of the lake not only appeal to the setting of the big boat, but also fit into the warm and peaceful atmosphere, adding much joys to life.

Activity plaza area is for the living area, space of which is reasonably used, empty in the building bottom to set up gym, leisure entertainment club. And leaves in forest is as designing elements, making the patterns on the bricks active and lovely. Playground for middle-age and elderly people to play Taiji is arranged in the plaza, it also can be used by the young to play table-tennis and badmiton, and be a fun fair for children, too. All people at all ages find somewhere the belong to when they are free. The tea bar area allows residents communicate with each other to build up a good relationship.

Private garden: townhouses are segregated by solid plants instead of walls, such devider line makes gardens keep subtle, better protecting residents private area. In terms of the garden designing, lawns are major, with megaphanerophytes, two or three micro phanerophytes at intervals, partly ornamented by herbal flavor plants, biref and easy for management.

11

12

13

14

1

顺驰·领海

开 发 商：顺驰置地达兴房地产开发有限公司
项目地址：大兴区 黄村西北端
建筑面积：住宅部分450 000 m²
景观设计：EDAW
建筑设计：加拿大UDS国际建筑事务所
容 积 率：1.41
绿 化 率：51%

2

　　顺驰·领海把法国普罗旺斯一年到头灿烂的纯金色阳光、难以抗拒的蔚蓝色诱惑和高尚的情调复制过来，成为让人心驰神往的第一千个理由。法国，是全世界公认的流行时尚样板；巴黎，更是将格调和品位做到发梢的楷模，是浪漫奢华的极乐世界。有人说巴黎人的眼睛都长在头顶上，他们相信自己的审美超过相信上帝。但是有个法国笑话说："想看看一个巴黎人真正妒嫉的样子吗？——告诉他你住在普罗旺斯吧！"全世界的人都羡慕巴黎人的生活，而巴黎人却在一心向往到普罗旺斯去度假。

　　处在阿尔卑斯山脉的庇护和蔚蓝海岸的悠闲生活之间，普罗旺斯有幸逃过汹涌的工业和商业大潮，把源自罗马时代和中世纪的宗教艺术以及贵族气质一直传承下来。塞尚、凡高、高更、毕

加索，这些毕生追求色彩和光线的大师都在这里找到了令他们燃烧激情和生命的精神物质。世界首富比尔·盖茨、威尔士王子查尔斯以及众多欧美名人富豪都置业在此，好像要寻找什么比财富和地位更重要的东西，就连英国足球大帅哥贝克汉姆和辣妹维多利亚也在去年硬挤进了"普罗旺斯族"的花名册。总之，在普罗旺斯，最多的是两类人：艺术家和富人，以及数千对来普罗旺斯度了一次假就辞掉大城市的工作举家搬迁的疯狂的夫妇。

　　顺驰·领海项目整个地段中央是3.8万平方米的"海"域，仔细琢磨一下就不难看出建筑师和开发商的用心——法国南部被称为"蓝色油漆桶"的地中海已经在中国北京大兴黄村1号地泛起了涟漪。在世界地图上找到地中海，然后找到法国普罗旺斯的位

置，在顺驰·领海项目总平面图相同位置上找到风情街，这两个地区的建筑及景观具有被卫星定位后逐行扫描过来的相同气质。设计在这里用的一招叫拼贴城市，精确复制。经典原版的普罗旺斯斜屋面、大波瓦、百叶窗和钟楼不仅为建筑师带来无限的设计乐趣，也会在地产品位的大战中占尽风头，将猎奇与惊艳的目光收集于一身。

　　在顺驰·领海随处流露着让人耳目一新的卖点和对传统问题的解决方案，就连一般设计中不能避免的交通空间——走廊，也被设计师做成有意思的东西了。设计始终用夹杂着对情调与品位的向往的设计手法去追求再现人们心灵最远端的认同，让异国的生活片断像戏剧场景一样被拉近和讴歌。

目前，北京的房地产似乎已经进入翻拍美加城市、瓜分欧洲文明、复制世界文化遗产的时代。楼盘广告上对于塔楼板楼是否户户朝阳的介绍慢慢被富于视觉冲击力的异国情调的巨幅画面所取代；每一个有"海归"身份的楼盘都会为设计带来对未来生活的无限憧憬；更多的二次以上（含二次）置业人群的选择正在从满足基本居住要求转变为选择品位和文化内涵的不容忽视的精神购买力，开发商和建筑师对于产品的定位和设计已经直接关系到设计的生活品质。但是，怎样使方鸿渐式的克莱登楼盘不光在案名上是个洋秀，不只做表面现象上的潮流跟随，而是要有真正拿来的东西，让开发商自信地对消费者说：你想过什么样的生活就可以选择什么样的楼盘，在选择地理位置、户型平面的同时还可以选择一下生活的品位。这是一个亟待解决的问题。在这种情况下，从规划设计方面来说，建筑师必须为地产开发商提供一种精神的产品，让他们能够用令人信服的建筑语言和崭新的生活方式来按耐住人们最不安份的想象力和荣耀之心。

1. 隔水相望
2. 建筑局部
3. 总规划鸟瞰图

3

4. 总平面图
5. 亲水区域
6. 远景

4

Project: north-west south of Huangcun Village, Daxing District.

Floor area(residence): 450 000 m^2

Landscape design: Canada UDS International Architecture Firm

Greening rate: 51%

Volume rate: 1.41

Repulse bay copy the glorious gold sunshine that shines throughout the year, the temptation of the irresistible sky blue, the elegant atmosphere, which have become the thousandth excuse that to be infatuated. France is a world-recognized fashion model, and Pair is even the model that put the style and taste into details, a romantic and luxury paradise. It is said that eyes of Parisians are overhead, and they believe in their own aesthetic than believing in God. However, there is a French joke saying that to people who really want to have a look at Parisian, just tell him that you live in Provence. People all over the world are all envy at living in Parisians, but Parisians only yarn for a holiday in Provence.

Under the protection of Alps and the leisurely pace of life along azure coast, Provence was fortunate enough to escape the surging tide of industry and commerce and religious art and the aristocratic temperament originated from the Roman era and medieval has been passed down. Cezanne, Van Gogh, Gauguin, Picasso, the life-long pursuit of a master of light and color are here to find a burning passion to make their life and the spirit of the material. Richest man in the world Bill Gates, the Prince of Wales Charles and many celebrities in Europe and America are rich in this home, like there are things more important than fortune and status in the world for them to seek, even the handsome British soccer great David Beckham and Spice Girl Victoria is also ranked into the roster "family Provence" last year. In short, there are two large types of people in Provence: the artists, the rich, as well as thousands of crazy couples who quit their jobs in big cities and move here after they have had a holiday in Provence.

There is a central "sea-domain" of 38,000 square meters' large in the whole area. And it is not hard to find the architects and the developers' purpose----Mediterranean in South France is known as the "blue paint", which has rippled gently in No, 1 Land, Daxing Huang Village, and Beijing China. You can find Mediterranean in the world map, then the

237

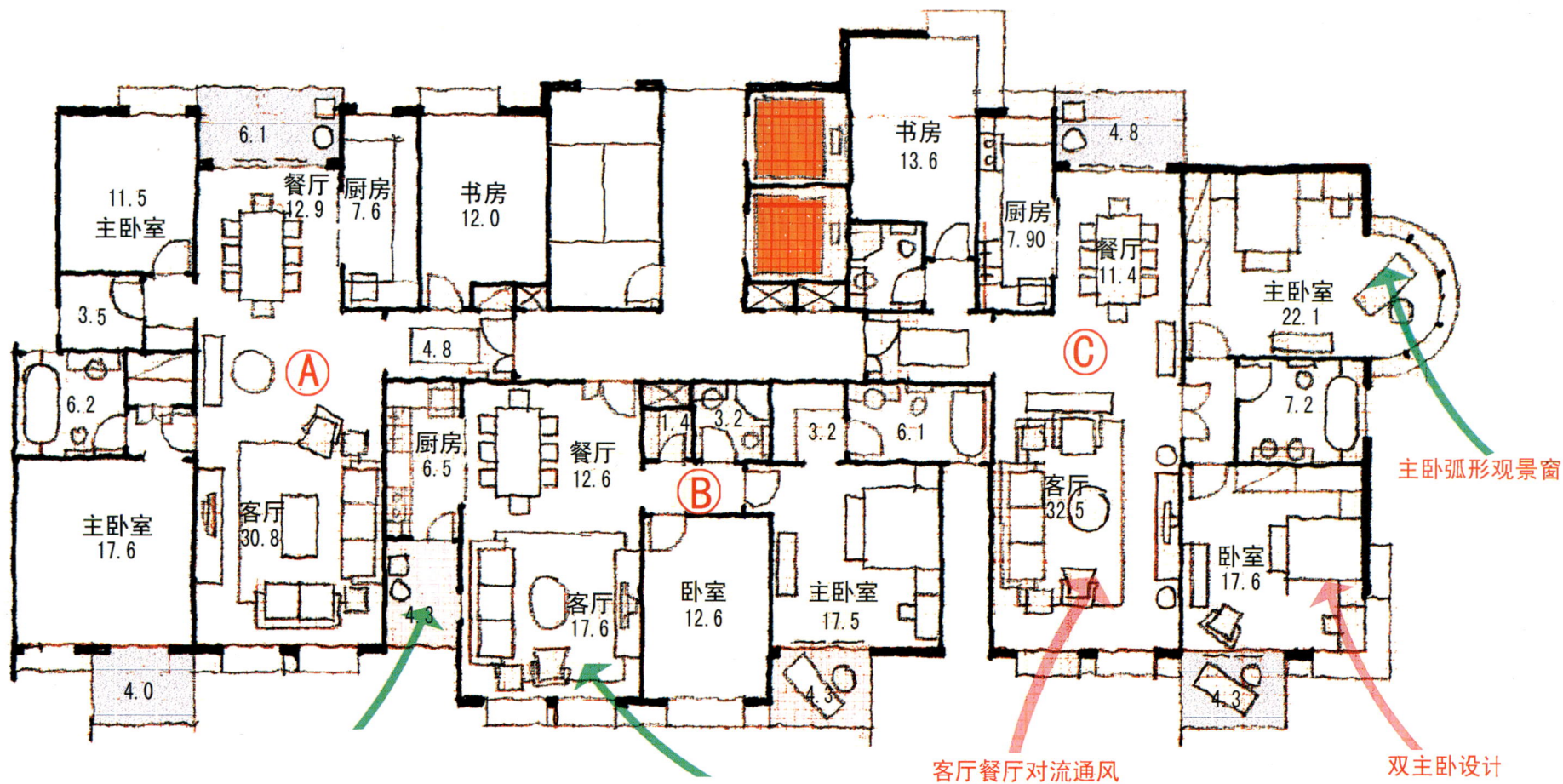

主卧弧形观景窗

客厅餐厅对流通风

双主卧设计

7．户型平面图

8．广场钟楼

9/10．广场

9

10

location of Provence, and in the same place in the overall planning of Repulse bay, you can find the street of styles, landscapes and architect. Construction and features of these two areas are in same temperament after scanning by the satellites. Such is called Collage City which is exactly copied. Classical original vision of Provence gets slanting roofs, big wave povoa, Venetian shades and bell tower, not only bring lots of joy for architects but also dominates in the invisible fight of ranking of production, attracting all eyes on it.

Everywhere in Repulse bay reveal a fresh selling point and the solutions to traditional problems, even the unavoidable transport space "corridors" in general design are regarded as the interesting things. Design have used a mixture of emotional appeal and tastes to pursuit and recur the agreement in the remotest soul, let the exotic fragments of life were drew near and eulogized

At present, Beijing's real estate seems to have entered times that copy American and Canadian cities, carve up the European civilization, and reproduce the world's cultural heritage .the introduction on Real estate ads whether every turret and slab-type apartment is face the sun or not are gradually replaced by huge screen of exotic atmosphere that is full of visual wallop. each of the real estate have identification of "returnees" will bring about the unlimited vision of future life, the choices of more home-buyers who buy over second time (including secondary) have changed from meeting the basic requirements of living into psychical purchasing powers that can not be ignored choosing culture and taste, there is no doubt that product positioning and design which developers and architects are directly concerned with the living quality. However, how to make the Calydon housing project that regard as the hero "Fang Hung-chine" in " Fortress Besieged" not only has a alien name, a trendy fellow in Appearance, but also give out something. Let the developers said to customers confidently:" life what you want to live is building what you can choose." While choosing the location, size of the plane, you can have the option of the quality of life. It is a problem demanding prompt solution. In such case, concerning layout, architects need to offer a kind of mental product that they can use convincing architect language and new lifestyle to calm down the honorable heart and fancy imagination.

11

12

13

11. 亲水区远眺
12. 隔水而居
13. 宅间园林

1

嘉兴巴黎都市

开　发　商：嘉兴市广源房地产开发有限公司
项目地址：嘉兴南湖区中环东路妇保院西侧
景观设计：杭州禾泽都林建筑景观设计有限公司
占地面积：253 104 m²
总建筑面积：800 000 m²
容　积　率：1.9
绿　化　率：49%

广源·巴黎都市位于风景秀丽的南湖之畔，嘉兴市政府与南湖区政府之间。项目秉承"建顶级楼盘，造特色产品，创嘉兴人骄傲"的开发目标，以总投资近20亿元的大手笔，分十二组团开发，汲取巴黎人文、自然、艺术、科学等诸元素，精心打造一座集生活、学习、休闲、娱乐、办公于一体的最适人居的法式品质建筑。

城市是人类在一定地域范围内营造的一个生存空间，建筑是人类高度文明和城市化的产物，住宅是建筑、艺术和技术在一定

城市空间与城市发展历史过程中的传承与共生。单从住宅规划与设计的角度来说，理想的城市住宅应该是应该把经济效益、环境效益和社会效益结合起来，以营造最佳居住环境、最好居住条件为中心，使区域规划与设计达到目标功能、环境功能、社会功能的要求，也可以简要地概述为理想的城市住宅应该是文化与绿化的结合。

项目背景

巴黎都市位于嘉兴的南湖板块，是嘉兴未来的行政、经济

文化中心，同时凭借优越的自然条件也被定义为嘉兴市环境高尚的休闲生活区。此地块曾以6.35亿元总价创嘉兴土地拍卖史上的新高。项目总投资20亿元，她秉承"建顶级楼盘，造特色产品，创嘉兴人骄傲"的开发目标，打造一个地标级建筑群体。

文化理念
文化定位

规划力图从法国巴黎的城市设计的角度出发，理性的解构巴黎城市空间，并吸纳巴黎的历史文化和建筑人文景观，彰显巴黎异国情调和历史文化内涵。在大空间形态上，巧妙地将埃菲尔铁塔以及塞纳河等法国元素融入到规划骨架中；在建筑艺术方面，将凯旋门、卢浮宫以及法国的休闲文化、咖啡色建筑色彩等法国建筑、人文、艺术等一一展现；在景观艺术方面充分利用建筑设计格局，突出空间景观，加强竖向设计，使整个小区高低错落，并紧紧围绕巴黎都市的空间概念，联想巴黎的生活"情境"，融入现代法国时尚、浪漫、自然、活力的都市风情，创造皇家宫廷巴黎生活意境，构筑都市特色生活。

规划空间文化

我们分析巴黎都市的城市空间形态并与先前先进的居住理念及实际的用地现状相结合下提出两区、一街、一心、二轴线的整体合理的空间组织布局。两区就是用地的南北两大区块，北区从西至东依次布置为居住用地及9班的幼儿园用地、48班的小学用地，购物中心用地，各用地布局紧凑合理，通过一街即香榭丽舍大街的过渡进入南区，大街仿造巴黎香榭丽舍的空间商业格局，成为休闲、娱乐、购物、旅游为一体的全新的商业形态；南区由一心、二轴线构成，也是整个项目的核心区域；一心即为西提岛（卢浮宫会所）一座人性化活动场所，内设结婚礼堂及健身、活动专用房，成为社区一个重要的集聚场所；二轴线就是十字相交的南北步行中央景观轴线及东西的塞纳河水轴。

南北中央步行轴线从凯旋门（南区的主入口）出发，穿越商业步行街到达协和广场，通过新桥到达西提岛广场（卢浮宫会所），从西提岛广场继续向南延续，穿越具有西方古典园林的布局的景观带——阳光大道，直至星河码头结束，这条轴线的空间序列布局犹如一曲动听的西方古典音乐，东西塞纳河水轴从中环

2

3

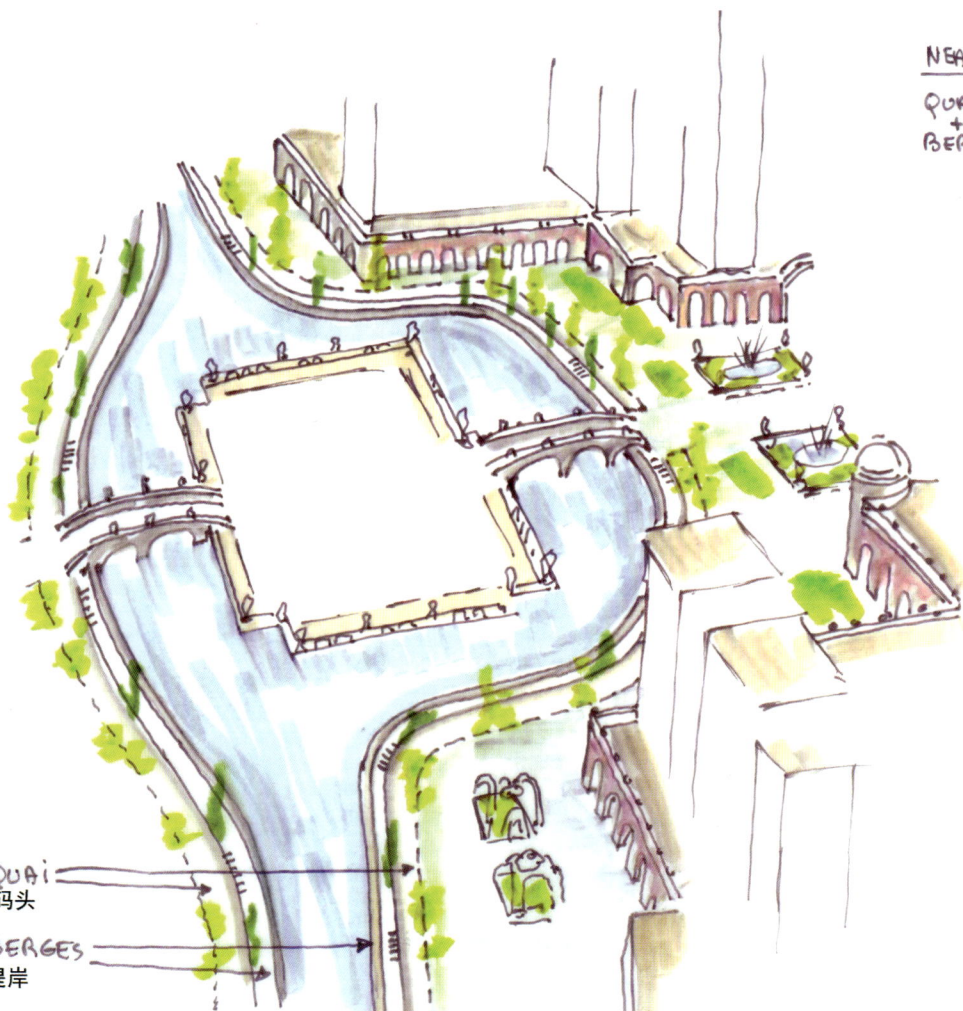

NEAR RiVER
QUAi
+
BERGES

QUAi
码头

BERGES
堤岸

4

东路（南区的次入口）进入沿着曲折的河流，布移景异，并在河的北侧设置左岸休闲街内设休闲酒吧、茶座、商业购物及一些景观小品，使人们仿佛身置于巴黎真正的塞纳河上，使塞纳河的优美风光再次在此淋漓尽致地体现。

居住空间文化

在小区的规划中，我们始终强调空间的交往、开放性，却有明确的领域特征空间，除了整个小区空间布局采用外围内散的布局手法外，同时还有意识地创造一个有丰富景观的空间序列，将整个室外空间划分为四个空间层次，即：公共空间——半公共空间——半私密空间——私密空间。公共空间即为贯穿南北的中央景观轴线及东西的塞纳河水轴，这是居住小区内居民可充分享

受的共享空间，尤其在塞纳河两岸，一侧为沿河景观绿带，可以静静地欣赏塞纳河的风光，另一侧则结合沿河的商业用房可以品茶、娱乐，以增加小聚居民的休闲去处。半公共空间是组团内的组团绿地，供组团内的居民使用，各组团绿地通过环形步行道，得以相互沟通。半私密空间是住宅楼之间的院落空间，其空间形状各异，也是居民最常去的地方。私密空间即住宅及住宅底层的庭院。小区的居民正是这些空间上相互交往、接触、熟悉、游乐，从而充分体现了"以人为本"的规划设计核心。

商业形态文化

嘉兴巴黎都市在商业形态上强调完整性、均好性、协调性，其商业主要由香榭丽舍大街、沿塞纳河及中央轴线入口处的步行

街和老佛爷购物中心组成；其中香榭丽舍大街主要通过道路、道路与商业建筑之间休闲人行道及商业建筑骑楼式的架空道廊这三者相互沟通，在大街的局部如中央轴线的主入口侧采用内街的形式，既扩大商业面，活跃商业氛围，又起到形态的变化；沿中央轴线入口处的步行街创造一个宜人尺度的休闲购物空间。

建筑艺术文化

构成建筑艺术的要素，包括建筑群体和单体的体形、内部和外部的空间组合、立面构图、细部处理、材料的色彩与质感以及光影和装饰的处理等等。住宅是生活的反映，巴黎都市试图从生活方式出发探寻居住需求。在户型方面根据小区品位与消费者需求，基本形成了单身贵族、两人世界、温馨家庭、豪宅（空中别

245

6

7

建筑是环境的科学与艺术

　　环境艺术是衡量理想城市住宅的一个关键要素。巴黎都市景观设计提出以居住休闲为主的花园式居住区的生活需求是景观设计的最大出发点，满足不同层次居民的需求，必须以人为本，使景观服务与人，构筑功能景观。组团内部分别抬高造坡的景致，把建筑遮掩在郁郁葱葱的乔木和灌木之间，营造丰富的天际线和地平线，局部建筑单元架空，增加景观和绿化的渗透。研究学习法国等欧洲园林的植物搭配手法，强调欧式造型与现代园林的搭配，并做到因地制宜，疏密有致，适当配以时令花卉。根据小区周边环境特点，对影响小区的"噪声"、"视线"采取"挡"的手法，对房前屋后的半公共空间采取"围"的手法。

文化与绿化的统一

　　巴黎都市吸取自然、文化、社会、经济等诸多元素，精心打造一座生活、学习、休闲、购物、娱乐、办公于一体的城市片段。该设计的成功之处主要体现在四大方面：运用巴黎的浪漫休闲文化、延续巴黎城市格局、商业文化脉络以及现代居住理念与自然环境、艺术景观相结合的总体布置形式、创造完善的城市配套，建筑类型的多样性及优美的住宅环境对城市景观的贡献，使小区更具识别性与个性特色。总之巴黎都市的设计从绿化、文化二者之关系入手，创造一个文化和绿化不断交融的人文绿色家园。

墅）等系列成熟产品；在建筑形态上强调建筑艺术与城市文脉的和谐，注重对传统建筑的传承；材料及构造的创新与应用也同样成为巴黎都市的一种建筑语言；建筑施工工艺上的创新与控制也成为巴黎都市设计师们参与讨论的一个热门话题。

浓郁的社区文化

　　在巴黎都市无论从巴黎公园到塞纳河，还是从协和广场到凯旋门；无论是香榭丽舍大街到老佛爷购物中心，还是从巴黎都市实验小学到卢浮宫会所；无论是河边漫步还是商场购物，无论是在社区里大草坪还是到中心广场，总能有您畅游的天地。透过卢浮宫里结婚礼堂依稀的听到远处传来的朗朗读书声，原来生活是这么美好。

绿化概念

　　自然是建筑的原始属性。

　　居住区的建设离不开土地，它与建筑基地地形、地貌、地物等密切相关。巴黎都市结合用地现状充分注意保护环境保证有足够的绿化面积，并要注意防止空气和水域的污染，应为居民创造出一个能健康地成长、愉快地生活的人居环境。尽量减少对人工能源的依赖，多利用自然能源，寻求适应于当地气候特点的建筑形式，利用建筑自身的气候调节能力创造宜人的小气候。创造了巴黎公园、塞纳河、长中港沿河生态景观走廊等一系列自然生态小环境。

8

Project location: West of Maternal and Child Care Service Centre, East Zhonghuan Road, Nanhu District, Jiaxing

Developer: Jiaxing Guangyuan Real Estate Development Co., Ltd.

Landscape design: Hesoms Architectural & Landscape Design Co., Ltd. (Hangzhou)

Floor space: 253 104 m^2

Gross building area: 800 000 m^2

Plot ratio: 1.9

Afforesting: 49%

Guangyuan•Paris City is located on the bank of Nanhu Lake, between the governments of Jiaxing City and Nanhu District. On the developing goal of "to build top-class buildings for sale, to make special product and to create pride of Jiaxiang citizens", 2 billion Yuan has been invested to elaborately construct French style architecture in 12 groups, and many elements have been taken into the project, such as Paris culture, nature, art and science, making it the most suitable for living with the functions of life, studying, leisure, entertainment and office.

City is a certain living space drew by human, architecture is the product of highly human civilization and urbanization and residences help architecture, arts and technology derivate and integrate together in such certain city space and its developing process. See from residential planning and designing, an ideal city residence, aimed to provide best living place and condition, should be a combination of economic effect, environmental effect and social effect. It can also be a combination of culture and greening.

Project background

Paris City is located in Nanhu in Jiaxing, a prospect administrative, economic and culture center of the city as well as a leisure residential area with its excellent natural condition. It covers an area of 23.3 hectare, which once created a highest price, 0.635 billion, in the history of Jianxing' land auction. This project takes a total investment of 2 billion, aimed to build a landmark construction group with the development goal of "to build top-class buildings for sale, to make special product and to create pride of Jiaxiang citizens".

Cultural rationale

Cultural orientation

Started from Paris' city design, we try our best to deconstruct Paris' city space reasonably and take its historical and architectural features to demonstrate foreign style and culture. Also in larger space shape, French elements such as Eiffel Tower and Seine River are taken into the planning skeleton. Then architecture, culture and art are all demonstrated by the Arch of Triumph, the Louvre. In terms of feature arts, the architectural layout is made full use of to show the space, and vertical design makes the whole community in picturesque disorder. Such houses revolving tightly around French city space

concept allow you to enjoy lively French life and its being fashioned, romantic, and natural as well as its vitality.

We analyze city space of French and combine it with advanced living rationale and real land using situation. Therefore, space layout of two blocks, one street, one center and two axes are put forward. "Two blocks" means two areas of the land, the south and the north ones.

The north one is arranged in order from west to east with residential land, 9 classes of kindergarten, 48 classes of primary school, and shopping center. Each land used are reasonably closed to each other, and by going through the Avenue Champs Elysees you come to the south block. The avenue learns from the Paris one to make a new commercial shape, combining leisure, entertainment, shopping and tourism as a whole. The south block consists of one center and

10

11

12

13

14

two axes, which is also a core area of the whole project. One center refers to Cite Island (the Louvre's chamber), a important humanized activity place with wedding assembly hall, places for fitting and activities. The second axis is the north-southward central walking landscape axis and west-eastward Seino River Axis.

Starting from the famous Arch of the Triumph (main entry of the south block), the north-south central walking axis reaches Xiehe Plaza through the commercial pedestrian street, gets to Cite Plaza (the Louvre Chamber) through the Xinqiao Bridge. Then it extends towards south from Cite plaza through the classic western garden landscape belt---- the Sunshine Avenue and ends at the Xinghe Harbor. Landscapes changes as you walk on the west-eastward Seino River axis, which enters the winding river from Middle Huandong Road (sub entry of block south). It will also make you feel like in the real Seino River, enjoying the beautiful scenery, by setting landscape ornaments such as leisure bars, tea houses and shops on the left bank, north side of the river.

Living space culture

In the plan of the community, we make an emphasis on interaction and openness of space from the outset, but there is clear space with region features. Except the way to build a community that are enclosed outside and free inside, we also make it a space full of landscapes. The whole outdoor space is divided into fore space layers, that is, public space, semi-public space, semi-private space and private space. The public space is the Seino River Axis which goes throughout the central landscape axis of the north and the south. People in the community can fully enjoy this space, especially on both bands of the river, one of which are greening landscape belt along the river where you can watch the scenery silently, and another are the houses for business, providing you places for tea and entertainment. Semi-public space is the block green lands provided for the residents. The green lands are connected to each by the ring pedestrian paths. Semi-private space is the courtyard space between the dwelling buildings. They are in different shapes and places where residents go the most often. Finally, private space are the residences and the courtyards in the bottom of them. Residents communicate, get familiar and have fu n with each other in these spaces, which demonstrates humanized designing core.

Commercial culture

We emphasized that in terms of business, Jiaxing Paris City must be complete, homogeneous and harmonious. It was made up of the Avenue Champs Elysees, the pedestrian street along the Seino River and the central axis, and Laofoye shopping center. The Avenue Champs Elysees is connected by the main roads, leisure pedestrian streets between the road and the commercial architectures and corridors of commercial verandah style. And part of the main street like the main entry of the central axis is in the for of inner streets which not only expand the commercial area, making commercial atmosphere active but also makes shape changes: there are leisure shopping space in proper scale along the pedestrian streets of the central axis.

Arts and culture of the architecture

Elements that format architectural art include outer shapes of the construction groups and individuals, internal and external space matches, elevation sketches, detail processing, colors and quality of the materials and process of light and shadow and decorations and so on. Residences reflect on life, and Paris City tries to search living demand from its lifestyle. According to the taste of the community and demand of the consumers, house types are classified into series of mature products for the single, childless people, warm family and luxurious houses (villas in the air). Then, harmony between architectural art and city culture are emphasized and we pay attention to passing on traditional architectures. Innovation and application for materials and formation also become an architectural language in Paris City, in the meantime, innovation and control for the crafts of construction is a hotly debated topic for designers of Paris City.

15

15. 建筑立面
16. 别墅实景

251

Strong community culture

Wherever in Paris City, from Paris Park to Seino River, or from Xiehe Plaza to the Arch of the Triumph; From the Avenue Champs Elysees to Laofoye Shopping center, or from Paris City Experimental Primary School to the Louvre Chamber; no matter you stroll along the river or go shopping in the center, no matter you are on the grass of the community or in the central plaza, you will always have your place to stay. And through the wedding assembly hall of the Louvre you can find a lovely life by listening to the vague sound of reading aloud.

Greening concept

Architecture is natural

Construction of the residential quarter cannot do without land, for it is closely related to the base terrain, land features and ground object of the architectures. So with the present situation of the land, Paris City pays full attention to the protection of the environment to ensure enough greening land as well as prevention of air and water pollution, providing the residents with a living environment which allows them

to live healthily and happily. Dependence on manmade energy is reduced as much as possible and natural energy is used instead to seek architectural form that fits local climate characteristics. A gorgeous climate inside the architecture is created by its ability of self-adjust. A series of natural and ecological small environment such as Paris Park, Seino River, and so on, are also built.

Architecture is science and art of the environment

Environmental art is a key factor to assent an ideal city. Landscape design of the Paris City provides garden living area to meet residents at different levels. Functional landscape is humanized-based. Landscapes inside the blocks are raised up to make the architectures stand among the dense shrubs and bushes and to create skyline and landline abundant. Part of the construction units are hollowed, adding to the landscapes and the greening. By studying the way of collocating plants from European countries such as France, we put emphasis on the collocation of European shape and modern garden, and meanwhile make full use of the land by planting proper flowers according to

different seasons. Noise and vision that affects the community is fenced off and means of "enclosure" is taken in the semi-public space of the front and back room.

Unity of culture and greening

Paris is a city section elaborately built combining life, study, leisure, shopping, entertainment and office as a whole with so many elements such as nature, culture, society and economic taken into it. The success of it can be concluded in four aspects: overall arranging form as well as perfect city amenities combined by natural environment and artistic features and the application of Paris' romantic and leisure culture, continuing city layout, commercial pulse and modern living ideal. Multiple architectural types and beautiful residential environment have made contribution to the city landscape, so the community is more recognizable and has more characteristics. In all, design of Paris City has created a cultural and green human homeland by taking care of the relationship between them two.

1

総用地面积 : 1 749 984 m²
総建筑面积 : 240 204 m²
容 积 率 : 1.2
绿 化 率 : 41.2%
総 户 数 : 1610
停 车 位 : 1214

1. 南区实景
2. 南区建筑立面
3. 鸟瞰图

2

深圳中海大山地

开 发 商：中海地产（深圳）有限公司
项 目 地 址：深圳市龙岗区横岗片区梧桐路北端
建 筑 设 计：城脉建筑设计（深圳）有限公司

中海大山地位于深圳横岗梧桐路与环城北路交汇处，振业城的北面，占地18万平米，住宅类型为双拼别墅、连排别墅、山地叠加别墅和多层住宅等，是深圳郊区大型、低密度TOWNHOUSE山居小城。分南北两区，南区的容积率非常低，只有0.7，主要以165—220平方米的连排别墅为主，另外还有叠加复式大约是180平方米，其中也有多层和带电梯的小高层。

深圳中海大山地项目是具有坡地特色的大型混合居住社区。规划中充分利用并整合原有地形地貌，将较为复杂的山体处理成对朝向及视线均有利的缓坡、台地，并将局部制高点作为景观要素，提高土地利用价值，同时确保场地内的土方平衡。用环状路与尽端路相结合的方式清晰简洁地解决了交通流线。住宅采用外部线状结构，内部组团式体系的布局，力求与场地有机地结合和

生长，在保证均好性的同时丰富与城市的界面。注重外部空间中公共空间、邻里空间与私家院落空间的等级划分及相互渗透和借景，创新的联院别墅布局提高了小户型联排住宅的用地效率。通过多种自然材料的组合及构成营造具有现代感、东方韵味及自然气息的居住建筑形象。

设计中将"山地"概念进行到底，着重刻画"山地规划"、"山地建筑"以及"山地环境"，强调人与自然的和谐对话。代表山地特色的"山、水、风、光"四大元素渗入建筑与环境设计的细节中，将生态主题具体化、精练化。山清水秀，风光旖旎的含义寓于其中，"山水"、"风光"成了风景、景观的代名词。形象地向客户传达了"隐、散、闲"的生活意境。

3

4

Project location: North end, Wutong Road, Henggang Area, Longgang District, Shenzhen.

Developer: China Overseas Holdings Limited (Shenzhen)

Architectural design: Shenzhen-based CityMark Architects and Engineers

Total plot area: 174 9984 m^2

Gross building area: 240 204 m^2

Plot ratio: 1.2

Afforesting rate: 41.2%

Total units: 1610

Parking space: 1214

5

China Overseas Dashandi is located in the intersection of Henggang Wutong Road and North Huancheng Road in Shenzhen, north of Zhenye City, and covers an area of 180,000 square meters. Being a large and low density mountain townhouse residential community, it has house types such as semi-detached houses, townhouses, mountain superimpose villa and multi-story residential houses and so on. It was divided into two parts, the north and the south ones, and the latter gets a low plot ratio of only 0.7 with townhouses of about 165 to 220 square meters large as its main house type. In addition, there are superimposed and duplex houses about 180 square meters large , multi-story buildings and small high rise residential building with elevators.

6

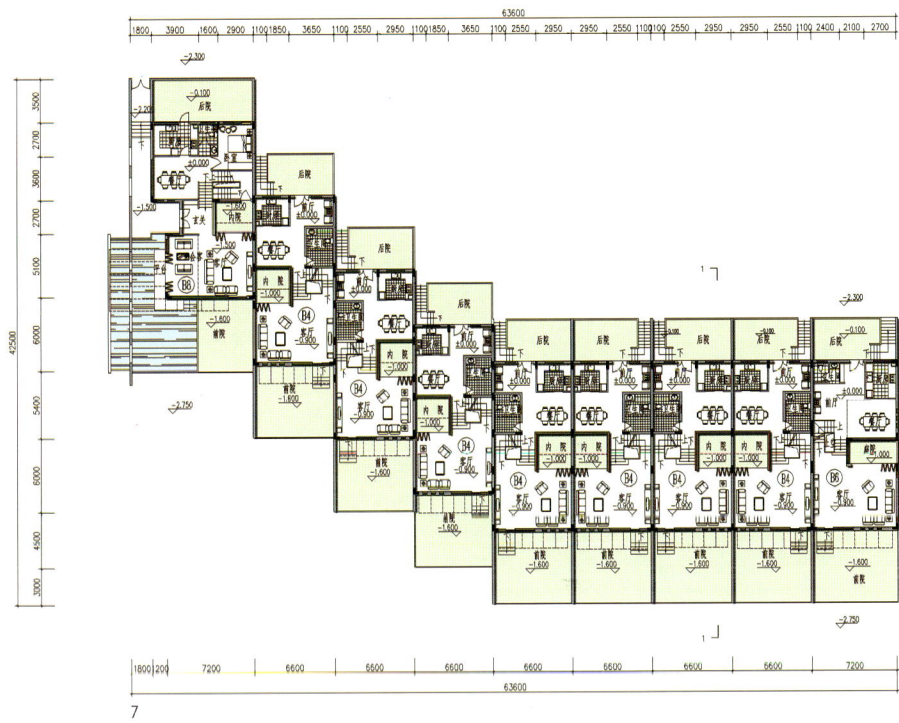

7

8

9

4．总平面图
5．北区高层效果图
6．坡地别墅效果图
7．八栋平面图
8．八栋立面图
9．别墅区景观

10

11

12

13

15

16

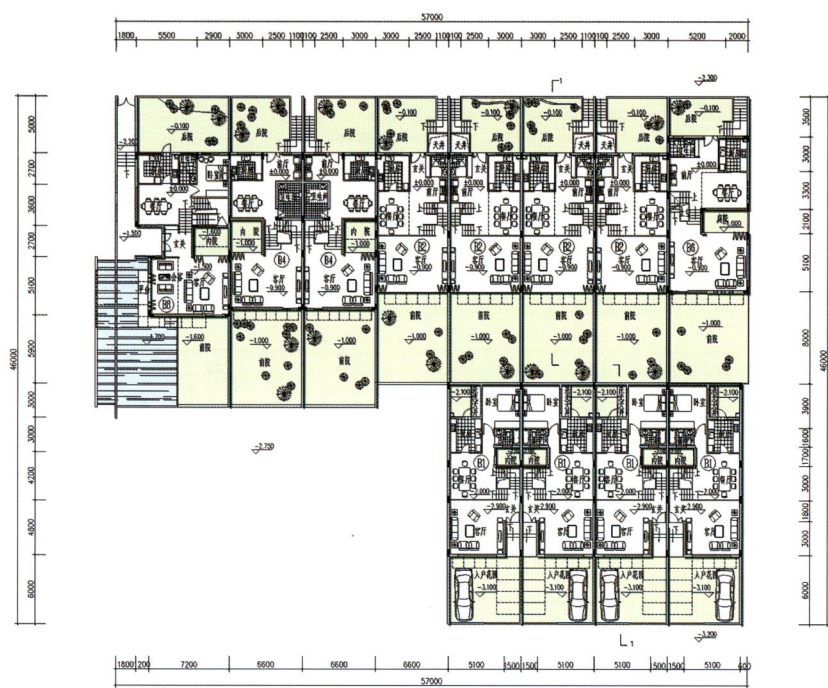

九栋一层平面图

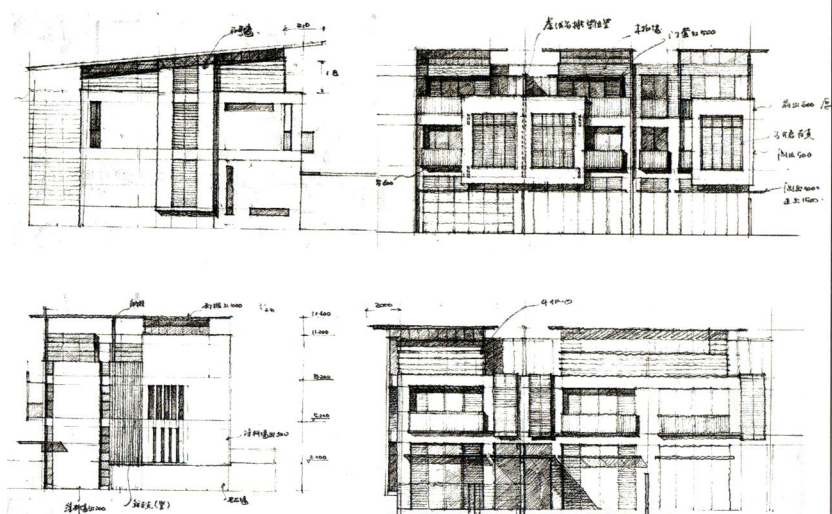

九栋立面构思草图图

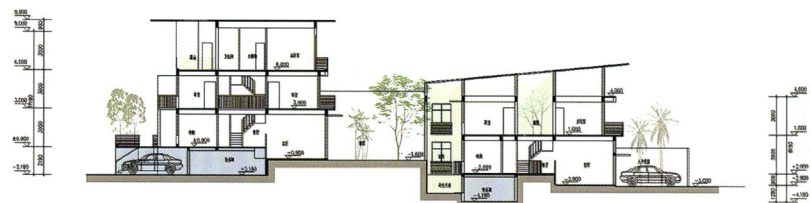

九栋1-1剖面图

九栋南立面图

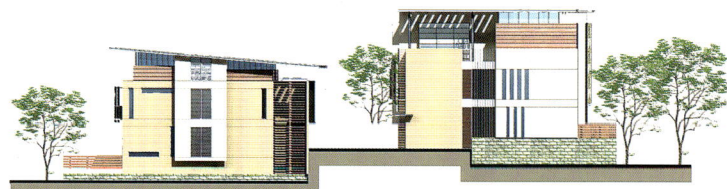

九栋东立面图

17

18

19

20

Project Shenzhen China Overseas Dashandi is a slope featured large mixed residential community. In the plan the original terrain features are made up and made full use of to promote the land using value and in the meantime to ensure the earthwork balance by making the comparative mountains gentle slopes and terrace which are both good for aspect and vision as well as taking part of the commanding point as landscape elements. The way of combining the ring road and culdesacs efficiently solve the flow of the traffic. External linear structure and layout of internal block pattern system are adopted in the residences in pursuit of the organic combination and further reach with the field, also of the homogeneous and enriching it interface with the city. Emphasis is laid on the classification of the public, neighborhood and private courtyard space of the external space, allowing them to penetrate and share landscapes with each other and the innovated townhouse layout has increased land using rate of small townhouses. With so many natural materials and structure, a modern, oriental and natural living architectural image has been created.

Concept of mountain terrain has been carried out throughout the design, which is laid emphasis upon to a mountain layout, mountain architecture and a mountain environment, and here people live in harmonious with nature. Four elements, mountain, water, wind and light which stand for mountain feature are used in the detailed design of the architecture and the environment, specifying the ecological theme. You can enjoy the green hills, clear waters and many beautiful scenic sights. Such four elements have become nouns which represent our landscape, vividly convey quiet, free and relaxed living feeling for our customers.

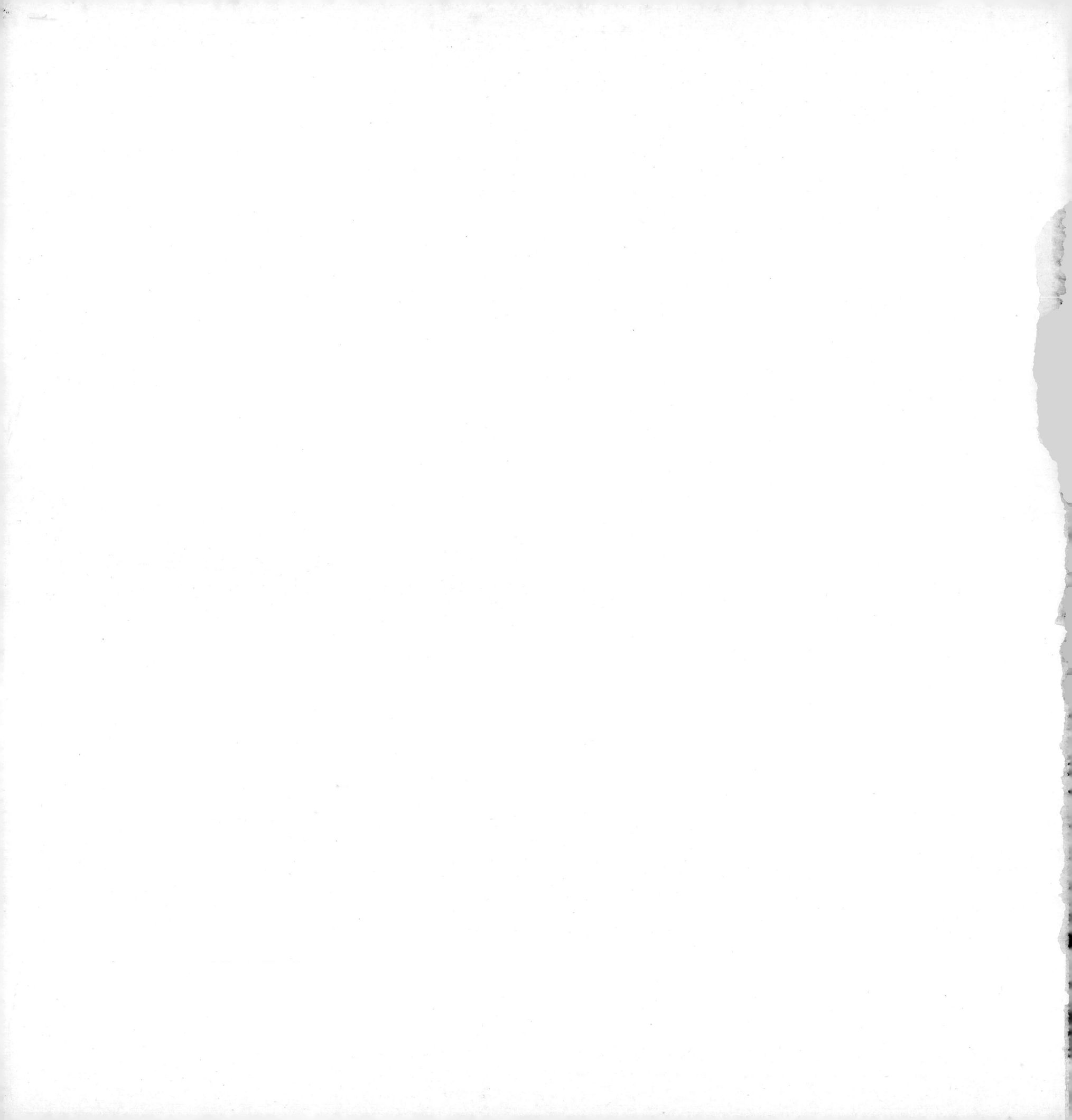